Featured Papers in Inorganic Materials 2024

Featured Papers in Inorganic Materials 2024

Guest Editors

Roberto Nisticò
Torben R. Jensen

MDPI

Basel • Beijing • Wuhan • Barcelona • Belgrade • Novi Sad • Cluj • Manchester

Guest Editors

Roberto Nisticò
Department of Materials
Science, U5
University of Milano-Bicocca
Milano
Italy

Torben R. Jensen
Department of Chemistry
and Interdisciplinary
Nanoscience Center
Aarhus University
Aarhus
Denmark

Editorial Office
MDPI AG
Grosspeteranlage 5
4052 Basel, Switzerland

This is a reprint of the Special Issue, published open access by the journal *Inorganics* (ISSN 2304-6740), freely accessible at: www.mdpi.com/journal/inorganics/special_issues/K9I97Q90IW.

For citation purposes, cite each article independently as indicated on the article page online and as indicated below:

Lastname, A.A.; Lastname, B.B. Article Title. *Journal Name* **Year**, *Volume Number*, Page Range.

ISBN 978-3-7258-3854-7 (Hbk)
ISBN 978-3-7258-3853-0 (PDF)
https://doi.org/10.3390/books978-3-7258-3853-0

Contents

About the Editors

Roberto Nisticò

Roberto Nistico (RN) is Associate Professor in General and Inorganic Chemistry at the University of Milano-Bicocca (Department of Materials Science, Italy). RN has published more than 90 papers in renowned international journals and 2 book chapters. Additionally, RN is a member of the Editorial Board of the journal *Inorganics* ("Inorganic Materials" Section). RN's research mainly focuses on several aspects at the interface between Inorganic Chemistry, Materials Science and Nanotechnology, always looking for novel and appealing technological solutions for a clean and sustainable future. RN's principal fields of interest are the development of magnet-responsive nanomaterials (i.e., iron oxides, ferrites) and other inorganic systems for the environmental remediation of contaminated (waste)water, photo(electro)catalysis, and energy applications.

Torben R. Jensen

Torben R. Jensen (TRJ) obtained a PhD degree in Materials Chemistry at the University of Southern Denmark, Odense, in 1999. In October 1998, he joined the Ris. Lab as a post-doc, where he changed research topics to biophysics and conducted surface diffraction and reflectivity experiments on lipids and enzymes. Much of his research was conducted at the synchrotron Hasylab, Hamburg. After 2 years, Oct. 2000, he became an Assistant Prof. at the Department of Chemistry, Aarhus University, and became associated with the Interdisciplinary Nanoscience Center (iNANO). Since then, he has assembled a unique and productive independent research group within energy materials science. His research achievements led to a Doctor of Science (D.Sc.) degree in 2014 and a promotion to Professor in 2016. Over the years, TRJ developed several new courses in chemistry and nanoscience and educated 16 post-docs, 19 PhDs, 44 MScs, and 61 BSc students and hosted about 45 guest professors and researchers at all levels. His research group synthesised a multitude of new metal hydrides by combining solvent-based methods, mechanochemistry, and solid–gas reactions. They also infiltrated materials in nanoporous scaffolds and conducted systematic studies of properties as a function of pore size and surface area to investigate their nanoeffects. This group now focuses on cationic conductivity in the solid state and on the development of novel battery materials. This group is a frequent user of large synchrotron X-ray and neutron facilities and has developed knowledge and new versatile in situ powder X-ray diffraction (PXD) sample environments and gas control systems for the investigation of solid–gas and solid–liquid reactions, also under extreme p,T conditions of up to p(H2) = 700 bar. TRJ has published >310 peer-reviewed papers.

inorganics

MDPI

Editorial

Featured Papers in Inorganic Materials 2024

Roberto Nisticò [1,*] and Torben R. Jensen [2,*]

1 Department of Materials Science, University of Milano-Bicocca, U5, INSTM, Via R. Cozzi 55, 20125 Milano, Italy
2 Interdisciplinary Nanoscience Center and Department of Chemistry, Aarhus University, DK-8000 Aarhus C, Denmark
* Correspondence: roberto.nistico@unimib.it (R.N.); trj@chem.au.dk (T.R.J.)

1. Introduction

After the impressive success of the Special Issue "10th Anniversary of *Inorganics*: Inorganic Materials", published in 2023 [1], the "Inorganic Materials" section launched the Special Issue "Featured Papers in Inorganic Materials 2024" as a complementary Special Issue, with the aim of continuing discussion on the recent advances in the field of functional inorganic materials for a green and sustainable future. In particular, the overarching aim of this Special Issue is to increase the knowledge of new emerging areas where the sustainable use of inorganic materials can play a pivotal role, such as catalysis [2–5], clean energy production and storage [6–9], environmental remediation [10–13], the development of value-added nanomaterials and nanocomposites (eventually from waste) [14–16], recycling [17,18], and biomedicine [19–22].

In particular, this Special Issue is composed of a collection of contributions touching different fundamental aspects strictly related to the storage and delivery of renewable energy (i.e., in particular by exploiting hydrogen as an energy source/carrier), the improvement of environmental clean-up approaches to remediate different contaminated media (e.g., air and liquid phases), and the development of inorganic materials with enhanced optical and thermal properties exploitable in many advanced technological fields (e.g., photovoltaics).

Prior to proceeding with the overview of individual contributions, the Guest Editors would like to thank all the Reviewers who spent their valuable time thoroughly reviewing and improving the quality of the articles published in this volume. Moreover, the Guest Editors sincerely thank all the Authors for choosing *Inorganics* and, in particular, the "Inorganic Materials" section, as the place to publish their valuable results.

Received: 18 March 2025
Revised: 28 March 2025
Accepted: 1 April 2025
Published: 2 April 2025

Citation: Nisticò, R.; Jensen, T.R. Featured Papers in Inorganic Materials 2024. *Inorganics* **2025**, *13*, 110. https://doi.org/10.3390/inorganics13040110

2. An Overview of the Published Articles

Overall, this Special Issue collected 10 original papers (9 research articles and 1 review) and received more than 13,000 views, thus, paving the way for further editions of this Special Issue. As expressed above, the scope of this collection covers the entire focus area where inorganic materials can play a key role in order to reach a sustainable future, and this is exemplified by the various topics covered by the 10 original papers published in this Special Issue. This section provides a brief overview of the contributions, organizing them into discreet subsections as reported in Table 1, including the following: (i) Storage and delivery of renewable energy: Hydrogen-based technology, (ii) Environmental remediation, and (iii) Development of inorganic materials with enhanced optical and thermal properties.

Table 1. Correlation between subsections and contributions collected in the present Special Issue.

Subsections	Contribution No.	Title
Storage and delivery of renewable energy: Hydrogen-based technology	1	Li, M.; Hu, Y.; Kong, H.; Huang, Q.; Chen, Y.; Yan, Y. A Study on the Volume Expansion of Vanadium-Based Alloy Powders and Compacts During Hydrogen Sorption. *Inorganics* **2024**, *12*, 318. https://doi.org/10.3390/inorganics12120318.
	2	Yang, Q.; Jia, X.; Qin, Z.; Ding, X.; Li, Y. Enhancements in Hydrogen Storage Properties of Magnesium Hydride Supported by Carbon Fiber: Effect of C–H Interactions. *Inorganics* **2024**, *12*, 273. https://doi.org/10.3390/inorganics12110273.
	3	Davis Cortina, M.; Romero de Terreros Aramburu, M.; Neves, A.M.; Hurtado, L.; Jepsen, J.; Ulmer, U. The Integration of Thermal Energy Storage Within Metal Hydride Systems: A Comprehensive Review. *Inorganics* **2024**, *12*, 313. https://doi.org/10.3390/inorganics12120313.
	4	Zhong, J.; Zhang, T.; Tian, J.; Gao, W.; Wang, Y. Nickel Foam-Supported FeP Encapsulated in N, P Co-Doped Carbon Matrix for Efficient Electrocatalytic Hydrogen Evolution. *Inorganics* **2024**, *12*, 291. https://doi.org/10.3390/inorganics12110291.
Environmental remediation	5	Wang, Y.; Yu, H.; Wang, H.; Chen, T. Hierarchically Porous Titanosilicate Hollow Spheres Containing TS-1 Zeolite Precursors for Oxidative Desulfurization. *Inorganics* **2025**, *13*, 37. https://doi.org/10.3390/inorganics13020037
	6	Sriram, G.; Baby, N.; Dhanabalan, K.; Arunpandian, M.; Selvakumar, K.; Sadhasivam, T.; Oh, T.H. Studies of Various Batch Adsorption Parameters for the Removal of Trypan Blue Using Ni-Zn-Bi-Layered Triple Hydroxide and Their Isotherm, Kinetics, and Removal Mechanism. *Inorganics* **2024**, *12*, 296. https://doi.org/10.3390/inorganics12110296.
	7	El Atti, O.; Hot, J.; Fajerwerg, K.; Lorber, C.; Lebeau, B.; Ryzhikov, A.; Kahn, M.; Collière, V.; Coppel, Y.; Ratel-Ramond, N.; Ménini, P.; Fau, P. Synthesis of TiO_2/SBA-15 Nanocomposites by Hydrolysis of Organometallic Ti Precursors for Photocatalytic NO Abatement. *Inorganics* **2024**, *12*, 183. https://doi.org/10.3390/inorganics12070183.
Development of inorganic materials with enhanced optical and thermal properties	8	Sassi, S.; Bouich, A.; Hajjaji, A.; Khezami, L.; Bessais, B.; Soucase, B.M. Cu-Doped TiO_2 Thin Films by Spin Coating: Investigation of Structural and Optical Properties. *Inorganics* **2024**, *12*, 188. https://doi.org/10.3390/inorganics12070188.
	9	Gáborová, K.; Hegedüs, M.; Levinský, P.; Mihok, F.; Matvija, M.; Knížek, K.; Milkovič, O.; Vatraľová, D.; Hejtmánek, J.; Saksl, K. Thermoelectric Characteristics of β-Ag_2Se_{1+x} Prepared via a Combined Rapid Mechano-Thermal Approach. *Inorganics* **2024**, *12*, 334. https://doi.org/10.3390/inorganics12120334.
	10	Costa Oliveira, F.d.A.; Galindo, J.; Rodríguez, J.; Cañadas, I.; Cruz Fernandes, J. Thermal Shock Resistance of Commercial Oxide-Bonded Silicon Carbide Reticulated Foams under Concentrated Solar Radiation at PSA: A Feasibility Study. *Inorganics* **2024**, *12*, 246. https://doi.org/10.3390/inorganics12090246.

2.1. *Storage and Delivery of Renewable Energy: Hydrogen-Based Technology*

To achieve energy independence, a stable supply of clean energy is essential. Hydrogen is an ideal energy carrier and technological solution for energy storage and transportation [23,24]. In this context, metal hydrides are very attractive materials as they can reversibly absorb hydrogen in large quantities under mild conditions, thus, offering a promising energy storage and carrier solution [25,26].

Li et al. (Contribution 1) studied the volume expansion of a V-based alloy (namely, $V_{61}Cr_{24}Ti_{12}Ce_3$, with a body-centered-cubic structure) after performing hydrogenation reactions, thus, testing its potential as a substrate material for the wall of a hydrogen storage tank. To monitor the in-situ volume expansion, the authors designed and constructed an expansion test apparatus. Results indicated an initial maximum reversible capacity of the V-based alloy of ca. 2.5 wt.%, with a lattice expansion of ca. 37.8% measured through XRD analysis. A powder bed made with the same V-based alloy showed a volume expansion ratio of 131% during the first hydrogen absorption cycle, and 40–45% in the following four cycles, whereas a compact bed, made of the V-alloy powders, organic silicone gel

(PDMS), and graphite flakes, showed a significantly smaller volume expansion ratio (i.e., 97% during the first cycle and 13% in the following cycles). These improved performances are due to the presence of both PDMS and graphite flakes, which do not affect the overall hydrogen absorption capacity of the V-alloys but speed up the absorption kinetics.

Another interesting technological solution is the one proposed by Yang et al. (Contribution 2), who prepared and studied the hydrogen storage performance of two MgH_2-C fiber-based cloth composites prepared by either dry ball milling or wet impregnation routes, obtaining distinct morphologies (namely, the formation of MgH_2 particles uniformly distributed on the surface of the C fibers in the direction of the long axis of the fibers, or the formation of MgH_2 particles diffusely distributed without any distinct aggregation phenomenon, respectively). Both composites show reduced preparation costs, and significantly improved kinetics compared to the bare MgH_2. The experimental results revealed that the occurrence of C-H bonding interactions involving the C fibers is responsible for the enhanced hydrogen absorption/desorption ability of the two composites.

Davis Cortina et al. (Contribution 3) reviewed the recent literature regarding metal hydride materials for hydrogen storage, focusing on their thermophysical, thermodynamic, and kinetic properties. Additionally, since thermal energy storage systems provide a means to enhance the energy efficiency and cost-effectiveness of metal hydride-based storage by effectively coupling thermal management with hydrogen storage processes, authors also explored thermal energy storage materials with particular attention to those that operate at temperatures compatible with the most widely studied metal hydride systems.

As previously noted in this introduction, hydrogen energy stands out with respect to other renewable energy sources due to its high energy density and clean, pollution-free characteristics. In this context, electrochemical water splitting to produce hydrogen using abundant non-noble metals as catalysts has become a research hotspot [27,28].

Zhong et al. (Contribution 4) reported the synthesis of metal–organic framework-derived N/P co-doped carbon-encapsulated FeP nanoparticles onto a nickel foam substrate and successfully tested these systems as catalysts for the hydrogen evolution reaction. The proposed substrates showed excellent activity in both acid (0.5 M H_2SO_4) and alkaline (1.0 M KOH) environments and stability with negligible decay over 48 h.

2.2. Environmental Remediation

During the combustion of gasoline and diesel, sulfur impurities produce sulfur oxides, which are harmful for the environment, thus, performing a preliminary oxidative desulfurization (ODS) treatment emerges as a very important reaction step [29]. In this context, Wang et al. (Contribution 5) investigated the synthesis of hierarchically porous titanosilicate hollow spheres from TS-1 zeolite precursor to obtain catalysts with uniformly incorporated tetrahedrally coordinated Ti sites. The synthesized materials demonstrated remarkable catalytic performance in ODS, achieving complete dibenzothiophene removal within 15 min and a high turnover frequency of up to 123 h^{-1} at 30 °C. This outstanding catalytic performance can be attributed to the tetrahedral coordination Ti of TS-1 sub-units and the hierarchically porous hollow structure of the material itself.

Furthermore, the need to remediate the water, air, and soil pollution caused by human activities has become urgent for society, not only from an environmental viewpoint, but also when considering the effects that environmental contaminants have on human health [30].

Hence, Sriram et al. (Contribution 6) synthesized a Ni-Zn-Bi-layered triple hydroxide via co-precipitation techniques and successfully tested this substrate for the removal of the dye Trypan blue from water via an adsorption mechanism. Adsorption tests indicate a maximum removal effectiveness of around 96.7% at natural pH (ca. 4.5–5.0), with maximum adsorption capacity of 5.3 $mg{\cdot}g^{-1}$ at dye concentrations ranging from 5 to 30 $mg{\cdot}L^{-1}$.

Interestingly, when combined with various anionic dye mixtures, selectivity studies showed a high selectivity for the removal of Trypan blue and the removal of other cationic dyes. Recyclability investigations revealed the notable removal of Trypan blue using 0.1 M NaOH for the desorption.

Lastly, El Atti et al. (Contribution 7) reported the synthesis of a TiO_2/mesoporous silica SBA-15 nanocomposite using an organometallic decoration method (based on Ti(III) amidinate) and compared it with an analogous system prepared via a Ti(IV) precursor. Photocatalytic performances of these two different nanocomposites were investigated for the abatement of NO under UV light irradiation in humidified air. Interestingly, the TiO_2/mesoporous silica SBA-15 nanocomposite prepared from the Ti(III) precursor demonstrated an NO abatement performance 40% more efficient than the reference photocatalyst TiO_2 P-25.

2.3. *Development of Inorganic Materials with Enhanced Optical and Thermal Properties*

Sassi et al. (Contribution 8) reported the synthesis of Cu-doped TiO_2 films (with Cu concentration in the 0–8% range) grown directly on FTO glass by means of a spin coating deposition. Morphological and structural characterization revealed the occurrence of small, spherical nanoparticles with a preferred TiO_2 anatase (101) orientation for all samples. Regarding the optical properties, the 2–4% Cu-doped sample showed high transmittance, thus, making these systems promising candidates for electron transport and potentially integrable in perovskite solar cells.

Gáborová et al. (Contribution 9) investigated the thermoelectric properties of Se-rich β-Ag_2Se synthesized via a solvent-free mechanochemical method followed by spark plasma sintering step. Importantly, single-phase Ag_2Se_{1+x} samples with varying Se content were produced. The increase in Se significantly influenced the material's thermoelectric performance, and the sample with nominal composition $Ag_2Se_{1.01}$ exhibited a high figure-of-merit ZT > 0.9 at ca. 111 °C, which is nearly six times higher than the reference sample (β-Ag_2Se).

Lastly, Costa Oliveira et al. (Contribution 10) investigated the thermal shock behavior of commercial oxide-bonded silicon carbide (ob-SiC) reticulated porous ceramic (RPC) foams, potentially exploitable as volumetric ceramic receivers for solar thermal electricity production. Foams were subjected to well-controlled temperature cycles (in the 800 to 1400 °C range), and the damage induced by thermal shock was quantified by performing crush tests. Results indicated that damage was critically dependent on both the bulk density and cell size. Interestingly, better thermal shock resistance was achieved by decreasing both the bulk density and cell size.

3. Conclusions

With this Special Issue "Featured Papers in Inorganic Materials 2024" published in the "Inorganics Materials" section and also published as a book, the Editors hope that the high quality of the contributions collected here will receive the visibility and attention they deserve. These would help readers increase their knowledge in the field of inorganic materials, and be a new source of inspiration for novel, focused investigations.

Acknowledgments: The Editors would like to thank all Authors, Reviewers, and the entire Editorial Staff of *Inorganics* who provided their new science, constructive recommendations, and assisted in the realization of the present Special Issue, respectively.

Conflicts of Interest: The authors declare no conflicts of interest.

References

1. Nisticò, R.; Idriss, H.; Carlos, L.; Aneggi, E.; Jensen, T.R. 10th Anniversary of *Inorganics*: Inorganic Materials. *Inorganics* **2024**, *12*, 62. [CrossRef]
2. Jacobs, P.A.; Dusselier, M.; Sels, B.F. Will Zeolite-Based Catalysis be as Relevant in Future Biorefineries as in Crude Oil Refineries? *Angew. Chem. Int. Ed.* **2014**, *53*, 8621–8626. [CrossRef]
3. Chakravorty, A.; Roy, S. A review of photocatalysis, basic principles, processes, and materials. *Sustain. Chem. Environ.* **2024**, *8*, 100155. [CrossRef]
4. Lei, Y.; Niu, Y.; Tang, X.; Yu, X.; Huang, X.; Lin, X.; Yi, H.; Zhao, S.; Jiang, J.; Zhang, J.; et al. Cu-based materials for electrocatalytic CO_2 to alcohols: Reaction mechanism, catalyst categories, and regulation strategies. *J. Energy Chem.* **2024**, *97*, 593–611. [CrossRef]
5. Chatenet, M.; Pollet, B.G.; Dekel, D.R.; Dionigi, D.; Deseure, J.; Millet, P.; Braatz, R.D.; Bazant, M.Z.; Eikerling, M.; Staffell, I.; et al. Water electrolysis: From textbook knowledge to the latest scientific strategies and industrial developments. *Chem. Soc. Rev.* **2022**, *51*, 4583–4762. [CrossRef]
6. Li, F.; Zeng, Q.; Li, J.; Hao, X.; Ho-Baille, A.; Tang, J.; Green, M.A. Emerging inorganic compound thin film photovoltaic materials: Progress, challenges and strategies. *Mater. Today* **2020**, *41*, 120–142. [CrossRef]
7. Oh, S.M.; Patil, S.B.; Jin, X.; Hwang, S.-J. Recent Applications of 2D Inorganic Nanosheets for Emerging Energy Storage System. *Chem. Eur. J.* **2018**, *24*, 4757–4773. [CrossRef]
8. Man, X.; Lu, H.; Xu, Q.; Wang, C.; Ling, Z. Review on the thermal property enhancement of inorganic salt hydrate phase change materials. *J. Energy Storage* **2023**, *72*, 108699. [CrossRef]
9. Vitillo, J.G. Magnesium-based systems for carbon dioxide capture, storage and recycling: From leaves to synthetic nanostructured materials. *RSC Adv.* **2015**, *5*, 36192–36239. [CrossRef]
10. Sarkar, S.; Das, R.; Choi, H.; Bhattacharjee, C. Involvement of process parameters and various modes of application of TiO_2 nanoparticles in heterogeneous photocatalysis of pharmaceutical wastes—A short review. *RSC Adv.* **2014**, *4*, 57250–57266. [CrossRef]
11. Cho, M.-S.; Younis, S.A.; Lee, C.S.; Li, X.; Kim, K.-H. The superior mineralization potential of a graphitic carbon nitride/titanium dioxide composite and its application in the construction of a portable photocatalytic air purification system against gaseous formaldehyde. *J. Mater. Chem. A* **2024**, *12*, 32239–32258. [CrossRef]
12. Sahu, D.; Pervez, S.; Karbhal, I.; Tamrakar, A.; Mishra, A.; Verma, S.R.; Deb, M.K.; Ghosh, K.K.; Pervez, Y.F.; Shrivas, K.; et al. Applications of different adsorbent materials for the removal of organic and inorganic contaminants from water and wastewater—A review. *Desalin. Water Treat.* **2024**, *317*, 100253. [CrossRef]
13. Polliotto, V.; Pomilla, F.R.; Maurino, V.; Marcì, G.; Bianco Prevot, A.; Nisticò, R.; Magnacca, G.; Paganini, M.C.; Ponce Robles, L.; Perez, L.; et al. Different approaches for the solar photocatalytic removal of micro-contaminants from aqueous environment: Titania vs. hybrid magnetic iron oxides. *Catal. Today* **2019**, *328*, 164–171. [CrossRef]
14. Li, Z.; Li, W.; You, J.; Huang, J.; Gan, R.; Guo, J.; Zhang, X. Critical secondary resource for porous ceramics: A review on recycling of inorganic solid wastes. *J. Eur. Ceram. Soc.* **2024**, *44*, 116781. [CrossRef]
15. Xu, C.; Nasrollahzadeh, M.; Selva, M.; Issaabadi, Z.; Luque, R. Waste-to-wealth: Biowaste valorization into valuable bio(nano)materials. *Chem. Soc. Rev.* **2019**, *48*, 4791–4822. [CrossRef]
16. Trovato, V.; Sfameni, S.; Ben Debabis, R.; Rando, G.; Rosace, G.; Malucelli, G.; Plutino, M.R. How to Address Flame-Retardant Technology on Cotton Fabrics by Using Functional Inorganic Sol-Gel Precursors and Nanofillers: Flammability Insights, Research Advances, and Sustainability Challenges. *Inorganics* **2023**, *11*, 306. [CrossRef]
17. Guo, X.; Zeng, M.; Yu, H.; Lin, F.; Li, J.; Wang, W.; Chen, G. Critical review for the potential analysis of material utilization from inorganic industrial solid waste. *J. Clean. Prod.* **2024**, *459*, 142457. [CrossRef]
18. Biswal, B.K.; Zhang, B.; Tran, P.T.M.; Zhang, J.; Balasubramanian, R. Recycling of spent lithium-ion batteries for a sustainable future: Recent advancements. *Chem. Soc. Rev.* **2024**, *53*, 5552–5592. [CrossRef]
19. Giner-Casares, J.; Henriksen-Lacey, M.; Coronado-Puchau, M.; Liz-Marzan, L.M. Inorganic nanoparticles for biomedicine: Where materials scientists meet medical research. *Mater. Today* **2016**, *19*, 19–28. [CrossRef]
20. Wang, X.; Zhong, X.; Li, J.; Liu, Z.; Cheng, L. Inorganic nanomaterials with rapid clearance for biomedical applications. *Chem. Soc. Rev.* **2021**, *50*, 8669–8742. [CrossRef]
21. Nisticò, R. A Comprehensive Study on the Applications of Clays into Advanced Technologies, with a Particular Attention on Biomedicine and Environmental Remediation. *Inorganics* **2022**, *10*, 40. [CrossRef]
22. Liu, C.; Xu, M.; Wang, Y.; Yin, Q.; Hu, J.; Chen, H.; Sun, Z.; Liu, C.; Li, X.; Zhou, W.; et al. Exploring the potential of hydroxyapatite-based materials in biomedicine: A comprehensive review. *Mater. Sci. Eng. R Rep.* **2024**, *161*, 100870. [CrossRef]
23. Moller, K.T.; Jensen, T.R.; Akiba, E.; Li, H.-w. Hydrogen—A sustainable energy carrier. *Prog. Nat. Sci. Mater. Int.* **2017**, *27*, 34–40. [CrossRef]
24. Sartbaeva, A.; Kuznetsov, V.L.; Wells, S.A.; Edwards, P.P. Hydrogen nexus in a sustainable energy future. *Energy Environ. Sci.* **2008**, *1*, 79–85. [CrossRef]

25. Von Colbe, J.B.; Ares, J.-R.; Barale, J.; Baricco, M.; Buckley, C.; Capurso, G.; Gallandat, N.; Grant, D.M.; Guzik, M.N.; Jacob, I.; et al. Application of hydrides in hydrogen storage and compression: Achievements, outlook and perspectives. *Int. J. Hydrogen Energy* **2019**, *44*, 7780–7808. [CrossRef]
26. Gebretatios, A.G.; Banat, F.; Cheng, C.K. A critical review of hydrogen storage: Toward the nanoconfinement of complex hydrides from the synthesis and characterization perspectives. *Sustain. Energy Fuels* **2024**, *8*, 5091–5130. [CrossRef]
27. Thakur, J.; Phogat, P.; Shreya. Catalyst design for efficient water splitting: A comprehensive review of challenges and opportunities. *Fuel* **2025**, *392*, 134954. [CrossRef]
28. Li, L.; Wang, P.; Shao, Q.; Huang, X. Metallic nanostructures with low dimensionality for electrochemical water splitting. *Chem. Soc. Rev.* **2020**, *49*, 3072–3106. [CrossRef]
29. Chen, L.; Yuan, Z.-Y. Design strategies of supported metal-based catalysts for efficient oxidative desulfurization of fuel. *J. Ind. Eng. Chem.* **2022**, *108*, 1–14. [CrossRef]
30. Shetty, S.S.; Deepthi, D.; Harshitha, S.; Sonkusare, S.; Naik, P.B.; Kumari, N.S.; Madhyastha, H. Environmental pollutants and their effects on human health. *Helyon* **2023**, *9*, e19496. [CrossRef]

inorganics

MDPI

Article

A Study on the Volume Expansion of Vanadium-Based Alloy Powders and Compacts During Hydrogen Sorption

Mojia Li [1], Yunfeng Hu [2], Hanyang Kong [1], Qiuwei Huang [1], Yusong Chen [1] and Yigang Yan [1,*]

[1] Institute of New Energy and Low-Carbon Technology, Sichuan University, Chengdu 610207, China; limojia@stu.scu.edu.cn (M.L.); hanyang_kong@163.com (H.K.); huangqiuwei2022@163.com (Q.H.); ty505606@outlook.com (Y.C.)

[2] Key Laboratory of Advanced Technologies of Materials, Ministry of Education, School of Materials Science and Engineering, Southwest Jiaotong University, Chengdu 610031, China; huyfeng@my.swjtu.edu.cn

* Correspondence: yigang.yan@scu.edu.cn

Abstract: Storing hydrogen in solid metal hydrides provides a safe and efficient storage approach. However, the large volume expansion of metal hydrides during hydrogen absorption imposes substantial stresses on the wall of a hydrogen storage tank. In this study, volume expansion behavior of a V-based hydrogen storage alloy, $V_{61}Cr_{24}Ti_{12}Ce_3$, with body centered cubic, was investigated using a self-developed in situ expansion testing device. The lattice expansion of the $V_{61}Cr_{24}Ti_{12}Ce_3$ alloy after full hydrogenation was determined to be 37.85% using X-ray diffraction(XRD). The powder bed, composed of alloy powder with an average size of 3.35 mm in diameter, displays a large volume expansion ratio of 131% at the first hydrogen absorption cycle and 40–45% in the following four cycles. The stable compact bed, made of alloy powders, organic silicone gel, and graphite flakes, shows significantly smaller volume expansion ratio, which is 97% at the first cycle and 21% at the second cycle, and stabilizes at 13% in the following cycles. Also, the compact bed shows similar hydrogen absorption capacity, but faster absorption kinetics compared to the powder bed.

Keywords: V-Ti-Cr alloy; volume expansion; metal hydride compact; organic silicone gel; graphite

Citation: Li, M.; Hu, Y.; Kong, H.; Huang, Q.; Chen, Y.; Yan, Y. A Study on the Volume Expansion of Vanadium-Based Alloy Powders and Compacts During Hydrogen Sorption. *Inorganics* **2024**, *12*, 318. https://doi.org/10.3390/inorganics12120318

Academic Editor: Rainer Niewa

Received: 24 November 2024
Accepted: 29 November 2024
Published: 8 December 2024

1. Introduction

To achieve the sustainable development of our society, a stable supply of environmentally harmless clean energy is essential. Renewable energies like solar and wind have great potential, but the time-dependent and geography-dependent variability limits their application as power resources [1,2]. Hydrogen, as an ideal energy carrier, offers a solution for energy storage and transportation [3]. Metal hydrides, which can reversibly absorb hydrogen in large quantity under mild conditions, have been attracting world-wide attention as the medium for hydrogen storage.

Metal hydrides are categorized into interstitial (e.g., VH_2 and $LaNi_5H_6$) and non-interstitial hydrides (e.g., MgH_2). Vanadium (V)-based alloy, with body centered cubic (BCC) structure, is noteworthy for its high hydrogen storage density and fast hydrogen sorption kinetics [4,5]. Different V-based alloy systems have been developed, including V-Ti-Cr, V-Ti-Cr-Fe, and so on, which exhibits reversible capacities of ~2.5 wt.% under ambient conditions (e.g., 0–80 °C at 10 MPa H_2) [4]. Furthermore, a big progress has been made in economical approaches to prepare V-based alloys, for example, from low-cost ferrovanadium master alloy or directly from oxides [4]. However, it is still a big distance from the practical application of vanadium-based alloys for hydrogen storage.

The hydrogen absorption process is accompanied by significant volume expansion, particularly in interstitial hydrides. For instance, the volume of $LaNi_5$ alloy can increase by about 20% after full hydrogenation [6], and the lattice of vanadium-based alloy can expand by approximately 40% [7]. Such large volume expansion would lead to the pulverization of alloy particles and reduce the heat and mass transfer efficiency within the tank [8], unavoidably imposing substantial stress on hydrogen storage tanks, potentially causing deformation or rupture [9,10]. Turning powders into compact after mixing with additives (such as binder and anti-expansion agent) could alleviate the impact of volume expansion of the metal hydrides [11–16]. For example, the $LaNi_5$ compact, made from 1 mm $LaNi_5$ powders and resin, shows very stable structure with only a small amount of powder detached after 100 hydrogen sorption cycles [17]. Tarasov [18] et al. employed a graphene–nickel composite to enhance the desorption reaction of magnesium hydride. While the graphene–nickel material facilitated the magnesium hydride desorption reaction, the graphene component was prohibitively expensive for large-scale implementation. Compacts of $Mg_{90}Ni_{10}$ powder and Ti-Mn-based powder after mixing with expanded graphite (ENG) could increase the thermal conductivity of the metal hydride bed and reduce the stress during the hydrogen sorption process [19]. There are also some studies on the preparation of membranes with hydrogen selective permeability using polymers mixed with metal hydrides [20]. However, few investigations focusing on the volume expanding behaviors of V-based hydrogen storage alloys during hydrogen sorption have been reported.

In the present study, we designed and constructed a dedicated expansion test apparatus to evaluate the in situ volume expansion of V-based alloy, $V_{61}Cr_{24}Ti_{12}Ce_3$ (at.%), during the hydrogen absorption process. Organic silicone gel, also known as PDMS, is a commercially available highly elastic polymer composed primarily of silicone oil and a curing agent. Its curing process is relatively straightforward, and it has been demonstrated to be non-toxic to V-Ti-Cr-Fe alloys [21]. So, the compact, composed of $V_{61}Cr_{24}Ti_{12}Ce_3$ powders, organic silica gel (PDMS) [22] and graphite flakes, was designed to reduce the volume expansion.

2. Results and Discussions

The pressure–composition–temperature (PCT) curves of the $V_{61}Cr_{24}Ti_{12}Ce_3$ alloy for hydrogen absorption (Figure 1a) and desorption (Figure 1b) were measured at 0 °C, 18 °C, and 50 °C, respectively. The hydrogen desorption plateau pressure (P_{eq}) of the $V_{61}Cr_{24}Ti_{12}Ce_3$ alloy at 18 °C reaches 4.6 bar, which meets with the requirement of entry pressure for the operation of fuel cell. Based on the desorption PCT curves in Figure 1b, the $V_{61}Cr_{24}Ti_{12}Ce_3$ alloy achieves a reversible hydrogen storage capacity of 2.51 wt.% at 50 °C within the pressure region of 0.01–10 MPa. The van't Hoff plots, obtained by plotting P_{eq} versus 1/T, are shown in Figure 1c. The enthalpy change (ΔH) and entropy change (ΔS) are derived and listed in Table 1.

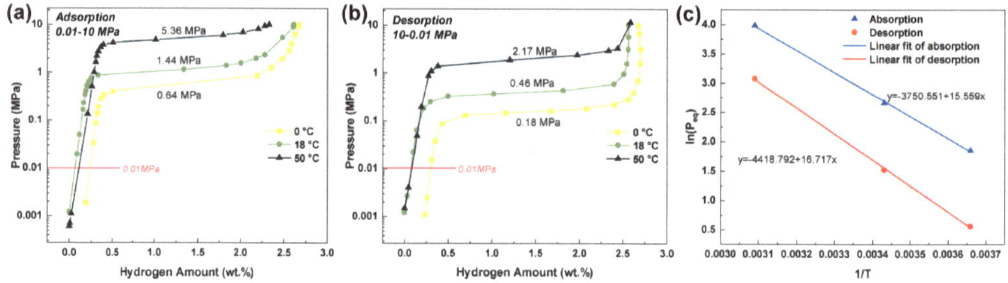

Figure 1. The PCT curves (**a**,**b**) and van't Hoff plots (**c**) of $V_{61}Cr_{24}Ti_{12}Ce_3$ alloy.

Table 1. The enthalpy change (ΔH) and the entropy change (ΔS) of the $V_{61}Cr_{24}Ti_{12}Ce_3$ alloy.

Properties	Hydrogen Absorption	Hydrogen Desorption
ΔH (kJ/mol H_2)	−31.18	36.74
ΔS (J/mol H_2)	−129.36	138.99

Figure 2a,b show the X-ray diffraction (XRD) patterns of the $V_{61}Cr_{24}Ti_{12}Ce_3$ alloy before and after hydrogenation, respectively. Before hydrogenation, the $V_{61}Cr_{24}Ti_{12}Ce_3$ alloy shows the typical BCC structure, with a = 0.3031 (1) Å and V = 0.02785 Å^3, as presented in Table 2. After hydrogenation, the minor $V_4H_{2.88}$-type BCT phase and the major VH_2-type FCC phase are observed, attributed to subsequent phase transition from BCC to BCT and from BCT to FCC during the hydrogen absorption process (Figure 2c). The BCT phase shows the lattice parameters of a = 0.3143 (0) Å and c = 0.3148 (3) Å, and lattice volume increases up to V = 0.03110 Å^3, with a lattice expansion ratio of 11.67% compared to the BCC phase. The FCC phase shows the lattice parameters of a = 0.4250 (3) Å, and lattice volume increases up to V = 0.07678 Å^3, with an expansion ratio of 23.44% compared to the BCT phase and an expansion ratio of 37.85% compared to the BCC phase.

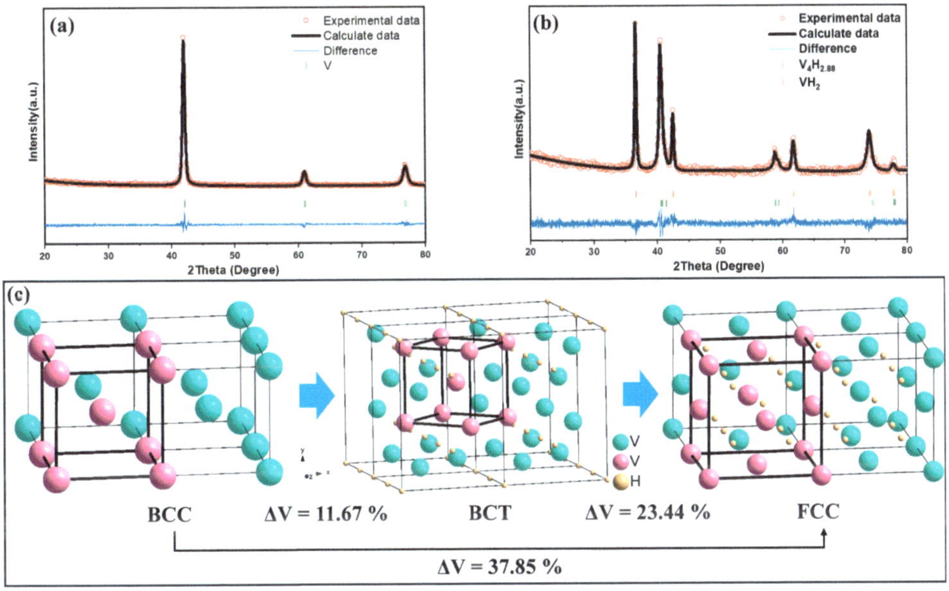

Figure 2. The XRD patterns of the $V_{61}Cr_{24}Ti_{12}Ce_3$ alloy before (**a**) and after (**b**) hydrogenation, and the schematic diagram of the $V_{61}Cr_{24}Ti_{12}Ce_3$ alloy volume expansion (**c**).

Table 2. Lattice constants BCC, BCT, and FCC phases in $V_{61}Cr_{24}Ti_{12}Ce_3$ alloy.

Phase	a (Å)	b (Å)	c (Å)	V (Å^3)
BCC	0.3031(1)	0.3031(1)	0.3031(1)	0.02785
BCT	0.3143(0)	0.3143(0)	0.3148(3)	0.03110
FCC	0.4250(3)	0.4250(3)	0.4250(3)	0.07678

Figure 3a,b show the hydrogen absorption plots of the pure powder bed and the compact bed with a diameter of 10 mm after five cycles. Both beds contain the same amount of alloy for testing. The first absorption reaction was conducted at 0 °C, and the rest of the absorption reactions were carried out at 20 °C. The powder bed shows the fast absorption kinetics at the first absorption cycle (Figure 3a), which reached saturation within 10 min with the absorption capacity of 3.5 wt.%. In the following four cycles, the powder bed could absorb hydrogen at ~2.0 wt.% within 10 min. In contrast, the compact bed shows slightly slower absorption kinetics at the first absorption cycle, which requires approximately 20 min to reach the saturation state with a capacity of 3.4 wt.%. In the subsequent four cycles, the powder bed displays much faster absorption kinetics with a hydrogen absorption amount of ~2.0 wt.% within 5 min. Furthermore, the compact bed still maintains its structure after five cycles, as shown in Figure 3c,d.

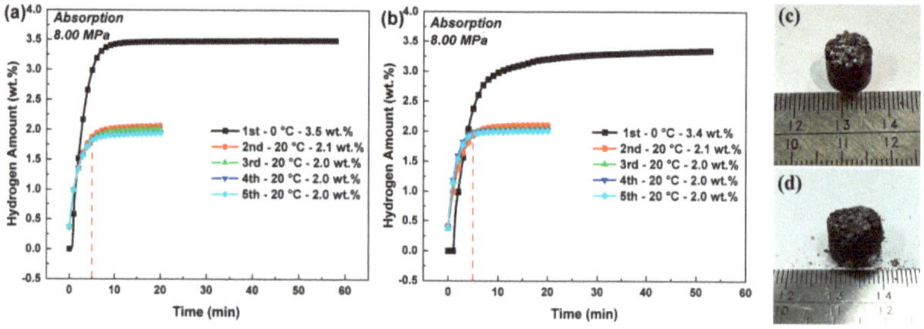

Figure 3. The hydrogen absorption plots of (**a**) the pure powder bed and (**b**) the compact bed within 5 cycles; the images of compact bed: (**c**) initial state and (**d**) after 5 cycles.

Figure 4a presents real-time data of position sensor for the $V_{61}Cr_{24}Ti_{12}Ce_3$ alloy powder bed during the five cycles of hydrogen absorption and desorption, which was measured using the self-made in situ expansion testing device shown in Figure 7. The time required for the powder bed to complete the hydrogen sorption reaction is 72, 49, 54, 58, and 59 min, for the first to fifth cycles, respectively. The powder bed was vibrated after the completion of each hydrogen absorption and desorption cycle. The volume expansion ratio in each cycle was calculated by comparing the height after hydrogen absorption of the current cycle with the height after hydrogen desorption in the initial cycle, as shown in Figure 4b. The alloy powder shows the highest expansion ratio of 131% at the first hydrogen absorption cycle, assigned to the phase evolution from BCC to FCC of $V_{61}Cr_{24}Ti_{12}Ce_3$. The expansion ratio decreases to 45% in the second cycle and stabilizes between 40% and 43% from the third to the fifth cycle. Figure 4c presents the particle size distribution of alloy powder after five hydrogen absorption/desorption cycles. The size of the majority of the powder decreases to below 0.355 mm, much smaller than the initial particle size of 3.35 mm.

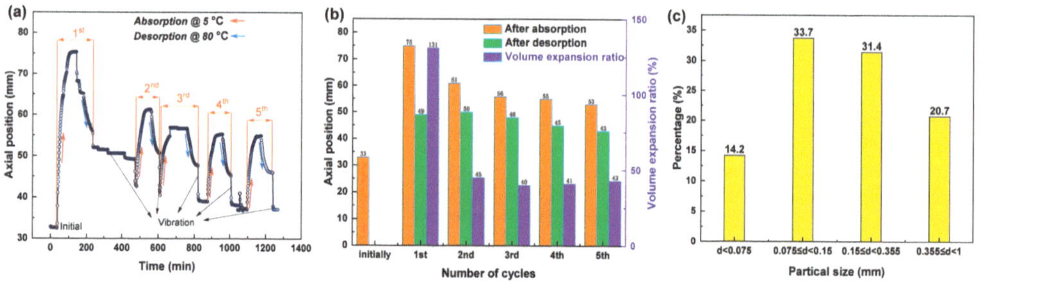

Figure 4. The height change (**a**) and volume expansion ratio (**b**) of the $V_{61}Cr_{24}Ti_{12}Ce_3$ alloy powder bed during the 5 cycles. The particle size (**c**) of $V_{61}Cr_{24}Ti_{12}Ce_3$ alloy powder after 5 cycles.

Figure 5a presents real-time data from the position sensor for compact bed during the five cycles. The compact bed was vibrated after the completion of each hydrogen absorption and desorption cycle. Figure 5b shows the corresponding positional information of the position sensor and the volume expansion ratio of each cycle. The expansion ratio of the first hydrogen absorption cycle is calculated to be 97%, a reduction of 34% compared to the alloy powder bed. The volume expansion ratio of the second circle decreases to 21% and stabilizes at 13% afterwards. Additionally, the durations required for the compact bed to complete the hydrogen sorption reaction are 91, 39, 44, 41, and 33 min for the first to the fifth cycle, respectively, which are significantly shortened compared to the values for the alloy powder bed in Figure 4a. Hence, the conversion of powder bed to compact bed accelerates the hydrogen absorption kinetics, probably attributed to improved effective thermal conductivity. Figure 5c,d show the initial state of the compact bed before and after five hydrogen sorption cycles, respectively. The shape of the compact bed was basically maintained.

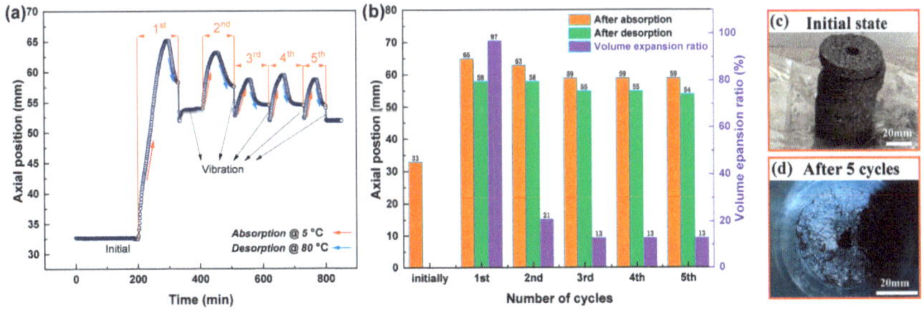

Figure 5. The height change (**a**) and volume expansion rate (**b**) of the compaction of $V_{61}Cr_{24}Ti_{12}Ce_3$ alloy powder during the 5 cycles. The state of composite hydrogen storage material before hydrogen absorption (**c**) and after 5 cycles (**d**).

For the two beds, the largest volume expansion was observed at the first cycle, which is attributed to lattice expansion during the phase transition of the $V_{61}Cr_{24}Ti_{12}Ce_3$ alloy from BCC to FCC structure. Since the second cycle, the volume expansion ratio of the powder bed almost stabilizes at 40–45%, accompanied with the phase transition from BCT to FCC. The compact bed shows a much smaller volume expansion ratio of 21% at the second cycle, which further decreases and stabilizes at 13% after the third cycle. The slight decrease in volume expansion ratio from the second to the third cycle may be attributed to continuous pulverization of $V_{61}Cr_{24}Ti_{12}Ce_3$ alloy powder or slow formation process of the stable silicone elastomers.

Note that after the completion of each cycle, the powder bed and compact bed were vibrated. The sensor positions after vibration are listed in Table 3. For the powder bed, sensor position after vibration decreases continuously from the first to the fifth cycle, implying that the bed became more compact after vibration probably owing to the pulverization of alloy powders. Differently, the sensor position of the compact bed changes slightly and keeps almost constant from the third to the fifth cycle, indicative of the formation of stable bed.

Table 3. The sensor position after vibration for each circle.

Sensor Movement Distance Caused by Vibration (mm)	1st	2nd	3rd	4th	5th
Powder bed	42.88	40.54	38.99	37.10	36.00
Compact bed	53.95	52.73	51.99	52.27	51.86

SEM and EDS analyses were conducted on the compact bed after five cycles, as illustrated in Figure 6. SEM and mapping results in Figure 6(a,a1,a2) reveal that graphite flakes and organosilicon gel are uniformly distributed on the surface of the alloy particles. Figure 6b–d demonstrate that the organic silica gel (PDMS) effectively binds the pulverized particles. Graphite has a lubricating effect, which could reduce the friction between particles [23]. Organic silica gel (PDMS) cured like ethylene-vinyl acetate copolymer is highly elastomeric, which not only secures the graphite flakes tightly to the alloy particles but also fixes the alloy particles in situ. Accordingly, the utilization of a combination of alloy particles, organic silica gel (PDMS), and graphite flakes in the form of a compact bed not only effectively mitigates the volume expansion of the alloy bed but also contributes to the immobilization of the pulverized alloy powders.

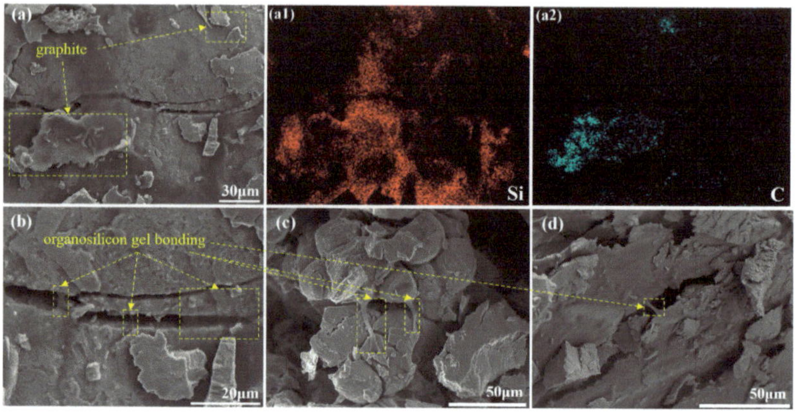

Figure 6. (**a–d**) The SEM images of the compact bed after 5 cycles and elemental mapping (**a1**,**a2**) of Si and C corresponding to figure (**a**).

3. Materials and Methods

The $V_{61}Cr_{24}Ti_{12}Ce_3$ alloy was prepared via floating melting with raw materials of V (purity: 99.9%), Ti (purity: 99.6%), Cr (purity: 99.6%) and Fe (purity: 99.9%). The as-cast $V_{61}Cr_{24}Ti_{12}Ce_3$ alloy was annealed at 1400 °C for 0.5 h under dynamic vacuum. The as-annealed alloy was crushed to a powder size of ~3.35 mm using a jaw crusher under nitrogen atmosphere.

The alloy powders were mixed with graphite flake (Macklin Biochemical Technology Co., Ltd., Shanghai, China) and organosilicon gel (Sylgard 184, Dow Corning, Midland, MI, America) with a mass ratio of 92%: 4%: 4%. The mixture was subjected to a pressing and molding process, with a pressure of 500 MPa maintained for a duration of one minute. Two types of compact bed, one with a diameter of 10 mm and another with an inner diameter of 10 mm and an outer diameter of 45 mm, were prepared. Prior to the commencement of cycling tests, the compact bed was stored at room temperature for $\geq$ 24 h in an Ar-circulated glove box to allow the silicone gel to fully cure and form an elastomer.

Alloy powders were loaded in a stainless-steel reactor for pressure–composition–temperature (PCT) measurement. To activate the alloy for hydrogen sorption, the alloy powders were evacuated for 1 h at room temperature, followed by the introduction of hydrogen at 7.5 MPa. The alloy powders could fast react with hydrogen at room temperature. The hydrogen absorption PCT curves were measured at 0, 18, and 50 °C. To measure the hydrogen desorption PCT curves, the alloy powders were fully hydrogenated at 0 °C under 7.5 MPa H_2, and subsequently, the temperature of the reactor was set to the targeting temperatures of 0, 18, and 50 °C, respectively. The measurement of hydrogen desorption PCT measurement started after the hydrogen pressure became stable. The X-ray diffraction (XRD) measurements were conducted using Cu Kα radiation (Rigaku, Tokyo, Japan), and the XRD data were analyzed using FullProf 2023 [24]. Microstructures was observed via scanning electron microscope (SEM, JSM-7800F, JEOL Ltd., Tokyo, Japan).

To investigate the actual expansion of the $V_{61}Cr_{24}Ti_{12}Ce_3$ alloy during the hydrogen absorption process, an in situ expansion testing device was designed and a corresponding testing platform was established, as depicted in Figure 7. The testing device contains the expansion test reactor, two water baths, and a flow and pressure controller. A steel plate together with a position sensor was placed on top of the sample. The sensor could measure the position of the steel plate, which reflects the volume change in the tested sample. Two water baths were maintained to offer cold and hot water for the hydrogen absorption and desorption processes, respectively. The hydrogen flows and pressures during the testing process could be controlled and monitored from the test platform which was equipped with flow meters and pressure sensors. Before testing, the alloy powder bed or compact bed was activated under vacuum for 2 h, followed by the introduction of 7.5 MPa H_2 at 5 °C. The hydrogen desorption was carried out under a dynamic vacuum at 80 °C. The hydrogen absorption and desorption cycle were repeated 5 times.

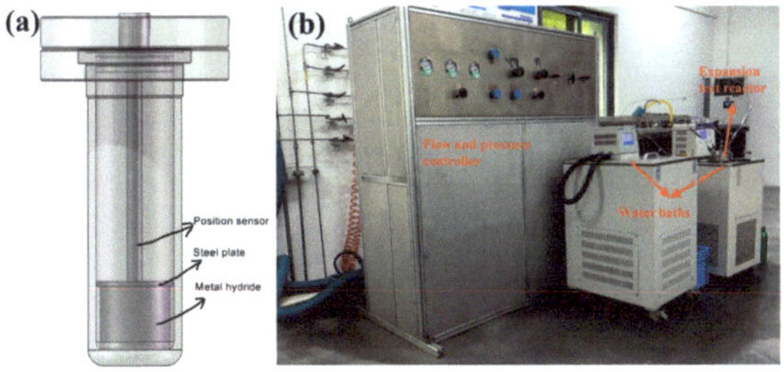

Figure 7. The schematic image of in situ expansion testing tank (**a**) and the test platform (**b**).

4. Conclusions

We have demonstrated the volume expansion behavior of the $V_{61}Cr_{24}Ti_{12}Ce_3$ alloy measured using the self-made volume expansion test device. Firstly, the maximum reversible capacity of $V_{61}Cr_{24}Ti_{12}Ce_3$ alloy is 2.51 wt.%, and the lattice expansion of $V_{61}Cr_{24}Ti_{12}Ce_3$ alloy in the saturated state of hydrogen absorption is found to be 37.85% and is analyzed via XRD test. The powder bed of the $V_{61}Cr_{24}Ti_{12}Ce_3$ alloy shows a large volume expansion ratio of 131% at the first hydrogen absorption cycle and 40~45% in the following four cycles. The compact bed, made of alloy powders, organic silicone gel (PDMS), and graphite flakes, shows a significantly smaller volume expansion ratio, which is 97% at the first cycle and 21% at the second cycle, and stabilizes at 13% in the following cycles. The highly elastic organic silicone gel (PDMS) serves to retain the graphite flakes on the alloy particles. Furthermore, the lubrication of the graphite flakes serves to reduce the interaction force between the powder particles. So, the addition of organic silicone gel and graphite flakes contributes to the formation of a stable bed of $V_{61}Cr_{24}Ti_{12}Ce_3$ alloy, which does not reduce the hydrogen absorption capacity but improves the hydrogen absorption kinetics after the second cycle.

Author Contributions: Data curation, Q.H. and H.K.; testing, Y.H. and Y.C.; writing—original draft preparation, M.L.; editing, Y.Y. All authors have read and agreed to the published version of the manuscript.

Funding: This study was financially supported by Sichuan Science and Technology Program (No. PG-PGFT-JFKF23-000009-0) and Xizang Autonomous Region Key R&D Program (XZ202401ZY0096).

Data Availability Statement: All data generated or analyzed during this study are included in this article.

Conflicts of Interest: The authors declare no conflicts of interest.

References

1. Chen, Z.; Ma, Z.; Zheng, J.; Li, X.; Akiba, E.; Li, H.W. Perspectives and challenges of hydrogen storage in solid-state hydrides. *Chin. J. Chem. Eng.* **2021**, *29*, 1–12. [CrossRef]
2. Dieterich, M.; Pohlmann, C.; Bürger, I.; Linder, M.; Röntzsch, L. Long-term cycle stability of metal hydride-graphite composites. *Int. J. Hydrogen Energy* **2015**, *40*, 16375–16382. [CrossRef]
3. Schlapbach, L.; Züttel, A. Hydrogen-storage materials for mobile applications. *Nature* **2001**, *414*, 353–358. [CrossRef]
4. Kong, H.Y.; Xie, Q.F.; Wu, C.L.; Wang, Y.; Chen, Y.G.; Li, H.W.; Yan, Y.G. Vanadium-based alloy for hydrogen storage: A review. *Rare Met.* **2024**, 6201–6232. [CrossRef]
5. Matsushita, M.; Monde, M.; Mitsutake, Y. Experimental formula for estimating porosity in a metal hydride packed bed. *Int. J. Hydrogen Energy* **2013**, *38*, 7056–7064. [CrossRef]
6. Cuscueta, D.J.; Silin, N.; Melnichuk, M. Stress reduction in a hydride container by the addition of a glidant agent. *Int. J. Hydrogen Energy* **2020**, *45*, 27452–27456. [CrossRef]

7. Luo, L.; Wu, C.; Yang, S.; Zhou, J.; Chen, Y.; Yang, F.; Xu, Y.; Liu, P. Decaying behaviors of $V_{40}(TiCr)_{51}Fe_8Mn$ hydrogen storage alloys with different particle sizes. *J. Alloys Compd.* **2015**, *645*, S178–S183. [CrossRef]
8. Herbrig, K.; Pohlmann, C.; Gondek, Ł.; Figiel, H.; Kardjilov, N.; Hilger, A.; Manke, I.; Banhart, J.; Kieback, B.; Röntzsch, L. Investigations of the structural stability of metal hydride composites by in-situ neutron imaging. *J. Power Sources* **2015**, *293*, 109–118. [CrossRef]
9. Wu, Z.; Yang, C.; Yan, Y.; Wang, Y.; Tang, X.; Chen, Y.; Li, J.; Wang, M.; Xie, Q.; Chen, Y.; et al. Effect of dehydrogenation depth on cyclic hydrogen desorp tion properties of $V_{40}Ti_{25.5}Cr_{26.5}Fe_8$ alloy. *J. Alloys Compd.* **2023**, *955*, 170036. [CrossRef]
10. Heubner, F.; Pohlmann, C.; Mauermann, S.; Kieback, B.; Röntzsch, L. Mechanical stresses originating from metal hydride composites during cyclic hydrogenation. *Int. J. Hydrogen Energy* **2015**, *40*, 10123–10130. [CrossRef]
11. Charlas, B.; Gillia, O.; Doremus, P.; Imbault, D. Experimental investigation of the swelling/shrinkage of a hydride bed in a cell during hydrogen absorption/desorption cycles. *Int. J. Hydrogen Energy* **2012**, *37*, 16031–16041. [CrossRef]
12. Wu, K.; Cai, B.; Fan, L.; Qin, L.; Chen, D.; Huang, Y. Stress measurement of $MlNi_{4.5}Cr_{0.45}Mn_{0.05}$ alloy during hydrogen absorption-desorption process in a cylindrical reactor. *Int. J. Hydrogen Energy* **2020**, *45*, 28175–28182. [CrossRef]
13. Gattia, D.M.; Jangir, M.; Jain, I.P. Behavior of Compacted Magnesium-Based Powders for Energy-Storage Applications. *Inorganics* **2020**, *8*, 54. [CrossRef]
14. Duan, W.; Du, J.; Wang, Z.; Niu, Y.; Huang, T.; Li, Z.; Pu, C.; Wu, Z. Strain variation on the reaction tank of high hydrogen content during hydrogen absorptiondesorption cycles. *Int. J. Hydrogen Energy* **2013**, *38*, 2347–2351. [CrossRef]
15. Gabriel, R.A.N.; Cesar, A.G.B.; Daniel, R.L.; Luiz, A.P. Polyetherimide-LaNi5 composite films for hydrogen storage applications. *Int. J. Hydrogen Energy* **2021**, *46*, 32767–32778. [CrossRef]
16. Shiraz, H.G.; Tavakoli, O. Investigation of graphene-based systems for hydrogen storage. *Renew. Sustain. Energy Rev.* **2017**, *74*, 104–109. [CrossRef]
17. Kazuya, K.; Yoshinori, K.; Hideaki, I. Development of large MH tank system for renewable energy storage. *Int. J. Hydrogen Energy* **2017**, *42*, 22475–22479. [CrossRef]
18. Tarasov, B.P.; Arbuzov, A.A.; Mozhzhuhin, S.A.; Volodin, A.A.; Fursikov, P.V. Composite materials with 2D graphene structures: Applications for hydrogen energetics and catalysis with hydrogen participation. *J. Struct. Chem.* **2018**, *59*, 830–838. [CrossRef]
19. Carsten, P.; Lars, R.; Siarhei, K.; Thomas, H.; Bernd, K. Magnesium alloy-graphite composites with tailored heat conduction properties for hydrogen storage applications. *Int. J. Hydrogen Energy* **2010**, *35*, 12829–12836. [CrossRef]
20. Strugova, D.V.; Zadorozhnyy, M.Y.; Berdonosova, E.A.; Yablokova, M.Y.; Konik, P.A.; Zheleznyi, M.V.; Semenov, D.V.; Milovzorov, G.S.; Padaki, M.; Kaloshkin, S.D.; et al. Novel process for preparation of metal-polymer composite membranes for hydrogen separation. *Int. J. Hydrogen Energy* **2018**, *43*, 12146–12152. [CrossRef]
21. Zheng, X.; Kong, H.Y.; Chu, D.S.; Hu, F.P.; Wang, Y.; Yan, Y.G.; Wu, C.L. Stress Reduction of a V-Based BCC Metal Hydride Bed Using Silicone Oil as a Glidant. *Inorganics* **2022**, *10*, 167. [CrossRef]
22. Marc, P.W.; Georgette, B.; Salieb, B.; Patrick, H. PDMS with designer functionalities-properties, modifications strategies, and applications. *Prog. Polym. Sci.* **2018**, *83*, 97–134. [CrossRef]
23. Warfsmann, J.; Puszkiel, J.A.; Passing, M.; Krause, P.S.; Wienken, E.; Taube, K.; Klassen, T.; Pistidda, C.; Jepsen, J. Applying wash coating techniques for swelling-induced stress reduction and thermal improvement in metal hydrides. *J. Alloys Compd.* **2023**, *950*, 169814. [CrossRef]
24. Rodríguez-Carvajal, J. Recent advances in magnetic structure determination by neutron powder diffraction. *Phys. B Condens. Matter* **1993**, *192*, 55–69. [CrossRef]

inorganics MDPI

Article

Enhancements in Hydrogen Storage Properties of Magnesium Hydride Supported by Carbon Fiber: Effect of C–H Interactions

Quan Yang [1], Xiansong Jia [1], Zhikang Qin [1], Xiaoli Ding [1,*] and Yongtao Li [1,2]

1 Key Laboratory of Green Fabrication and Surface Technology of Advanced Metal Materials of Ministry of Education, School of Materials Science and Engineering, Anhui University of Technology, Maanshan 243002, China; yq20001214@163.com (Q.Y.); jiaxiansongyj@163.com (X.J.); ahut.zhikangqin@gmail.com (Z.Q.); toni-li@163.com (Y.L.)
2 Key Laboratory of Efficient Conversion and Solid-State Storage of Hydrogen & Electricity of Anhui Province, Anhui University of Technology, Maanshan 243002, China
* Correspondence: dingxiaoli@ahut.edu.cn

Abstract: Carbon-based materials with excellent catalytic activity provide new ideas for the development of magnesium-based hydrogen storage. C-H bonding interactions may play a key role in performance improvement. In this work, we comprehensively compare the magnesium-carbon cloth composites (CC) prepared by method of dry ball milling and wet impregnation. The results were that the hydrogen release activation energy (E_a) of MgH_2@CC composites prepared by wet immersion method was 175.1 ± 19.5 kJ·mol^{-1}, which was lower than that of pure MgH_2 (E_a = 213.9 ± 6.4 kJ·mol^{-1}), and the activation energy of MgH_2-CC composites prepared by ball milling method was 137.3 ± 8.7 kJ·mol^{-1}, which provided better results. The kinetic enhancement should be attributed to C-H interactions. The presence of carbon carriers and electron transfer to reduce the activation energy of Mg-H bond fracture. These results will provide further insights into the promotion of hydrogen ab-/desorption from metal hydrides.

Keywords: hydrogen storage; magnesium-carbon composites; preparation methods; carbon-hydrogen bond

Citation: Yang, Q.; Jia, X.; Qin, Z.; Ding, X.; Li, Y. Enhancements in Hydrogen Storage Properties of Magnesium Hydride Supported by Carbon Fiber: Effect of C–H Interactions. *Inorganics* **2024**, *12*, 273. https://doi.org/10.3390/inorganics12110273

Academic Editor: Maurizio Peruzzini

Received: 29 August 2024
Revised: 20 October 2024
Accepted: 22 October 2024
Published: 23 October 2024

1. Introduction

Energy is a potential driver of human production and social activity, and the depletion of fossil fuels and global climate change have prompted the search for clean energy carriers [1–3]. As a clean, efficient and abundant secondary energy source, hydrogen is considered one of the ideal energy carriers for storing and transferring energy in the future [4–8]. However, gaseous and liquid hydrogen storage have limitations in terms of safety and storage density for various applications [9–12]. Solid-state hydrogen storage, especially in the form of metal hydrides, offers the most compact and safe technology for storing hydrogen, and among the various solid-state hydrogen storage materials, MgH_2 is valuable for research due to its high hydrogen storage capacity (7.6%) and abundant magnesium metal resources (2.3%) [1,13–19]. However, the high thermodynamic stability and slow reaction kinetics of MgH_2 make it difficult to meet the requirements of practical applications [20–23].

Nanosizing [13,24–27] has been shown to be one of the effective ways to improve the thermodynamics/kinetics of the MgH_2 reaction, and chemical reduction and hydrogenation [28–31] have attracted interest due to their low cost and portable operating conditions [28,32]. Liu et al. [33] used carbon aerogel (CA) as a precursor, wet impregnated with dibutyl magnesium solution and hydrogenated to prepare Mg-CA composites. The average particle size of the Mg nanoparticles was 19.3 nm, and the enthalpies of hydrogenation and dehydrogenation were −65.1 and 68.8 kJ·mol^{-1}. However, the hydrogen uptake kinetics of the Mg–CA composites were slower than those of the LaMg-containing ones prepared by the arc plasma method, and the kinetics need to be further improved.

In recent decades, a variety of novel carbon isomers have been synthesized, such as nanotubes [26,34], nanofibers [28,35], graphene [36,37], fullerenes [38,39], Mxenes [40,41], and carbon nanospheres [42,43]. Due to carbon's light weight, large specific surface area, thermal stability, load-bearing capacity, and carbon atom substitutability, the carbon composites studied those all contribute to enhance the kinetics of hydrogen uptake and/or desorption [44]. Carbon-based materials have been investigated as catalytic materials with the promise of further enhancing the MgH_2 kinetics. Andrey Lakhnik et al. [45] prepared MgH_2/C composites of elemental magnesium with graphite powder using high-energy ball milling method with particle sizes predominantly in the range of 10–30 μm. The materials have a minimum activation energy for hydrogen absorption of 65 $kJ \cdot mol^{-1}$ and a hydrogen capacity of ~4.3 wt%, which gives them an excellent kinetic performance. It was also found that the activation energy of the absorption process monotonically decreases with the increase of the graphite powder surface area during sample synthesis. Carbon composites also have excellent cycling performance, as they affect hydrogen storage kinetics primarily by preventing particle agglomeration and improving thermal conductivity [46]. R.M. Carson et al. [47] found that Mg-5 wt% C retains a capacity of 5.5 wt% H_2 after 1000 cycles and is suitable for most commercial applications.

The surface of carbon fiber has a high degree of surface defects and abundant growth sites, which can be used as a good growth carrier. Meanwhile, carbon fibers can improve cycling performance and inhibit MgH_2 agglomeration due to their hydrogen storage absorption/desorption cycling stability [8,48–50]. Herein, carbon fiber materials were selected to be combined with Mg-based hydrogen storage, and two different preparation processes, wet impregnation and dry ball milling, were designed in order to investigate the effect of the preparation processes on the hydrogen storage properties of magnesium-carbon fiber composites. It was found that the two preparation processes had a significant effect on the enhanced hydrogen storage kinetic properties of MgH_2. The introduction of carbon fiber carriers caused the carbon-hydrogen bond to weaken the interaction between hydrogen atoms and magnesium atoms, thus promoting the migration and release of hydrogen atoms.

2. Results and Discussion

2.1. Structural Features and Microstructures Induced by Different Processes

2.1.1. Structural Features of MgH_2@CC and MgH_2-CC Composites

We chose carbon fiber materials combined with magnesium-based hydrogen storage and designed two different preparation processes, wet impregnation and mechanical milling, as shown in Figure 1, with details as discussed in experimental sections. First, in order to investigate the formation process of MgH_2 on the carbon cloth, XRD tests were performed on the samples. The crystal structure analysis of MgH_2@CC and MgH_2-CC composites by XRD is shown in Figure 2a. The MgH_2@CC composite shows the presence of two broadened diffusion diffraction peaks near 25.2° and 43.1° that correspond to the characteristic diffraction peaks for carbon. The impregnated and vacuum-treated samples showed characteristic peaks of $MgBu_2$ near 12.4°, 19.2°, and 21.1°, respectively, indicating that the heptane solution had been completely removed and $MgBu_2$ was successfully loaded on the surface of the carbon cloth. The disappearance of the $MgBu_2$ phase and the appearance of the MgH_2 phase on the surface of the carbon cloth after the hydrogenolysis reaction imply that the hydrogenolysis reaction is complete. At the same time, no other heterogeneous phases appeared, suggesting that the dibutyl magnesium was completely transformed into MgH_2.

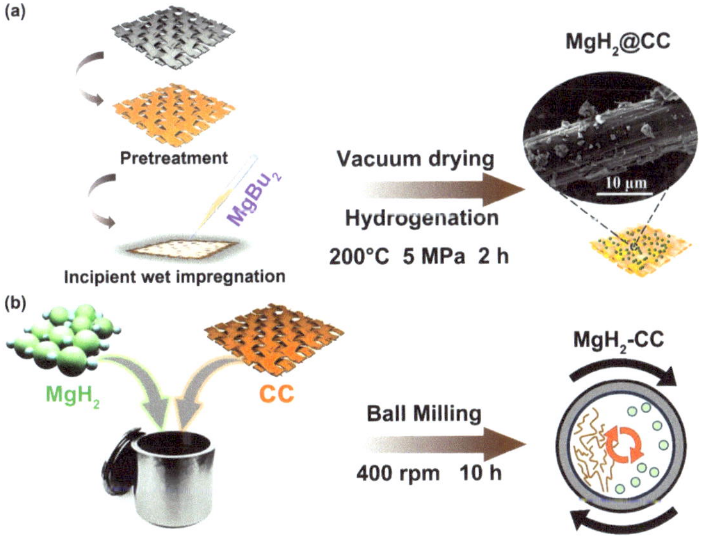

Figure 1. Schematic illustration of preparation process of (**a**) MgH_2@CC and (**b**) MgH_2-CC.

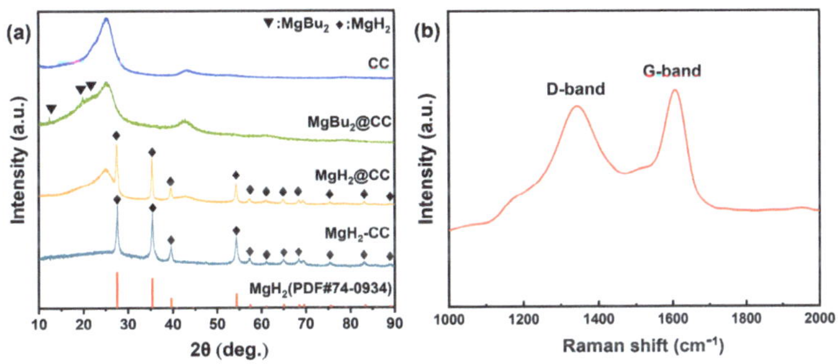

Figure 2. (**a**) XRD patterns of MgH_2@CC and MgH_2-CC composites and (**b**) Raman pattern of MgH_2-CC composites.

In addition, the MgH_2-CC composites prepared by ball milling showed characteristic peaks belonging to MgH_2 at 27.8°, 35.6°, and 39.7°, but no more obvious carbon peaks were detected, which may be attributed to the fact that the carbon fibers were uniformly dispersed in the MgH_2 matrix after ball milling. In order to demonstrate the successful loading of carbon fibers with MgH_2 in the MgH_2-CC composites, Raman tests were carried out on the MgH_2-CC composites. As shown in Figure 2b, typical Raman peaks belonging to the D-band and G-band of the graphite phase, respectively, appeared at wave numbers of 1350 cm^{-1} and 1590 cm^{-1}. D-band represents the sp3 defects in the carbon and the G-band represents the E2g vibration of the carbon due to the sp2 hybridization. The results also confirmed that the carbon fibers were successfully loaded with MgH_2 in the MgH_2-CC composites. The samples were then analyzed for particle size. The Scheller formula can be used to calculate the crystal size from the XRD data:

$$D = \frac{K\lambda}{\beta \cos\theta} \quad (1)$$

where D is the grain size perpendicular to the observed grain surface; K is a constant, usually taking the value of 0.9; λ is the wavelength of X-rays, β is the radian of the half-height width of the sample's diffraction peak, and θ is the diffraction angle. According to the calculation results, the average grain size of MgH_2@CC is 35.75 nm, the average grain size of MgH_2-CC is 20.01 nm, and the average grain size of commercial MgH_2 is 181.56 nm, which are both improved.

2.1.2. Distinct Morphologies of MgH_2@CC and MgH_2-CC Composites

The microstructure of MgH_2@CC composites obtained by hydrogenolysis of organic magnesium was observed by scanning electron microscopy and is shown in Figure 3a–c. It can be clearly observed that the MgH_2 particles are uniformly distributed on the surface of the carbon fibers, and the high-density defects on the surface of the carbon cloth provide the growth sites for MgH_2, which grows along the direction of the rod-shaped carbon fibers. It was also observed that some of the MgH_2 on the surface of the carbon fibers was deposited due to physical changes during the impregnation process, but the MgH_2 particles on the surface still showed a diffuse distribution. The microstructure of MgH_2-CC composites obtained by ball milling is shown in Figure 3d–f. It can be clearly seen in the image that the MgH_2 particles and carbon fibers are diffusely distributed on the surface of the conductive adhesive, and the particles do not show an obvious aggregation phenomenon. Compared with MgH_2@CC composites, the particles are more uniformly dispersed, and it is easier to form a uniform carbon fiber loading system by the ball milling method, while the dispersed carbon after ball milling can inhibit the agglomeration of MgH_2 particles.

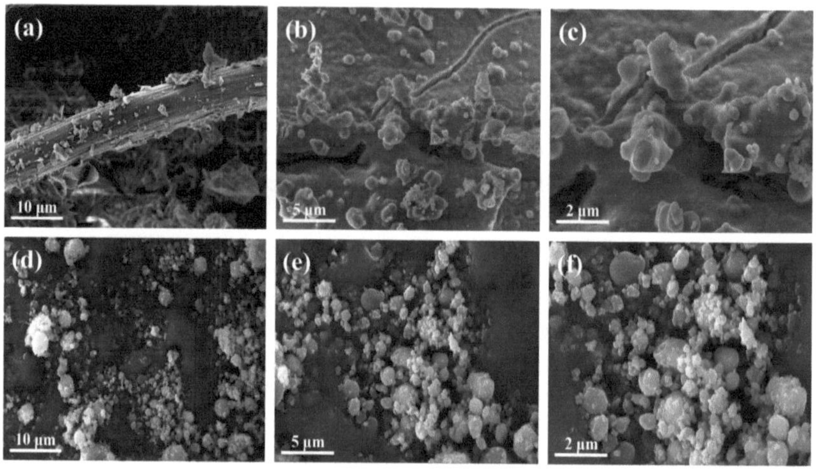

Figure 3. SEM images of (**a–c**) MgH_2@CC composites and (**d–f**) MgH_2-CC composites.

2.2. *Improved Hydrogen Storage Performance*

2.2.1. Kinetic Properties of MgH_2@CC and MgH_2-CC Composites

In order to investigate the kinetic properties of MgH_2@CC composites obtained after hydrolysis of organic magnesium, the composites were heated to 500 °C, with a heating rate of 3 °C/min, 5 °C/min, 8 °C/min, and 10 °C/min, and their DSC curves were tested. As a comparison, the DSC curves of commercially pure MgH_2 and BM-MgH_2 were tested under the same conditions. For the dehydrogenation, the exothermic decomposition of $MgH_2 \rightarrow Mg$ is accompanied by the exothermic process during the temperature rise stage, and the corresponding exothermic peaks will appear in the DSC curve. As shown in Figure 4, the peak exothermic temperatures of MgH_2-CC, MgH_2@CC, BM-MgH_2, and pure MgH_2 are 332 °C, 358 °C, 371 °C, and 449 °C, respectively. The peak hydrogen release temperatures of MgH_2-CC and MgH_2@CC composites were reduced by 117 °C and 91 °C, respectively, compared with that of pure MgH_2, indicating that the introduction of carbon fiber carriers can improve the kinetic properties of MgH_2 and reduce the peak hydrogen release temperature significantly.

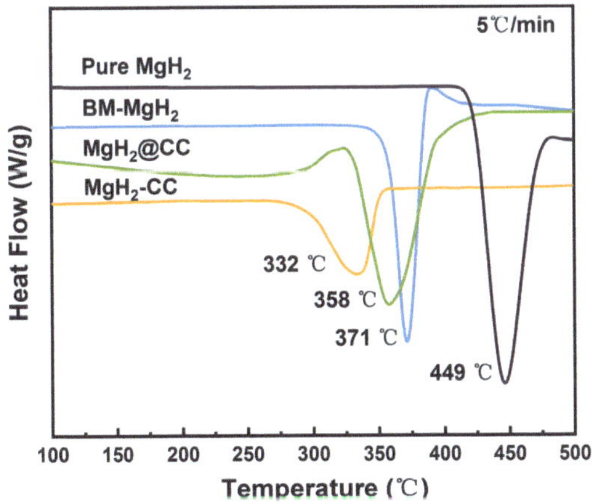

Figure 4. DSC curves of pure MgH_2, BM-MgH_2, MgH_2@CC and MgH_2-CC.

To further investigate the hydrogen release properties of MgH_2@CC and MgH_2-CC composites, Figure 5a–h shows the DSC curves of Pure MgH_2, BM-MgH_2, MgH_2@CC, and MgH_2-CC. It can be observed that both MgH_2@CC and MgH_2-CC show a significant decrease in the peak heat uptake temperature at different heating rates. The exothermic activation energy is also an important measure of the kinetic performance, and in general, the exothermic activation energy can be calculated based on the peak temperature of heat absorption and the rate of warming in combination with the Kissinger equation:

$$\ln\left(\frac{\beta}{T_P^2}\right) = A - \frac{E_a}{RT_P} \quad (2)$$

where β is the rate of temperature increase, T_p is the peak temperature of heat absorption, and A is a linear constant. The peak heat absorption temperatures of MgH_2@CC were 348.4 °C, 358.6 °C, 364.5 °C, and 372.1 °C at 3 °C/min, 5 °C/min, 8 °C/min, and 10 °C/min, respectively, and the scatter plots of $\ln(\beta/T_p{}^2)$ versus $1000/T_p$ were made subsequently, as shown in Figure 5b. Based on the slope of the straight line in the fitting result, the hydrogen release activation energy of MgH_2@CC can be obtained as ~175.1 kJ·mol^{-1}, which is 38.8 kJ·mol^{-1} lower than that of commercially pure MgH_2 (213.9 kJ·mol^{-1}), suggesting that the MgH_2@CC composites obtained by hydrolysis of organomagnesium improve their hydrogen release properties under the synergistic effect of nanoeffects and the carbon fiber carrier. Meanwhile, the hydrogen release properties of MgH_2-CC composites and BM-MgH_2 were investigated, and Figure 5e–h shows the DSC curves of MgH_2-CC composites and BM-MgH_2 and the straight line of $\ln(\beta/T_p{}^2)$ fitted to $1000/T_P$. It can be observed that the exothermic peak temperatures of MgH_2-CC composites are 321.3 °C, 332.8 °C, 341.5 °C, and 346.6 °C, respectively, and the activation energy of hydrogen release is ~137.3 kJ·mol^{-1}, which is also significantly improved compared with that of the ball-milled MgH_2 samples (~176.4 kJ·mol^{-1}).

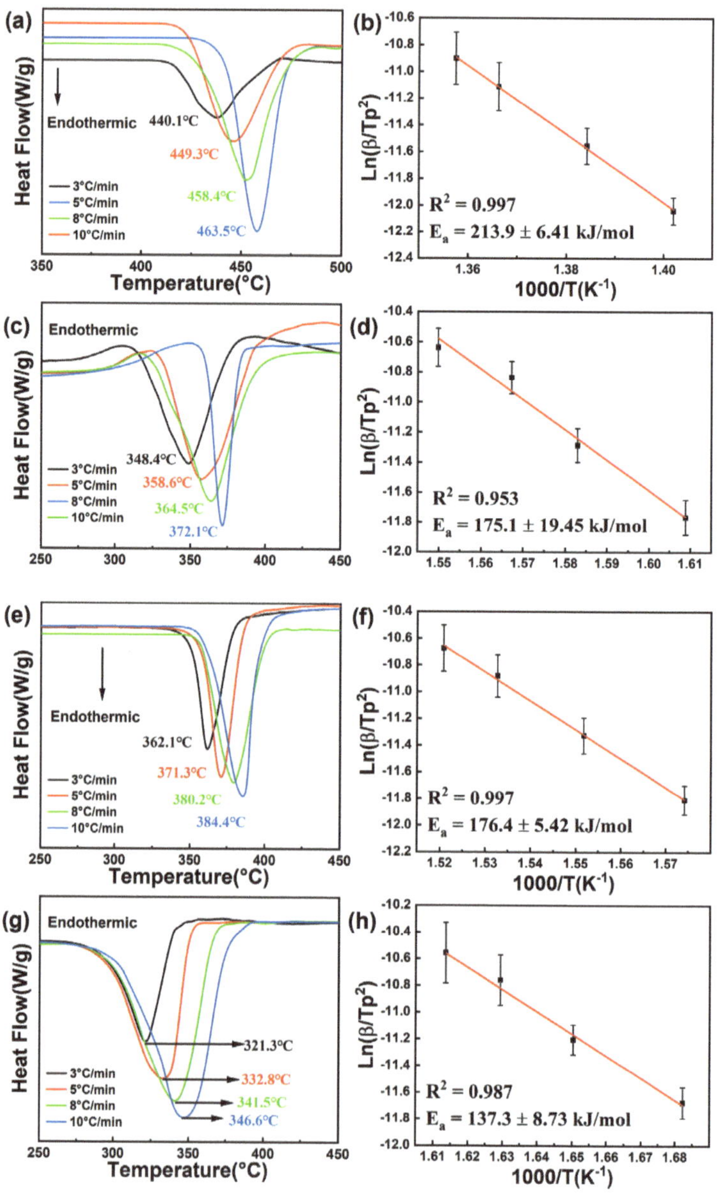

Figure 5. DSC curves and Arrhenius equation of (**a**,**b**) pure MgH_2, (**c**,**d**) MgH_2@CC, (**e**,**f**) BM-MgH_2 and (**g**,**h**) MgH_2-CC.

2.2.2. Thermodynamic Properties of MgH_2@CC and MgH_2-CC Composites

Thermodynamic tests were based on the equilibrium pressure of the desorption platform in PCI at different temperatures. The enthalpy (ΔH) and entropy (ΔS) of deuterium desorption were determined according to the van't Hoff equation:

$$\ln\left(\frac{P_e}{P_0}\right) = \left(\frac{\Delta H}{RT}\right) - \left(\frac{\Delta S}{R}\right) \tag{3}$$

where P_e is the equilibrium desorption pressure, P_0 is the standard atmospheric pressure, R is the gas constant, and T is the temperature in Kelvin.

To investigate the thermodynamic properties of MgH_2@CC and MgH_2-CC composites, as shown in Figure 6a–d, P-C-T tests were performed at 275 °C, 300 °C, 325 °C, and 350 °C, and the hydrogen absorption and release plateau pressures were obtained for the corresponding temperatures. The plateau pressures (P_{eq}) at different temperatures are shown in Table 1, and each plateau pressure is positively correlated with the temperature; ΔH and ΔS were calculated according to the Van't Hoff Equation (3). Based on the slope and intercept of the straight line in the fitting results, the enthalpy changes of hydrogen release (ΔH_{des}) and entropy change of hydrogen release (ΔS_{des}) of MgH_2@CC composites are 73.5 kJ·mol^{-1} and 140.5 J·mol^{-1}·K^{-1}, respectively, and the enthalpy change of hydrogen absorption (ΔH_{abs}) and entropy change of hydrogen absorption (ΔS_{abs}) were −71.7 kJ·mol^{-1} and −132.2 J·mol^{-1}·K^{-1}, respectively. The enthalpy changes of hydrogen release ΔH_{des} and entropy change of hydrogen release ΔS_{des} of MgH_2-CC composites were 73.0 kJ·mol^{-1} and 140.5 J·mol^{-1}·K^{-1}, respectively, and the enthalpy change of hydrogen absorption ΔH_{abs} and entropy change of hydrogen absorption ΔS_{abs} were −72.8 kJ·mol^{-1} and −136.2 J·mol^{-1}·K^{-1}, respectively. Compared with the commercially pure MgH_2 (ΔH_{des} = 76 kJ/mol^{-1} and ΔS_{des} = 134.1 J·mol^{-1}·K^{-1}), the thermodynamic properties have not been improved, indicating that reducing the size of the MgH_2 particle size and preventing MgH_2 particle aggregation is more likely to be reflected in the improvement of kinetic properties.

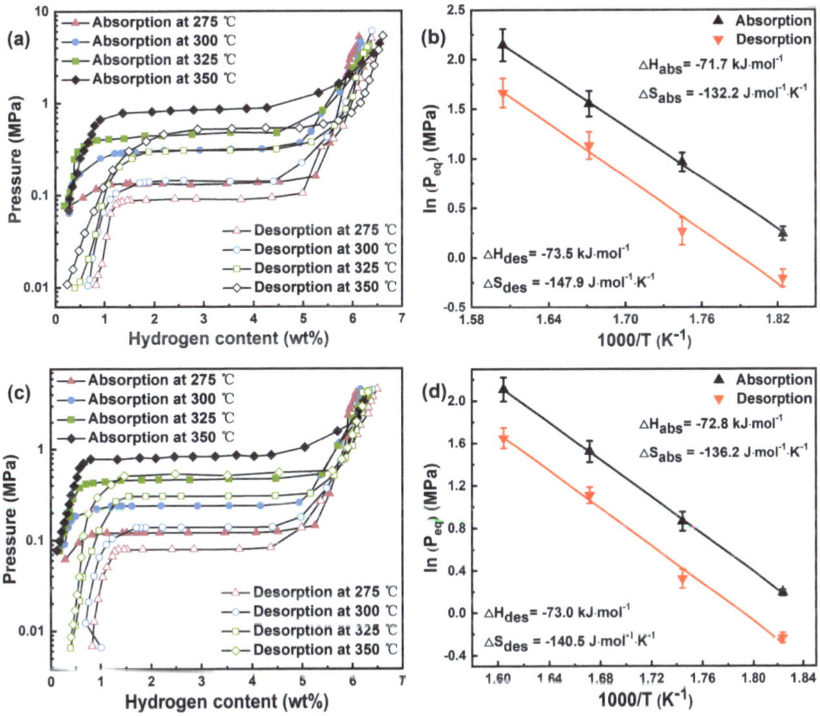

Figure 6. P-C-T curves and Van't Hoff equation of (**a**,**b**) MgH_2@CC composites and (**c**,**d**) MgH_2-CC composites.

Table 1. Hydrogen absorption and desorption plateau of MgH_2@CC and MgH_2-CC composite under different temperature.

Temperature (°C)	Absorption Plateau (MPa)		Desorption Plateau (MPa)	
	MgH_2@CC	**MgH_2-CC**	**MgH_2@CC**	**MgH_2-CC**
350	0.853	0.823	0.526	
325	0.472	0.458	0.310	0.303
300	0.262	0.237	0.130	0.138
275	0.127	0.121	0.081	0.079

2.3. Mechanism Understanding

In order to investigate the reason for the enhanced hydrogen release properties of MgH_2@CC composites, Fourier Transform Infrared Spectroscopy (FTIR) was conducted to analyze the material; the results are shown in Figure 7a. Among them, the peaks in the 1150–1500 cm^{-1} region correspond to the Mg-H stretching bands, the peaks in the 759–1000 cm^{-1} region correspond to the Mg-H bending bands, and the peaks in the 1450–1550 cm^{-1} region correspond to the O-H bonding, which means that the material undergoes partial oxidation. The peaks in the 2850–2960 cm^{-1} region correspond to the C-H bonding, which means that the carbon fiber matrix interacts with MgH_2 at the hydrogen release stage; this implies that the carbon fiber matrix interacts with MgH_2 during the hydrogen release phase [51–53]. When the hydrogen absorption reaction occurs, the carbon in magnesium-carbon composites adsorbs hydrogen from H_2 atmosphere or MgH_2 to form sp^2 C-H and sp^3 C-H bands [54]. When the hydrogen release reaction occurs, the carbon becomes a carrier for H atom transfer due to the C-H bond [55,56], the high temperature energy will drive the C-H bond to break, and the broken carbon bond will act as a carrier for H transfer and reduce the activation energy of the Mg-H bond, which will promote MgH_2 to overcome the hydrogen release energy barrier, similar to the photocatalytic Ni/TiO_2 composites, where the electrons are excited by the visible light and transferred from Ni to Ti, which drives the photocatalytic hydrogen production [57]. The activation energy is reduced, which improves the hydrogen storage performance of MgH_2@CC composites. In contrast, in MgH_2-CC composites, Mg has more binding sites with C. The C-H bonds and high-energy ball mills synergize with each other to provide more reaction sites and energy for hydrogen uptake/desorption, and thus produce better catalytic effects. The schematic diagram of the desorption enhancement mechanism of magnesium-carbon composites is shown in Figure 7b.

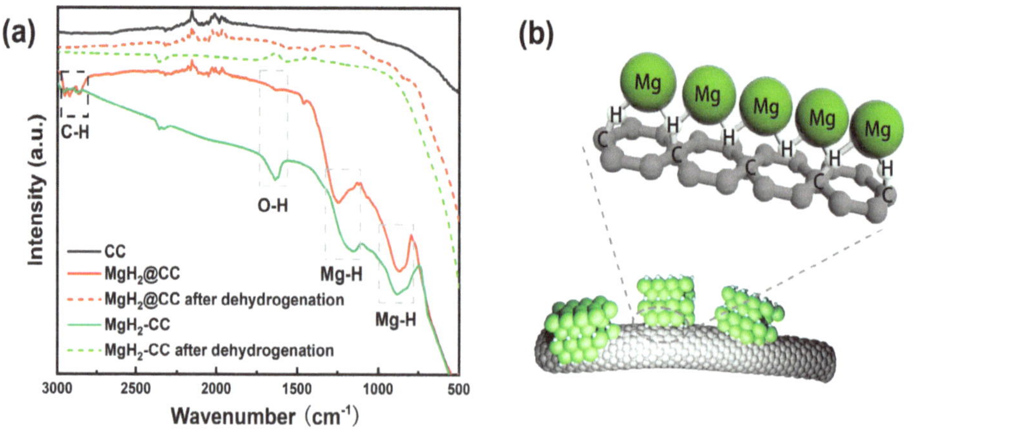

Figure 7. (**a**) FTIR images of MgH_2@CC composite and MgH_2-CC composites and (**b**) Schematic diagram of the enhanced mechanism for desorption of magnesium-carbon composites.

3. Experimental

3.1. Material Preparation

Carbon fiber cloth (CDD, 200 × 200 mm, Yongfu Electrochemical, Nanajing, China) was purchased and further processed. Dibutyl magnesium (1.0 mol/L in heptane, Aladdin, solution), MgH_2 (>95% from Mg Power Technology Co., Ltd., Shanghai, China), and anhydrous ethanol (C_2H_5OH, AR, Anhui Tiandi High Purity, Anqing, Anhui) were purchased and directly used as received without further purification. First, as shown in Figure 1a, the carbon fiber cloth was pretreated. The carbon fiber cloth was rinsed with ethanol, dried in an oven, and then placed in a stainless-steel autoclave and heated at 300 °C under 2 MPa hydrogen atmosphere for 3 h to remove hydrophilic groups and other impurities. In a sealed glove box filled with argon gas, weigh 0.43 g of pretreated carbon fiber cloth into a beaker and draw up a total of 3 mL of dibutyl magnesium solution using a syringe with a volume capacity of 5 mL. The absorbed solution was added to the surface of the carbon fiber cloth with 0.5 mL drops at a time, and after the carbon fiber cloth was completely impregnated, another drop was added until 3 mL was added. Subsequently, the beaker was placed in a vacuum chamber and vacuum dried for 12 h to remove the heptane solvent to obtain $MgBu_2$@CC. Finally, the sample was placed in a stainless-steel sample chamber and kept at 5 MPa hydrogen pressure and 200 °C for 2 h. The final sample obtained was named as MgH_2@CC composite. Subsequently, the sample was vacuum-dried in a vacuum oven at 80 °C for 6 h to obtain the carbon fiber-loaded precursor composite. The equation of the reaction was as follows:

$$(C_4H_9)_2Mg + 2H_2 \rightarrow MgH_2 + 2C_4H_9 \tag{4}$$

The loading of MgH_2 in MgH_2@CC can be calculated to be ~24.2 wt% from the before and after mass difference.

MgH_2-CC composites were prepared as shown in Figure 1b. In an argon glove box, pure MgH_2 and carbon fiber cloth were weighed according to the mass percentage of 75:25 and put into a stainless-steel ball milling jar with a ball to material ratio of 40:1. The stainless-steel balls were DECO-304-B, with diameters of 5 mm, 6 mm, 8 mm, and 10 mm, respectively. The ball milling time was 10 h. The ball milling speed was 400 rpm, and the interval between each 30-min revolution of the ball mill was 5 min. Ball milled MgH_2 (BM-MgH_2) samples were prepared in the same way described above.

3.2. Material Characterization

The phase composition of the composites was analysed by using an X-ray diffractometer (XRD, Mini Flex 600, Rigaku, Japan) with Cu Kα radiation (λ = 0.154056 nm) at 40 kV and 15 mA. The 2θ angle ranged from 10° to 90°with increments of 0.02°; samples were placed in a hermetically sealed XRD sample stage and sealed with polymer tape. The morphologies and microstructures of the samples were examined by using scanning electron microscopy (SEM, MIRA3 XMU, Tescan). We analysed the elemental distribution in the local region of the samples by EDS (Xplore30, OXFORD) attached to SEM, with magnification rates ranging from 200 to 5000. For SEM samples, ethanol or organic magnesium was used as the dispersant, and a small amount of the solution was taken and dropped on the conductive adhesive after sonication. SEM tests were performed at magnifications of 5 K, 10 K, and 20 K, with a 5 kv energy. Thermogravimetric analysis (TG, DTG-60H, SHIMADZU) was performed in the range of 30~700 °C with a heating rate of 10 °C/min to verify the loaded composition, Weigh about 5 mg of the sample in the glove box, place it in an aluminium crucible with a lid to seal the aluminium crucible and then take it out and put it into the DTG-60H instrument for testing. Maintain the argon gas environment at 0.2 Mpa during the testing process with a flow rate of 300 mL/min, and determine the peak value by selecting the beginning and end intervals of the melting peak in the DSC test. Raman spectra (HORIBA LabRAM HR Evolution spectrometer) were measured and processed with a laser wavelength of 532 nm. Fourier Transform Infrared Spectroscopy (FTIR) was conducted to analyse the material with a detection wavelength range of 500~3000 cm^{-1}. The samples were mixed with KBr and then pelletized in the argon-filled glove box.

The thermodynamic properties were evaluated by measuring the pressure−composition isotherms (PCI) on an automatic Sievert-type apparatus. Prior to the PCI measurements, we weighed 0.1–0.5 g of sample and heated it at 300 °C under vacuum conditions and activated it through two hydrogen sorption cycles consisting of a 0.5 h absorption under 3.0 MPa of H_2 gas and a 0.5 h desorption under vacuum. All the tests were performed under initial pressures of 0.0001 MPa for desorption and 5 MPa hydrogen pressure for absorption.

4. Conclusions

In summary, two carbon fiber-loaded MgH_2 hydrogen storage composites were successfully prepared by hydrogenation of organic magnesium and ball milling. Both the obtained MgH_2@CC and MgH_2-CC composites show significantly improved kinetics, with the MgH_2-CC composite being better in terms of preparation cost and kinetics. Our experimental results reveal that C-H bonding interactions in the two composite hydrogen storage materials are responsible for promoted hydrogen absorption/desorption from hydrides. These results provide further insights into promoting the hydrogen ab-/desorption of metal hydrides. Unfortunately, the thermodynamics were not improved, and we will work on the improvement of thermodynamic properties in subsequent work.

Author Contributions: Conceptualization, Q.Y. and X.D.; Methodology, X.J.; Validation, Z.Q.; Formal analysis, Q.Y.; Investigation, X.J.; Writing—original draft, Q.Y.; Writing—review & editing, X.D.; Visualization, Z.Q.; Supervision, Y.L.; Project administration, Y.L. All authors have read and agreed to the published version of the manuscript.

Funding: This work was financially supported by the National Natural Science Foundation of China (No. 52471027), the Natural Science Foundation of Anhui Province (No. 2308085QB50), the Natural Science Foundation of Anhui Provincial Education Department (Nos. KJ2021A0393, 2022AH020033 and 2022AH010025).

Data Availability Statement: Data is contained within the article.

Conflicts of Interest: The authors declare no conflicts of interest.

References

1. Zhang, J.; Li, Z.; Wu, Y.; Guo, X.; Ye, J.; Yuan, B.; Wang, S.; Jiang, L. Recent Advances on the Thermal Destabilization of Mg-Based Hydrogen Storage Materials. *RSC Adv.* **2019**, *9*, 408–428. [CrossRef] [PubMed]
2. Bolarin, J.A.; Zou, R.; Li, Z.; Munyentwali, A.; Zhang, Z.; Cao, H. Recent Path to Ultrafine Mg/MgH_2 Synthesis for Sustainable Hydrogen Storage. *Int. J. Hydrogen Energy* **2024**, *52*, 251–274. [CrossRef]
3. Yang, F.; Wang, J.; Zhang, Y.; Wu, Z.; Zhang, Z.; Zhao, F.; Huot, J.; Grobivć Novaković, J.; Novaković, N. Recent Progress on the Development of High Entropy Alloys (HEAs) for Solid Hydrogen Storage: A Review. *Int. J. Hydrogen Energy* **2022**, *47*, 11236–11249. [CrossRef]
4. Baum, Z.J.; Diaz, L.L.; Konovalova, T.; Zhou, Q.A. Materials Research Directions Toward a Green Hydrogen Economy: A Review. *ACS Omega* **2022**, *7*, 32908–32935. [CrossRef]
5. Lu, Y.; Kim, H.; Sakaki, K.; Hayashi, S.; Jimura, K.; Asano, K. Destabilizing the Dehydrogenation Thermodynamics of Magnesium Hydride by Utilizing the Immiscibility of Mn with Mg. *Inorg. Chem.* **2019**, *58*, 14600–14607. [CrossRef]
6. Qin, Z.-K.; He, L.-Q.; Ding, X.-L.; Si, T.-Z.; Cui, P.; Li, H.-W.; Li, Y.-T. Liquid Channels Built-In Solid Magnesium Hydrides for Boosting Hydrogen Sorption. *Inorganics* **2023**, *11*, 216. [CrossRef]
7. Chen, Y.; Habibullah; Xia, G.; Jin, C.; Wang, Y.; Yan, Y.; Chen, Y.; Gong, X.; Lai, Y.; Wu, C. Hydrogen Storage Properties of Economical Graphene Materials Modified by Non-Precious Metal Nickel and Low-Content Palladium. *Inorganics* **2023**, *11*, 251. [CrossRef]
8. Xu, Y.; Zhou, Y.; Li, Y.; Ding, Z. Carbon-Based Materials for Mg-Based Solid-State Hydrogen Storage Strategies. *Int. J. Hydrogen Energy* **2024**, *69*, 645–659. [CrossRef]
9. Floriano, R.; Zepon, G.; Edalati, K.; Fontana, G.L.B.G.; Mohammadi, A.; Ma, Z.; Li, H.-W.; Contieri, R.J. Hydrogen Storage in TiZrNbFeNi High Entropy Alloys, Designed by Thermodynamic Calculations. *Int. J. Hydrogen Energy* **2020**, *45*, 33759–33770. [CrossRef]
10. Mohammadi, A.; Ikeda, Y.; Edalati, P.; Mito, M.; Grabowski, B.; Li, H.-W.; Edalati, K. High-Entropy Hydrides for Fast and Reversible Hydrogen Storage at Room Temperature: Binding-Energy Engineering via First-Principles Calculations and Experiments. *Acta Mater.* **2022**, *236*, 118117. [CrossRef]

11. Cao, W.; Ding, X.; Chen, R.; Zhang, J.; Zhang, Y.; Guo, J.; Fu, H. Dual Tuning of the De-/Hydrogenation Thermodynamics and Kinetics of the Mg–Ni Alloy by Introducing the Ag–H Bond: Enhanced Hydrogen Storage Properties at Moderate Temperatures. *J. Mater. Chem. A* **2023**, *11*, 20761–20773. [CrossRef]
12. Panigrahi, P.K.; Chandu, B.; Motapothula, M.R.; Puvvada, N. Potential Benefits, Challenges and Perspectives of Various Methods and Materials Used for Hydrogen Storage. *Energy Fuels* **2024**, *38*, 2630–2653. [CrossRef]
13. Zhou, C.; Zhang, J.; Bowman, R.C.; Fang, Z.Z. Roles of Ti-Based Catalysts on Magnesium Hydride and Its Hydrogen Storage Properties. *Inorganics* **2021**, *9*, 36. [CrossRef]
14. Zhou, C.; Fang, Z.Z.; Lu, J.; Zhang, X. Thermodynamic and Kinetic Destabilization of Magnesium Hydride Using Mg–In Solid Solution Alloys. *J. Am. Chem. Soc.* **2013**, *135*, 10982–10985. [CrossRef] [PubMed]
15. Kefi, C.; Huot, J. Entropy-Enthalpy Compensation in Ti-V-Mn-Cr BCC Alloys Used as Hydrogen Storage Materials. *Inorganics* **2023**, *11*, 479. [CrossRef]
16. Zhang, Q.; Fang, M.; Si, T.; Fang, F.; Sun, D.; Ouyang, L.; Zhu, M. Phase Stability, Structural Transition, and Hydrogen Absorption−Desorption Features of the Polymorphic La_4MgNi_{19} Compound. *J. Phys. Chem. C* **2010**, *114*, 11686–11692. [CrossRef]
17. Zhou, C.; Peng, Y.; Zhang, Q. Growth Kinetics of MgH_2 Nanocrystallites Prepared by Ball Milling. *J. Mater. Sci. Technol.* **2020**, *50*, 178–183. [CrossRef]
18. Zhang, Q.A.; Liu, D.D.; Wang, Q.Q.; Fang, F.; Sun, D.L.; Ouyang, L.Z.; Zhu, M. Superior Hydrogen Storage Kinetics of $Mg_{12}YNi$ Alloy with a Long-Period Stacking Ordered Phase. *Scr. Mater.* **2011**, *65*, 233–236. [CrossRef]
19. Ding, X.; Li, Y.; Fang, F.; Sun, D.; Zhang, Q. Hydrogen-Induced Magnesium–Zirconium Interfacial Coupling: Enabling Fast Hydrogen Sorption at Lower Temperatures. *J. Mater. Chem. A* **2017**, *5*, 5067–5076. [CrossRef]
20. Zhang, K.; Chang, Y.; Lei, J.; Chen, J.; Si, T.; Ding, X.; Cui, P.; Li, H.-W.; Zhang, Q.; Li, Y. Synergy of inside Doped metals–Outside Coated Graphene to Enhance Hydrogen Storage in Magnesium-Based Alloys. *J. Magnes. Alloys* **2024**, *12*, 2462–2471. [CrossRef]
21. Aguey-Zinsou, K.-F.; Ares-Fernández, J.-R. Hydrogen in Magnesium: New Perspectives toward Functional Stores. *Energy Environ. Sci.* **2010**, *3*, 526. [CrossRef]
22. Lotoskyy, M.; Denys, R.; Yartys, V.A.; Eriksen, J.; Goh, J.; Nyamsi, S.N.; Sita, C.; Cummings, F. An Outstanding Effect of Graphite in Nano-MgH_2–TiH_2 on Hydrogen Storage Performance. *J. Mater. Chem. A* **2018**, *6*, 10740–10754. [CrossRef]
23. Yan, S.; Wei, L.; Gong, Y.; Yang, K. Enhanced Hydrogen Storage Properties of Magnesium Hydride by Multifunctional Carbon-Based Materials: A Review. *Int. J. Hydrogen Energy* **2024**, *55*, 521–541. [CrossRef]
24. Zhang, X.L.; Liu, Y.F.; Zhang, X.; Hu, J.J.; Gao, M.X.; Pan, H.G. Empowering Hydrogen Storage Performance of MgH_2 by Nanoengineering and Nanocatalysis. *Mater. Today Nano* **2020**, *9*, 100064. [CrossRef]
25. Zhang, X.; Liu, Y.; Ren, Z.; Zhang, X.; Hu, J.; Huang, Z.; Lu, Y.; Gao, M.; Pan, H. Realizing 6.7 Wt% Reversible Storage of Hydrogen at Ambient Temperature with Non-Confined Ultrafine Magnesium Hydrides. *Energy Environ. Sci.* **2021**, *14*, 2302–2313. [CrossRef]
26. Liu, M.; Zhao, S.; Xiao, X.; Chen, M.; Sun, C.; Yao, Z.; Hu, Z.; Chen, L. Novel 1D Carbon Nanotubes Uniformly Wrapped Nanoscale MgH_2 for Efficient Hydrogen Storage Cycling Performances with Extreme High Gravimetric and Volumetric Capacities. *Nano Energy* **2019**, *61*, 540–549. [CrossRef]
27. El-Eskandarany, M.S.; Ali, N.; Al-Ajmi, F.; Banyan, M.; Al-Duweesh, A.A. Hydrogen Storage Behavior and Performance of Multiple Cold-Rolled MgH_2/Nb_2O_5 Nanocomposite Powders. *Processes* **2022**, *10*, 1017. [CrossRef]
28. Maafa, I.M.; Zouli, N.; Abutaleb, A.; Yousef, A.; Qudsieh, I.Y.; Matar, S.M.; Adam, A.S.M.; El-Halwany, M.M. In Situ Preparation of 2D Co-B Nanosheets@1D TiO_2 Nanofibers as a Catalyst for Hydrogen Production from Sodium Borohydride. *Inorganics* **2023**, *11*, 342. [CrossRef]
29. Mehrabi, M.; Reyhani, A.; Parvin, P.; Mortazavi, S.Z. Surface Structural Alteration of Multi-Walled Carbon Nanotubes Decorated by Nickel Nanoparticles Based on Laser Ablation/Chemical Reduction Methods to Enhance Hydrogen Storage Properties. *Int. J. Hydrogen Energy* **2019**, *44*, 3812–3823. [CrossRef]
30. Mehrabi, M.; Parvin, P.; Reyhani, A.; Mortazavi, S.Z. Hybrid Laser Ablation and Chemical Reduction to Synthesize Ni/Pd Nanoparticles Decorated Multi-Wall Carbon Nanotubes for Effective Enhancement of Hydrogen Storage. *Int. J. Hydrogen Energy* **2018**, *43*, 12211–12221. [CrossRef]
31. Li, Z.-Y.; Sun, L.-X.; Xu, F.; Luo, Y.-M.; Xia, Y.-P.; Wei, S.; Zhang, C.-C.; Cheng, R.-G.; Ye, C.-F.; Liu, M.-Y.; et al. Modulated Noble Metal/2D MOF Heterostructures for Improved Hydrogen Storage of MgH_2. *Rare Met.* **2024**, *43*, 1672–1685. [CrossRef]
32. Zhang, Q.; Huang, Y.; Ma, T.; Li, K.; Ye, F.; Wang, X.; Jiao, L.; Yuan, H.; Wang, Y. Facile Synthesis of Small MgH_2 Nanoparticles Confined in Different Carbon Materials for Hydrogen Storage. *J. Alloys Compd.* **2020**, *825*, 153953. [CrossRef]
33. Liu, Y.; Zou, J.; Zeng, X.; Wu, X.; Tian, H.; Ding, W.; Wang, J.; Walter, A. Study on Hydrogen Storage Properties of Mg Nanoparticles Confined in Carbon Aerogels. *Int. J. Hydrogen Energy* **2013**, *38*, 5302–5308. [CrossRef]
34. Satawara, A.M.; Shaikh, G.A.; Gupta, S.K.; Andriotis, A.N.; Menon, M.; Gajjar, P.N. An Ab-Initio Analysis of the Hydrogen Storage Behaviour of V Doped Si_2BN Nanotube. *Int. J. Hydrogen Energy* **2024**, *52*, 1560–1567. [CrossRef]
35. Abutaleb, A. Synthesis of Copper/Sulfur Co-Doped TiO_2-Carbon Nanofibers as Catalysts for H_2 Production via $NaBH_4$ Hydrolysis. *Inorganics* **2023**, *11*, 352. [CrossRef]
36. Zeng, L.; Qing, P.; Cai, F.; Huang, X.; Liu, H.; Lan, Z.; Guo, J. Enhanced Hydrogen Storage Properties of MgH_2 Using a Ni and TiO_2 Co-Doped Reduced Graphene Oxide Nanocomposite as a Catalyst. *Front. Chem.* **2020**, *8*, 207. [CrossRef]
37. Boateng, E.; Thiruppathi, A.R.; Hung, C.-K.; Chow, D.; Sridhar, D.; Chen, A. Functionalization of Graphene-Based Nanomaterials for Energy and Hydrogen Storage. *Electrochim. Acta* **2023**, *452*, 142340. [CrossRef]

38. Mahamiya, V.; Shukla, A.; Chakraborty, B. Scandium Decorated C24 Fullerene as High Capacity Reversible Hydrogen Storage Material: Insights from Density Functional Theory Simulations. *Appl. Surf. Sci.* **2022**, *573*, 151389. [CrossRef]
39. EL-Barbary, A.A.; Shabi, A.H. Stone-Wales Defective C60 Fullerene for Hydrogen Storage. *Int. J. Hydrogen Energy* **2024**, *71*, 155–164. [CrossRef]
40. Kumar, P.; Singh, S.; Hashmi, S.A.R.; Kim, K.-H. MXenes: Emerging 2D Materials for Hydrogen Storage. *Nano Energy* **2021**, *85*, 105989. [CrossRef]
41. Huang, T.; Huang, X.; Hu, C.; Wang, J.; Liu, H.; Xu, H.; Sun, F.; Ma, Z.; Zou, J.; Ding, W. MOF-Derived Ni Nanoparticles Dispersed on Monolayer MXene as Catalyst for Improved Hydrogen Storage Kinetics of MgH_2. *Chem. Eng. J.* **2021**, *421*, 127851. [CrossRef]
42. Su, Y.; Yang, M.; Wang, X.; Tian, F.; Jia, H.; Li, L. Synthesis of Ni Nanoparticles Embedded Porous Mo_2C Nanospheres for Enhancing the Electrochemical Hydrogen Storage Properties of Co_2B Material. *J. Alloys Compd.* **2023**, *954*, 170143. [CrossRef]
43. Tseng, F.-G.; Bhalothia, D.; Lo, K.-H.; Syu, C.-H.; Chen, Y.-C.; Sihag, A.; Wang, C.-W.; Chen, H.-Y.T.; Chen, T.-Y. Glucose-Based Highly-Porous Activated Carbon Nanospheres (g-ACNSs) for High Capacity Hydrogen Storage. *Energy Adv.* **2024**, *3*, 1283–1292. [CrossRef]
44. Alsabawi, K.; Webb, T.A.; Gray, E.; Webb, C.J. The Effect of C60 Additive on Magnesium Hydride for Hydrogen Storage. *Int. J. Hydrogen Energy* **2015**, *40*, 10508–10515. [CrossRef]
45. Lakhnik, A.; Kirian, I.; Rud, A. Comparative Analysis of Hydrogen Uptake and Release Kinetics in MgH_2/C Composites Synthesized Using Varied Surface Areas Graphite Powders. *Appl. Phys. A* **2024**, *130*, 283. [CrossRef]
46. Guo, S.; Yu, Z.; Li, Y.; Fu, Y.; Zhang, Z.; Han, S. Preparation of Mg-Mg_2Ni/C Composite and Its Excellent Hydrogen Storage Properties. *J. Alloys Compd.* **2024**, *976*, 173035. [CrossRef]
47. Carson, R.M.; Ellis, B.L.; Long, F.; Persaud, S.Y. Structural Evolution of a Mg–C Composite over 1000 H_2 Storage Cycles. *Int. J. Hydrogen Energy* **2024**, *51*, 676–687. [CrossRef]
48. Huang, Y.; Li, M.; Liang, T.; Zhou, Y.; Guan, P.; Zhou, L.; Hu, L.; Wan, T.; Chu, D. Structural Optimization and Electrocatalytic Hydrogen Production Performance of Carbon-Based Composites: A Mini-Review. *Carbon Trends* **2024**, *15*, 100363. [CrossRef]
49. Qiao, Z.; Bian, K.; Ding, C.; Zhao, Y. Recent Progress of Carbon-Fiber-Based Electrode Materials for Energy Storage. *Diam. Relat. Mater.* **2023**, *138*, 110208. [CrossRef]
50. Fang, B.; Li, L.; Guo, J.; Qin, Y.; Wei, Y.; Zhang, J.; He, C.; Chen, Y. Development of a Novel Co-Doped Carbon Fiber from Textile Waste as Catalyst for the Highly Efficient Degradation of Organic Pollutants: The Key Role of C–O–Co Bond. *Chem. Eng. J.* **2024**, *498*, 155494. [CrossRef]
51. Mishra, M.K.; Ghalsasi, P.; Deo, M.N.; Bhatt, H.; Poswal, H.K.; Ghosh, S.; Ganguly, S. In Situ High Pressure Study of an Elastic Crystal by FTIR Spectroscopy. *CrystEngComm* **2017**, *19*, 7083–7087. [CrossRef]
52. Amaral, M.M.; Yukuhiro, V.Y.; Vicentini, R.; Peterlevitz, A.C.; Da Silva, L.M.; Fernandez, P.; Zanin, H. Direct Observation of the CO_2 Formation and C–H Consumption of Carbon Electrode in an Aqueous Neutral Electrolyte Supercapacitor by in-Situ FTIR and Raman. *J. Energy Chem.* **2022**, *71*, 488–496. [CrossRef]
53. Song, Y.; Zhang, T.; Zhou, S.; Liu, P.; Yu, H.; Han, Z. Confinement of Mg Nanoparticles by Bituminous Coal and Associated Synergistic Hydrogen Storage Effect. *J. Mater. Eng Perform* **2020**, *29*, 760–768. [CrossRef]
54. Peng, C.; Li, Y.; Zhang, Q. Enhanced Hydrogen Desorption Properties of MgH_2 by Highly Dispersed Ni: The Role of in-Situ Hydrogenolysis of Nickelocene in Ball Milling Process. *J. Alloys Compd.* **2022**, *900*, 163547. [CrossRef]
55. Salamone, M.; Martin, T.; Milan, M.; Costas, M.; Bietti, M. Electronic and Torsional Effects on Hydrogen Atom Transfer from Aliphatic C–H Bonds: A Kinetic Evaluation via Reaction with the Cumyloxyl Radical. *J. Org. Chem.* **2017**, *82*, 13542–13549. [CrossRef]
56. Capaldo, L.; Ravelli, D.; Fagnoni, M. Direct Photocatalyzed Hydrogen Atom Transfer (HAT) for Aliphatic C–H Bonds Elaboration. *Chem. Rev.* **2022**, *122*, 1875–1924. [CrossRef]
57. Fan, L.; Long, J.; Gu, Q.; Huang, H.; Lin, H.; Wang, X. Single-Site Nickel-Grafted Anatase TiO_2 for Hydrogen Production: Toward Understanding the Nature of Visible-Light Photocatalysis. *J. Catal.* **2014**, *320*, 147–159. [CrossRef]

inorganics

MDPI

Review

The Integration of Thermal Energy Storage Within Metal Hydride Systems: A Comprehensive Review

Matias Davis Cortina [1,2,†], Manuel Romero de Terreros Aramburu [1,2,†], Andre Martins Neves [3,4,†], Lourdes Hurtado [2], Julian Jepsen [3,4] and Ulrich Ulmer [1,5,6,*]

[1] Department of Process Engineering, Technische Hochschule Nürnberg Georg Simon Ohm, Keßlerplatz 12, 90489 Nuremberg, Germany; a2175856@correo.uia.mx (M.D.C.)
[2] Chemical Engineering Department, Universidad Iberoamericana Ciudad de México, Prolongación Paseo de la Reforma 880, Mexico City 01229, Mexico; lourdes.hurtado@ibero.mx
[3] Institute of Hydrogen Technology, Helmholtz-Zentrum Hereon, Max-Planck-Straße 1, 21502 Geesthacht, Germany
[4] Chair of Applied Materials Engineering, Helmut Schmidt University/University of the German Federal Armed Forces Hamburg, Holstenhofweg 85, 22043 Hamburg, Germany; jepsen@hsu-hh.de (J.J.)
[5] Institute of Applied Hydrogen Research, Electro- and Thermochemical Energy Systems (H2Ohm), Technische Hochschule Nürnberg Georg Simon Ohm, Keßlerplatz 12, 90489 Nuremberg, Germany
[6] Energie Campus Nürnberg, Fürther Strasse 250, 90429 Nürnberg, Germany
* Correspondence: ulrich.ulmer@th-nuernberg.de
† These authors contributed equally to this work.

Abstract: Hydrogen storage technologies are key enablers for the development of low-emission, sustainable energy supply chains, primarily due to the versatility of hydrogen as a clean energy carrier. Hydrogen can be utilized in both stationary and mobile power applications, and as a low-environmental-impact energy source for various industrial sectors, provided it is produced from renewable resources. However, efficient hydrogen storage remains a significant technical challenge. Conventional storage methods, such as compressed and liquefied hydrogen, suffer from energy losses and limited gravimetric and volumetric energy densities, highlighting the need for innovative storage solutions. One promising approach is hydrogen storage in metal hydrides, which offers advantages such as high storage capacities and flexibility in the temperature and pressure conditions required for hydrogen uptake and release, depending on the chosen material. However, these systems necessitate the careful management of the heat generated and absorbed during hydrogen absorption and desorption processes. Thermal energy storage (TES) systems provide a means to enhance the energy efficiency and cost-effectiveness of metal hydride-based storage by effectively coupling thermal management with hydrogen storage processes. This review introduces metal hydride materials for hydrogen storage, focusing on their thermophysical, thermodynamic, and kinetic properties. Additionally, it explores TES materials, including sensible, latent, and thermochemical energy storage options, with emphasis on those that operate at temperatures compatible with widely studied hydride systems. A detailed analysis of notable metal hydride–TES coupled systems from the literature is provided. Finally, the review assesses potential future developments in the field, offering guidance for researchers and engineers in advancing innovative and efficient hydrogen energy systems.

Keywords: hydrogen storage; thermal energy storage; metal hydrides

Citation: Davis Cortina, M.; Romero de Terreros Aramburu, M.; Neves, A.M.; Hurtado, L.; Jepsen, J.; Ulmer, U. The Integration of Thermal Energy Storage Within Metal Hydride Systems: A Comprehensive Review. *Inorganics* **2024**, *12*, 313. https://doi.org/10.3390/inorganics12120313

Academic Editors: Christian Julien and Hicham Idriss

Received: 15 September 2024
Revised: 11 November 2024
Accepted: 21 November 2024
Published: 3 December 2024

1. Introduction

One of the main challenges we currently face as a modern, technology-dependent society is guaranteeing universal access to our energy supplies, and ensuring their reliability as the global population continues to increase. The issue of climate change and the urgency to mitigate our overall carbon footprint have also imposed new restrictions on how we can continue to expand the energy infrastructure worldwide, raising the bar for our required efforts.

The world's overall demand for electricity grew by 2.2% in 2023, mainly driven by emerging economies such as India and China. However, demand is expected to rise faster over the next two years, averaging 3.4% growth annually [1]. Projections indicate that energy consumption in the industrial sector will grow between 9% and 62%, and between 8% and 41% in the transportation sector from 2022 to 2050 [2]. Until recently, the majority of our newly installed energy infrastructure was reliant on fossil fuels such as petrol, coal, and natural gas, but zero-carbon technologies (including their storage) are quickly picking up the pace. They are expected to account for up to 81% to 95% of the new global generating capacity installed in the period between 2022 and 2050, with hydrogen emerging as an extremely attractive candidate energy carrier molecule. Hydrogen, when combusted or combined with oxygen in an electrochemical reaction, results only in water as a by-product, avoiding the production of any greenhouse gases during its consumption. On top of that, provided that the hydrogen is produced from renewable energy sources, it can become a completely non-polluting energy carrier during its entire life cycle. Molecular hydrogen is also the lightest known element and has the highest gravimetric energy density of all known substances, with a lower heating value (LHV) of approximately 120 $kJ \cdot g^{-1}$, three times higher than gasoline [3].

Despite its benefits, hydrogen still faces major bottlenecks impeding its widespread usage, with its transportation and storage presenting critical opportunities for improvement [4]. Hydrogen can currently be stored both through physical means, e.g., as a compressed gas, in liquefied form, and by cryo/cold compression, or it can be stored in materials and chemical bonds (e.g., adsorbents, ammonia, liquid organic carriers, and metal hydrides) through physisorption and chemisorption [5]. Even with the wide array of available options, most still present significant challenges to overcome. Physical storage methods require high-energy inputs, wasting a considerable part of the hydrogen's energy potential. Hydrogen compression to 750 bar uses approximately 15% of its lower heating value, and cooling hydrogen down to -253 °C required for storage as a liquid consumes between 20 and 50% of its LHV [6]. For material-based storage options, the main drawbacks are the need for separation and storage of carrier molecules, the need for additional carrier molecule cracking or hydrogen extraction processes, and the potential health and environmental safety concerns during events of leakages [6,7]. It is in this context that metal hydrides (MHs) emerge as a promising option for energy storage in electronics, vehicles, and renewable energy systems. Metal hydrides have been known to store hydrogen at larger densities than that of liquid hydrogen [8], are considered reusable solutions (as they can store and release hydrogen multiple times), and their properties can be tailored to meet the needs of various applications with different capacity and temporal requirements [9,10]. Additionally, their ability to release hydrogen only under specific pressure/temperature conditions and their high thermodynamic stability makes them a safer alternative than other known solutions [10]. For metal hydrides to become a commercially viable solution, however, there is still much work to be done, particularly in terms of heat management [9]. The reaction for the storage (hydride formation) and release (hydride dissociation) of hydrogen in a metal hydride is known to produce or consume heat, respectively. The heat involved in these reactions can range from less than 10 $kJ \cdot (mol\ H_2)^{-1}$ to over 200 $kJ \cdot (mol\ H_2)^{-1}$, depending on the chemical nature and composition of the MH [11]. Both the removal and provision of heat from and to the MH can pose technical challenges and impact the economics of a MH hydrogen storage system. In particular, identifying a viable source to provide the heat for hydride dissociation, or finding a way to reduce the amount of heat demanded by the hydride, has thus presented itself as a major problem to be solved. Various strategies have been devised, including the use of waste heat from fuel cells [12–14] as well as the improvement of metal hydride properties themselves. The latter can be achieved through the use of preparation methods such as mechanical alloying [15], destabilization of the metal hydride alloy [16], and catalysts usage to overcome the reaction's energy barriers [17], or the addition of high thermal conductivity materials [9].

Another promising approach to reduce the amount of external heat required to release hydrogen from metal hydrides is to couple them with thermal energy storage (TES) materials. The basic idea behind this concept is to store the heat produced by the MH during the hydrogen absorption process in a TES material. Later, this heat can be reused by the MH to release its stored hydrogen. By applying this concept, a MH–TES storage system with high thermal efficiency, good economic performance, and sustainability can be developed. The focus of this review article is therefore to explore the current developments in the field of utilizing metal hydride hydrogen storage systems in combination with thermal energy storage systems. The article is structured in the following way: first, an overview of the main working principles of metal hydride and thermal energy storage materials are presented; second, the technical concepts of facilitating heat exchange in metal hydride reactor systems are presented; third, the current developments in coupling metal hydride systems with TES systems are reviewed; and fourth, some future developments are proposed, which are necessary to improve the technical maturity of MH–TES systems. This review provides a reference point that can aid engineers and scientists to design MH–TES systems and provide valuable information for future developments in hydrogen storage technologies.

2. Metal Hydrides

2.1. Overview

The term 'metal hydride' refers to any kind of compound formed between metal and hydrogen atoms, in which these are chemically bound [9]. Materials of the metal hydride family have many modern technological applications, including hydrogen purification, isotope separation, acting as heat pumps, hydrogen compression, and—most importantly—hydrogen storage [18]. The fact that metal hydrides can be used for hydrogen storage is due to their ability to reversibly absorb and release H_2 gas in what are known as 'hydriding' and 'dehydriding' reactions, respectively [9]. These reactions are governed by the following general equation:

$$M_{(s)} + \frac{x}{2} H_{2\ (g)} \Leftrightarrow MH_{x\ (s)} + Q$$

in which M is a metal-hydride-forming compound and Q is the amount of heat liberated during the reaction [15]. The conditions under which hydrating or dehydrating reactions are favored depend not only on the system's temperature, but also on the H_2-partial pressure at any given time [19]. Generally, the hydriding reaction proceeds exothermically, while the dehydriding reaction is endothermic. The absolute value of the heat liberated/required during reaction depends on the composition of the hydride-forming material [6,15,20].

2.2. Classification

Metal hydrides can currently be classified into one of three main families: the binary (or elemental) hydrides, the ternary (or intermetallic) hydrides, and the complex hydrides [10,21].

Binary metal hydrides (AH_x) are the chemically simplest form a metal hydride can adopt, as hydrogen bonds to only one element [6]. It is known that most of the 91 naturally occurring elements are able to hydride under the right conditions [11], but not all of them do so in the same way. As a result of its particular electronic configuration, hydrogen ($1s^1$) is able to behave differently with individual elements, forming ionic hydrides with alkaline metals, covalent compounds with group 4a and 5a metals [18], and metallic bonds with transition metals, including the rare earth and actinide series [8] (Figure 1). While most of these compounds form hydrides under conditions that are too extreme to be of interest for practical applications [11], MgH_2 has been identified as a worthy candidate because of its favorable hydrogen-release properties and good reversibility at moderate conditions (with a hydrogen storage capacity of 7.6 wt.%) [22].

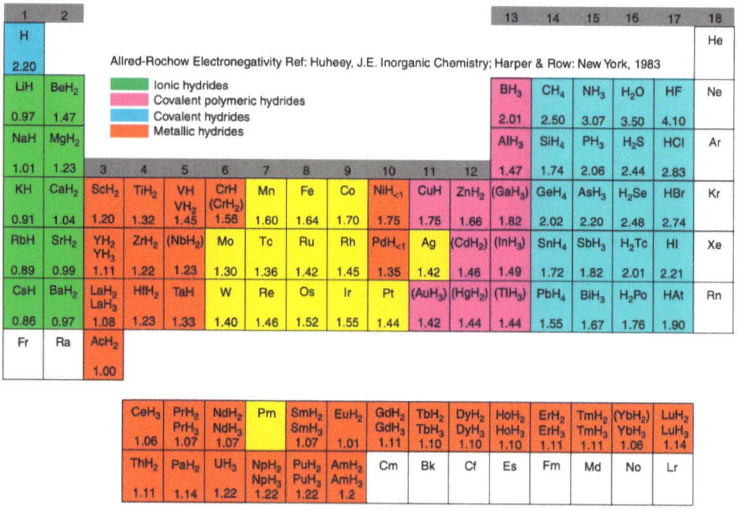

Figure 1. Binary hydrides and their bond types. Prepared by Züttel et al. [23] using data from Huheey et al. [24]. "Reprinted with permission from Ref. [23]. 2009, Elsevier".

Ternary metal hydrides ($A_nB_mH_x$), on the other hand, are the result of reacting hydrogen with an alloy of two distinct metals (A and B). Because of the fact that upon absorption, hydrogen atoms occupy interstitial sites on the parent lattice, these ternary compounds are often referred to as 'interstitial hydrides' (along with a few binary MHs that behave the same) [18]. Alloys offer great benefits over binary hydrides in terms of versatility, stemming from the possibility of combining different metal behaviors and thermodynamic properties within a single compound, and thus opening the door for selective design [11]. Their biggest advantage so far is that they have demonstrated good hydrogen absorption rates and kinetics at near ambient conditions, while eliminating the need for thermal activation in most cases [6,10]. As a general rule, ternary hydrides are mixtures of stable (A) and unstable (B) hydride-forming metals, resulting in more complex structural configurations than binary hydrides [8]. They can be broadly classified according to their stoichiometry in one of five groups: AB, AB_2, A_2B, AB_5, and BCC solid solutions [8,25]. Other additional configurations (e.g., AB_3, A_2B_7, and A_6B_{23}) have been known to exhibit reversible hydrogen absorption behavior; however, they have either never really been considered as promising materials for practical applications, or only until very recently (as is the case for AB_3 and A_2B_7 alloys) [11,26]. Some of the most well-known and studied examples of ternary metal hydrides are TiFe (AB-type) and $LaNi_5$ (AB_5-type), due to their remarkable hydriding properties at room temperature [27].

The complex metal hydride group stands out from the rest, as in these cases hydrogen atoms covalently bond with transition metals to form anionic coordination compounds, that are then stabilized by certain Group IA or IIA elements [11]. These materials can be divided into transition-metal complex hydrides (such as Mg_2FeH_6 and Mg_2NiH_4 Mg_2CoH_5), borohydrides (like $LiBH_4$), alanates (like $NaAlH_4$), and amides (like $LiNH_2$ and $Mg(NH_2)_2$) [11,28]. Complex metal hydrides have been known to possess the highest theoretical hydrogen absorption capacities among all existing metal hydrides (18.5 wt.% in the case of lithium borohydride $LiBH_4$ [29]). However, high thermodynamic stabilities, kinetic barriers, and multi-step formation reactions with potentially irreversible side reactions limit their practical applications [20,30–32]. Additionally, their formation and decomposition reactions require a certain degree of metal atom diffusion, resulting in slow kinetics compared to interstitial hydrides [11]. Among the transition-metal complex hydrides, Mg_2FeH_6 is one of the most studied due to its very high volumetric hydrogen storage capacity (150 $g{\cdot}L^{-1}$) and good cycling properties, although the system's reversibility has been shown to depend strongly on the experimental conditions [33]. Additionally, the slow diffusion rate of Fe still leads to high temperature and pressure conditions being required for hydride formation [34]. $LiBH_4$ is another example of a widely studied complex metal hydride. However, because of the high stability of LiH [35], the desorption reaction of $LiBH_4$ will yield LiH and B, reducing its practical reversible capacity to around 13.8 wt.% [35,36]. Even then, the rehydrogenation of LiH + B still requires very harsh conditions (155 bar and 600 °C) [37].

One approach to improve the reversibility of some of these complex metal hydrides are the so-called reactive hydride composites (RHCs), mixtures of complex metal hydrides with more traditional hydride forms (binary/ternary), that can react together exothermically to release hydrogen and form a new, more stable, compound [38]. For example, through the mixture of $LiBH_4$ and MgH_2 in the stoichiometric proportion of 2:1, the so-called Li-RHC can be produced [36]. While the absorption enthalpies for $LiBH_4$ and MgH_2 are, respectively, 67 and 76 $kJ \cdot (mol\ H_2)^{-1}$, it has been found that the theoretical value for this composite is reduced to 46 kJ/mol H_2 [39]. Said reduction is caused by the formation of MgB_2 upon dehydrogenation, which is much more stable than the products of the individual decomposition reactions of MgH_2 and $LiBH_4$ (Mg and B, respectively) [40]. While this composite has a high theoretical gravimetric capacity of 11.4 wt.%, the involved reactions are rather complex. The absorption reaction proceeds as a one-step reaction [41], but desorption proceeds in two steps, with the fast decomposition of MgH_2 followed up by a slow reaction for the formation of MgB_2 and LiH as H_2 is released [42].

2.3. Thermodynamics and Kinetics

To evaluate the potential coupling of metal hydrides with TES materials, it is first necessary to understand the thermodynamic and kinetic phenomena that govern their ability to store hydrogen. All chemical compounds store energy in their bonds, and the amount of energy stored can change whenever said compound undergoes a structural reconfiguration. For hydrogen uptake by metals, the net energy change between the reactants and the products over the course of the reaction is generally negative (exothermic reaction), meaning that the system releases excess energy and a more stable compound is formed [43]. This behavior is illustrated in Figure 2.

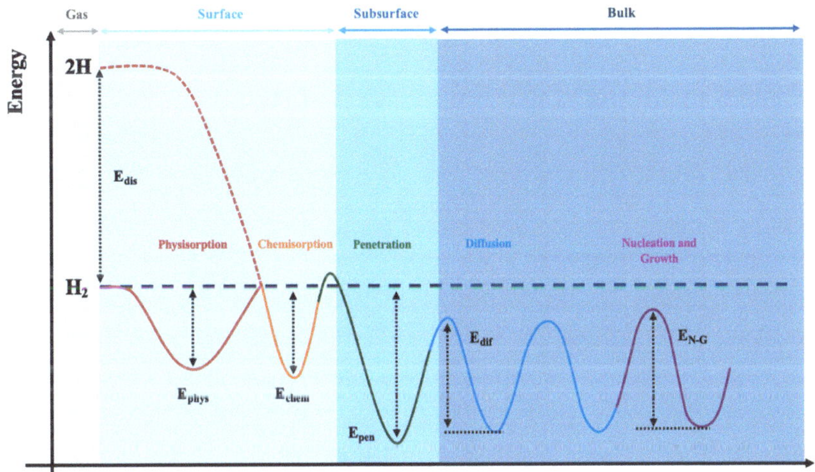

Figure 2. Potential energies of hydrogen during absorption by metals, and its mechanism. "Adapted with permission from Ref. [44]. 2010, Royal Society of Chemistry".

As shown in Figure 2, metal hydriding occurs via the following sequence of steps: physisorption of molecular hydrogen, dissociation of the hydrogen molecule and coordination with metal atoms (chemisorption), hydrogen penetration and diffusion into the host lattice, and finally, hydride nucleation and phase progression. These steps will be briefly explained, as they will provide us with valuable insights on how MH performance can be improved for practical applications.

Following the pressure-induced mass transport of hydrogen towards the metal, the H_2 molecules adhere to the metal surface as a result of Van der Waals forces, in a process known as physisorption [15,45]. At the surface, hydrogen is then chemisorbed into the metal: the molecules split into individual hydrogen atoms at specific dissociation sites, and consequently coordinate with the exposed metal atoms [45]. These sites have been found to be various physical defects in the metal surface, and in the case of alloys, clusters of the least stable-hydride forming metal (B) [15,45,46]. Once individual hydrogen atoms have been made available, those with high energies penetrate into the metal, to occupy octahedral and tetrahedral interstitial sites and form a solid-solution phase, otherwise known as the α-phase (or low-concentration phase). Hydrogenation then proceeds with further diffusion of hydrogen atoms into the metal's bulk structure, a process that can occur at rates comparable to those for ions in aqueous solutions, one of the many reasons why metal hydrides are so effective at storing hydrogen. By the time enough hydrogen has diffused, and the concentration of hydrogen reaches its saturation point, a homogeneous higher concentration phase (commonly known as β-phase) begins to nucleate. This new phase is what is ultimately, and formally, known as the 'metal hydride' [45,47,48]. The nucleation of the metal hydride phase usually follows the hydrogen concentration gradient in metals, preferring zones with the lowest activation energy and paths of higher diffusion, such as the metal surface and grain boundaries [49]. Dehydriding reactions can be thought to proceed through these mechanisms in the reversed order [50].

Pressure-composition-isotherms (PCIs) are recorded to describe the thermodynamics of hydrogen sorption, as shown in Figure 3a.

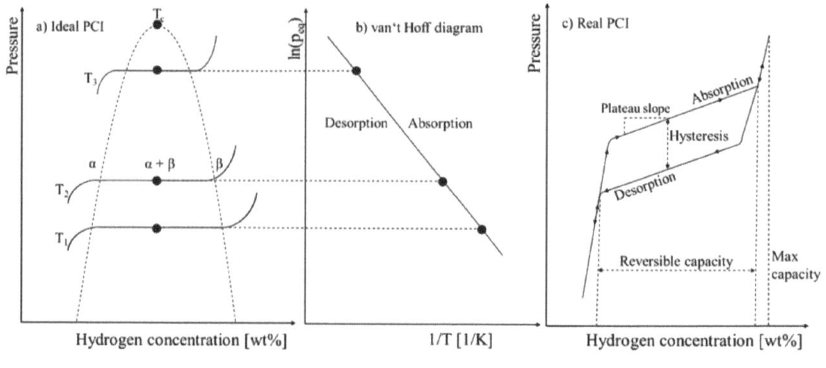

Figure 3. Examples of (**a**) a pressure-composition isotherm (PCI) graph for an ideal system; (**b**) the corresponding Van't Hoff line; (**c**) and a PCI graph for a real system. "Reprinted with permission from Ref. [10]. 2023, Elsevier".

Three main regions are easily identifiable on the ideal PCI graph and separated by a dotted curve (the biphasic zone, inside the dotted curve; the α-phase, to the left of the biphasic zone; and the β-phase, to the right of the biphasic zone). For each of the isotherm (solid) lines, the following pattern is observed from left to right, according to the correct sequence of steps for hydride formation. In the low hydrogen concentration region (α-phase), hydrogen concentration in the metal (C_H) can be found to increase along with the square root of the partial pressure of hydrogen (P_{H2}), a relationship that can be described by Sievert's Law (Equation (1)), where K_H is Sievert's constant [47,51]:

$$C_H = K_H \sqrt{P_{H_2}} \quad (1)$$

By the time the dotted phase boundary curve is reached, the system is at its saturation point, and the β-phase begins to nucleate. Here, the system attains its equilibrium pressure (P_{eq}), and a pressure plateau appears. Hydrogen concentration can now increase with no changes in the partial pressure of hydrogen. The width of the biphasic zone is therefore referred to as the 'reversible hydrogen storage capacity', which is also visible in Figure 3c. While the system is in equilibrium, the pressure values are, however, still dependent on temperature conditions. Said relationship can be expressed through the Van't Hoff equation (Equation (2)):

$$In\left(\frac{P_{eq}}{P_0}\right) = \frac{\Delta H}{R}\left(\frac{1}{T}\right) + \frac{\Delta S}{R} \tag{2}$$

where P_0 is the reference pressure, T is the system temperature, ΔH and ΔS are the net enthalpy and entropy changes in the reaction, respectively, and R is the ideal gas constant [52]. This relationship can be found in graphical form in Figure 3b. Figure 3c shows a schematic representation of a real PCT curve. The presence of deviations from an ideally flat plateau become evident in the form of a plateau slope and hysteresis. Plateau slopes are generally attributed to compositional inhomogeneities, and the relaxation of the metal matrix that results from lattice expansion during metal hydriding [8]. Hysteresis, on the other hand, mainly refers to the difference in equilibrium pressures between absorption and desorption plateaus [53], with hydrogenation occurring at a lower equilibrium pressure. There are, however, further implications of hysteresis: differences in the saturation point of the α-phase during absorption and desorption become present (solvus hysteresis), as well as differences in the formation and decomposition temperatures of metal hydrides (thermal hysteresis) [54]. All of these hysteresis-related behaviors have been attributed to the creation of additional energy barriers by either the plastic deformation of the metal lattice (Flanagan–Clewley theory) [55], or by the elastic strains exerted on it (Schwarz–Khachaturyan theory) [56] during hydrogen absorption.

The kinetic behavior of gas–solid reactions, which include MH–H_2 reactions, can be described by three main components, as shown in Equation (3):

$$\frac{d\alpha}{dt} = K(T) \times F(P) \times G(\alpha) \tag{3}$$

in which alpha (α) is the reacted fraction, $K(T)$ is the temperature dependency, $F(P)$ is the pressure dependency, and $G(\alpha)$ is associated with the morphological changes experienced by the metal during the course of the reaction [50]. Temperature has a very important impact on the kinetic behavior of hydration/dehydration reactions, and its influence on the overall reaction rate can be described through the Arrhenius equation (Equation (4)):

$$K(T) = A{\cdot}exp\left(\frac{-E_a}{R{\cdot}T}\right) \tag{4}$$

where A is the pre-exponential factor, E_a is the activation energy, and R is the ideal gas constant [50]. The pressure dependency $F(P)$ term is associated with the driving force of the chemical reaction (in this case, pressure), and is typically represented by an empirical mathematical relationship between the system's pressure and the hydride's equilibrium pressure P_{eq} at the operating temperature (T) [41,42]. These expressions usually differ for hydriding and dehydriding reactions. Finally, it is worth mentioning that the morphological changes in the metal hydride during hydrogen absorption/desorption can be accounted for by the assumptions made in specific mathematical models for gas–solid reactions, whose integral form can be used to derive an expression for $G(\alpha)$. Some models that have been developed and proved experimentally throughout the years include nucleation and growth models, geometrical contracting models, diffusion models, and autocatalytic models. A few of them are shown in Figure 4. These, along with their variations have been thoroughly reviewed and described in the works of J. A. Puszkiel [50], Wang and Suda [57], and Pang and Li [58].

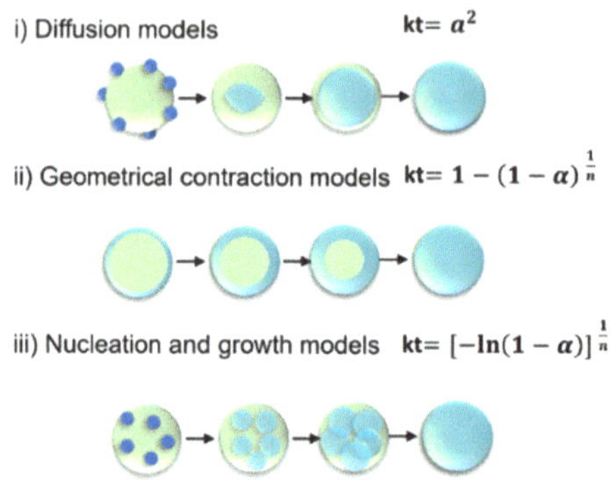

Figure 4. Illustration of some common kinetic models for hydrogen absorption by metals, and their mathematical expressions in integral form. "Reprinted with permission from Ref. [17]. 2022, John Wiley & Sons".

Another aspect that should be mentioned briefly is that both hydriding and dehydriding reactions, like all multi-step chemical reactions, have what is considered a 'rate-limiting step'. This term refers to the slowest step in a sequence of chemical reactions and the overall reaction cannot occur faster than the speed of its rate-limiting step [59]. For the absorption/desorption of hydrogen in metals, it is widely agreed that the rate-limiting step can be any of the following, depending on the specific metal and the experimental conditions: a surface related process such as H_2 dissociation and penetration, the diffusion of H_2 within the metal, or a phase transformation step (metal-metal hydride or between hydride phases) [49]. The rate-limiting step is considered within the $G(\alpha)$ term in the overall reaction rate expression [48], as it is part of the assumptions made for the different mathematical gas–solid models mentioned in the last paragraph. Having said this, the determination of the rate-limiting step for specific metal hydrides therefore plays an important role in selecting specific methods, and modifying process conditions, to improve the overall reaction kinetics.

One of the most common methods to tailor reaction kinetics in metal hydrides includes the addition of catalysts. Oxides and halides of multivalent transition metals have been extensively studied and proven as viable options [17]. Catalysts have been found to facilitate hydrogen dissociation at the gas–metal interface, reduce the energy barrier for hydrogen diffusion into the bulk metal, and act as nucleation centers for both the hydrogenated and dehydrogenated metal phases [60,61]. Some typical heterogeneous catalysts used to enhance the kinetics of binary and complex hydrides are Ti, Co, Fe, V, Nb, as well as their oxides and halides, while metals such as Cr, Mn, Fe and Ni are preferred for interstitial MHs [17,62]. These additives significantly increase the hydrogenation and dehydrogenation rates in MH. For example, Zhang et al. found that adding 10 wt.% Mn_3O_4 nanoparticles to MgH_2 can decrease the apparent activation energy for H_2 absorption from 72.5 to 34.4 $kJ\cdot(mol\ H_2)^{-1}$ and reduce the temperature required to desorb 6 wt.% H_2 by approximately 100 °C [63]. Furthermore, Ali et al. showed that upon doping $LiAlH_4$ with 10 wt.% $MgFe_2O_4$, 4 wt.% H_2 could be successfully desorbed after 40 min at 90 °C and 1 bar(a), compared to no desorption at all in the same conditions for pristine $LiAlH_4$ [64]. Similarly to catalyst addition, partial metal atom substitution in intermetallic hydride matrices has also been proven not only to speed up reaction kinetics, but to also lower pressure plateaus and improve cyclability [10,65]. Substitution can either be conducted with a single element or multiple elements, with alkali-earth metals, transition metals, p-block elements, rare-earth metals and even non-metals having been studied as candidates [65].

While understanding the reaction kinetics of gas–solid interactions in hydride systems plays a pivotal role in overcoming some material limitations, it is well known that as the characteristic dimensions of hydride-based H_2 storage systems increase, the hydrogen release rate will depend much more on the heat transfer than on the intrinsic kinetics of the hydride [66]. Therefore, considering the high amount of heat associated with gas–solid reactions in hydride materials and the typical low heat conductivity of metal hydrides, managing the heat release or consumption during hydriding/dehydriding reactions is a critical factor to consider if adequate reaction rates are to be maintained. To provide a reference point on the thermodynamics of selected metal hydriding and dehydriding reactions, Table 1 presents a summary of prominent metal hydride materials and reactive hydride composites.

Table 1. Absorption and desorption enthalpies and entropies for select metal hydrides.

Material	Absorption		Desorption		Other	References
	Enthalpy [kJ mol^{-1}H$_2$]	Entropy [J mol^{-1}H$_2$K^{-1}]	Enthalpy [kJ mol^{-1}H$_2$]	Entropy [J mol^{-1}H$_2$K^{-1}]	Hysteresis [⊘]	
MgH_2	−74.6	−134.8	74.6 *	134.8 *	N/A	[67,68]
TiFe (L)	−24.3	−100	27.4	103	N/A	[69]
TiFe (U)	N/A	N/A	N/A	N/A	N/A	[69]
$TiFe_{0.85}Mn_{0.05}$ (L)	−27.8	−99	30.6	103	N/A	[69]
$TiFe_{0.85}Mn_{0.05}$ (U)	−32.5	−121	35.2	126	N/A	[69]
ZrNi	−68.2	−125.4	73.6	127.6	N/A	[70]
$TiMn_{1.5}$	N/A	N/A	28.7	114	0.93	[11]
$TiCr_{1.8}$	N/A	N/A	20.2	111	0.11	[11]
Hydralloy C5	−22.69	−97.2	27.83	109.90	N/A	[71]
Mg_2Ni	−57.47	−94.94	61.26	99.26	N/A	[72]
$LaNi_5$	−28.4	−102.3	28.3	100.2	N/A	[73]
$CaNi_5$	N/A	N/A	31.9	101	0.16	[11]
$MmNi_5$	N/A	N/A	21.1	97	1.65	[11]
$CeNi_5$	−17.0	−105.8	22.2	111.0	N/A	[74]
$NaAlH_4$ (Step 1)	−35.1	−118.1	38.4	126.3	N/A	[75,76]
$NaAlH_4$ (Step 2)	−46.1	−123.8	47.6	126.1	N/A	[75]
$NaBH_4$	N/A	N/A	108	133	N/A	[77]
Mg_2FeH_6	−66	−124	67 (Step 1) 80 (Step 2)	123 (Step 1) 137 (Step 2)	N/A	[78]
Li_3N ($LiNH_2$-LiH)	−66.1	−118	66.6	120	N/A	[79,80]
Li-RHC	−34	−70	76 (Step 1) 61 (Step 2)	110 (Step 1) 107 (Step 2)	N/A	[41,42,81]

L = lower plateau; U = upper plateau. * Values correspond to MgH_2 absorption, but were presented here like this since it is known that the hysteresis for MgH_2 is negligible [68].

If heat management strategies are not sufficiently well designed, the temperature changes resulting from heat accumulation/extraction could shift the equilibrium pressure sufficiently to considerably alter hydrogen absorption/release rates [8], or even invert the direction of the process completely [82]. Hence, due to the importance of heat and mass transport for the performance of hydride-based H_2 storage systems, the next section is dedicated to covering the physical and thermophysical properties of these materials.

2.4. Physical and Thermophysical Properties

Some of the most important properties for the practical application of MHs as hydrogen storage materials are their volumetric and gravimetric energy storage densities. As previously mentioned, every metal hydride can only store a limited amount of hydrogen. Gravimetric and volumetric storage densities refer to the hydrogen stored per unit mass and volume, respectively.For convenience purposes, when comparing hydrides to other energy storage technologies, these capacities are often converted to energy densities using the LHV of hydrogen, and consequently become known as gravimetric energy density or volumetric energy density, respectively [83]. Excelling both in terms of gravimetric and volumetric energy densities has been identified as a critical factor to competitively introduce metal hydrides, especially in mobile applications [6]. In terms of stationary storage applications, volumetric energy densities tend to be of much greater importance than gravimetric densities, given the need to minimize costs associated with land use [27]. Having said this, it is worth mentioning that the volumetric energy densities of metal hydrides can be further improved, with metal-hydride powder compaction into pellets already being a proven and commonly used method [84–87]. Additionally, Safyari et al. have recently identified that exposing Mg–Ni based alloys to small amounts of oxygen on the second absorption cycle leads to improved storage capacities (and therefore volumetric energy densities), but more research is needed on the topic [88].

The effective heat conductivity of the MH bed is another property upon which hydrogen uptake and release depends, given that it can limit the rate at which heat is transferred within the hydride, and therefore limit the rate at which the absorption and desorption reactions occur [66,82,89]. The effective heat conductivity is not only a function of the individual thermal conductivities of both the metal and the interstitial hydrogen atoms, but also of pressure, temperature, reacted hydrogen fraction, porosity, and particle size [90,91]. Reference values for the effective thermal conductivities of some common metal hydride powders are provided in Table 2.

Table 2. Effective thermal conductivities of some selected metal hydride powders under hydrogen atmosphere.

Material	Effective Thermal Conductivity [W/(m·K)]	Pressure [bar]	Temperature [°C]	Reference
Mg	0.64–1.24	0.97–29.61	25–410	[92]
Mg_2Ni	0.35–0.75	1–50	35–200	[93]
$LaNi_{4.7}Al_{0.3}$	0.878	1	20	[94]
$TiMn_{1.5}$	0.2–1.3	1–50	21	[95]
$NaAlH_4$	0.46–0.55	1	25	[96]

Increasing either the pressure or the temperature of the system has been found to increase the effective heat conductivity of the hydride bed [97]. Regarding the reacted hydrogen fraction, it has been observed that the effective heat conductivity will diminish as the reacted fraction increases; this happens because adding more hydrogen atoms into the metal lattice blocks the ability of free electrons to conduct heat. Similarly, a smaller particle size in the hydride bed leads to larger surface areas, and therefore an increase in thermal contact resistance, which brings down the heat conductivity. However, smaller powder particles may agglomerate and sinter to form a denser structure during hydriding/dehydriding cycles, leading to a rise in thermal conductivity throughout the hydride's lifetime [82]. The agglomeration of metal hydride particles can also be used to explain the relationship between porosity and conductivity. As the particles agglomerate, the reactor bed becomes less porous, and there is less room for hydrogen gas (whose heat conductivity is smaller than that of the bulk metal) in a given volume of hydride [98]. Therefore, in less porous beds, the effective thermal conductivity's value approaches that of the bulk metal, which is greater. It is relevant to mention that several strategies have been identified to avoid these changes that occur during cycling, or to generally increase the effective thermal conductivity of the MH bed. These consist of the addition of high-conductivity materials, mechanical compaction, or the combination of both. Some of the more widely applied high-conductivity materials include carbon-based materials (graphite powder, carbon nanotubes, carbon bars, and expanded graphite) [9], metal foams and powders (aluminum, nickel, copper) [82,99], or copper mesh [9]. None of these materials will react with the metal hydride powders, and they will fill the void space in the hydride bed, increasing contact between particles to create more effective heat conduction channels [9,82].

Microstructural aspects of metal hydrides such as their crystallinity must also be taken into account for hydride-based H_2 storage systems, as they can provide crucial insights on how to maximize their storage capacities and for the selection of adequate preparation methods for a specific metal hydride. It is known that most conventional metal hydrides are crystalline phases; however, materials with varying degrees of crystallinity have been investigated as potential hydrogen storage candidates. Amorphous metal hydrides have been reported to possess much larger hydrogenation capacities, faster kinetics, and lower dehydrogenation temperatures as a consequence of the increased availability of interstitial sites resulting from their disordered structure [100]. However, the absence of a pressure plateau, poor structural stability, and reversibility are still challenges that need to be overcome [11,100]. In a similar manner, quasicrystals have been shown to possess a greater number of interstitial sites than crystalline metal alloys, increasing their capacity to store H_2; TiZr-based quasicrystals have been identified as good candidates for H_2 storage due to their low thermodynamic stability [101,102].

Another important behavior that must be considered for practical applications of metal hydrides is the lattice expansion of metals resulting from the occupation of interstitial sites by hydrogen and the formation of the β-phase. Upon repeated hydrogenation and dehydrogenation, the constant expansion–contraction of the metal hydride lattice (up to an additional 25% of the original size) can lead to cracking in metal hydrides and their subsequent pulverization, a phenomenon also known as decrepitation [8,9]. If not properly managed, decrepitation can result in the metal hydride powder sinking to the bottom of the reactor and, in combination with the expansion–contraction cycles, induce enough stress on the reactor walls to be able to deform or even destroy them [11,103]. Various techniques have been long known to mitigate the consequences of volume expansion. Some relevant examples are strengthening container walls, lubricating the reactor bed with non-volatile oils such as silicon, adding aluminum powder or fiber, or partially substituting the metal matrixes with other metallic atoms [103–105].

2.5. Hydride Based H_2 Storage Systems

Metal hydrides present several advantages in comparison with more traditional H_2 storage methods. To begin with, hydrides are able to store hydrogen at a very wide range of temperatures and pressures (including ambient conditions), whereas other alternatives have very specific requirements to successfully store it. For instance, some methods such as liquified hydrogen storage and the use of some solid adsorbents (organic polymer networks and metal–organic frameworks) require cryogenic temperatures of −253 °C and ca. −196 °C [106,107], respectively), while gaseous hydrogen storage involves elevated pressures that typically range from 350 to 750 bar, and cryo-compression needs the combination of both high pressures and very low temperatures [106]. Besides the fact that maintaining these temperatures and pressures translates to increased operational costs, the use of specialized tanks (with reinforced walls, extreme insulation, and lightweight materials) is usually necessary to account for such harsh storage conditions [106,108], meaning a larger capital investment is required. Metal hydrides also excel in terms of volumetric hydrogen storage densities, outcompeting both compressed and liquified hydrogen [10]. Their performance is only rivaled by some chemical hydrogen carriers, such as liquid organic hydrogen carriers (LOHCs) or ammonia [109], but it has been found that metal hydrides can be more energy efficient than both of these options [110]. Regarding safety, metal hydrides are advantageous over other H_2 storage alternatives mainly because of the endothermic nature of their desorption reactions, and their high bond stability [106]. Having an endothermic hydrogen release reaction allows for the careful control of H_2 gas exposure to the tank surroundings, as it is only released when heat is supplied. Additionally, a high bonding energy means that desorption will not occur without external heat supply. Therefore, spontaneous H_2 leaks that might lead to accidents, and even tank depletion (as is the case in passive boil-off for liquid hydrogen storage [111]) can be avoided. The possibility of operating metal hydride storage tanks at near-ambient pressure conditions (depending on the selected metal hydride) also eliminates the risk of explosions or jet fire in the case of a tank rupture, as sometimes occurs with gaseous hydrogen tanks [112]. Finally, metal hydrides are promising options for H_2 storage because of their compositional and structural diversity. By choosing specific alloys, preparation methods and additives, their properties can be tailored to suit very particular operating requirements.

Despite their advantages, metal hydrides still present some drawbacks, such as their high weights, high material costs, poor reversibility, poor reaction kinetics, and the requirement for high desorption temperatures [106]. Nonetheless, the benefits of using hydrides for H_2 storage outweigh their disadvantages, as can be seen by their gradual incorporation into existing energy systems.

The potential of MH reactors for energy systems has already been demonstrated by real world applications, mainly in the transportation and energy storage sectors. The German and Italian navies have implemented hydride-based power systems in submarines [6], and pilot studies with mini-passenger trains [113,114], forklifts [115], and canal boats [116] have been successfully carried out. Road-bound applications, however, are still very scarce as no metal hydride has managed to accomplish all the targets set by the DOE [117] for on-board H_2 storage yet. In terms of stationary energy storage, off-grid hydrogen energy storage systems for buildings, hotels, and small communities [118], as well as auxiliary power units [119,120], have been tested at a pilot scale. Furthermore, the use of metal hydrides in heating/cooling systems and as thermal energy storage systems has received a lot of attention lately, with prototypes for heat transformers, heat pumps, refrigeration systems, steam production systems, and CSP energy storage having already been built and tested [121]. Despite the wide variety of current implementations, most of the other lesser-known applications of hydrides that might benefit from thermal energy storage coupling (such as hydrogen purification) are still in the experimental phase. In-depth reviews of the current developments can be found in the scientific literature [6,122,123]. The following paragraphs will address some key design aspects of MH H_2 storage systems, for consideration when coupling with TES solutions.

The first crucial factor is the geometry of the hydride reactor. A proper reactor design allows for the control of heat and mass transfer in a hydride bed, and thus of the metal hydride's performance in practical applications [9,124]. Reactor geometries can be mainly divided into the following categories, as shown in Figure 5: tubular, tank, planar, disk, and spherical.

Although there are some existing studies on planar reactors with a rectangular geometry [125,126], these are not widely adopted. Their planar shape does not allow for a uniform distribution of mechanical stresses generated by the gas pressure, thus making their use with high operating pressures not viable [127]. Cylindrical tubular and tank-type reactors, on the other hand, are the most widely employed configurations, given that they facilitate the equal distribution of heat and mass transfer within the bed, tolerate high operating pressures/stresses, and are relatively easy to manufacture [124,127,128]. Tubular reactors typically employ a 'central artery' to feed hydrogen into the hydride bed, while tank reactors have a larger volume, and involve multiple arteries or surrounding outer filters for hydrogen diffusion [9,127,129]. It is worth mentioning that despite both tubular- and tank-type reactors commonly involving circular cross-sections, they are not limited to it; elliptical configurations have been proposed for the former [130], while the latter can come in cubic shapes [9,129]. Additionally, one advantage that tubular reactors present is that the individual units can be treated as modules, and bundled up whenever there is a demand for more hydrogen storage capacity [131,132]. Disk reactors are composed of a flat metal hydride bed, with hydrogen flowing axially in and out of the reactor, and a large surface area enables efficient heat exchange and therefore remarkable hydrogen uptake/release rates [124,129]. Recent progress has shed light on their true potential, outperforming even multi-tubular reactors, particularly in modular annulus configurations [133–135].

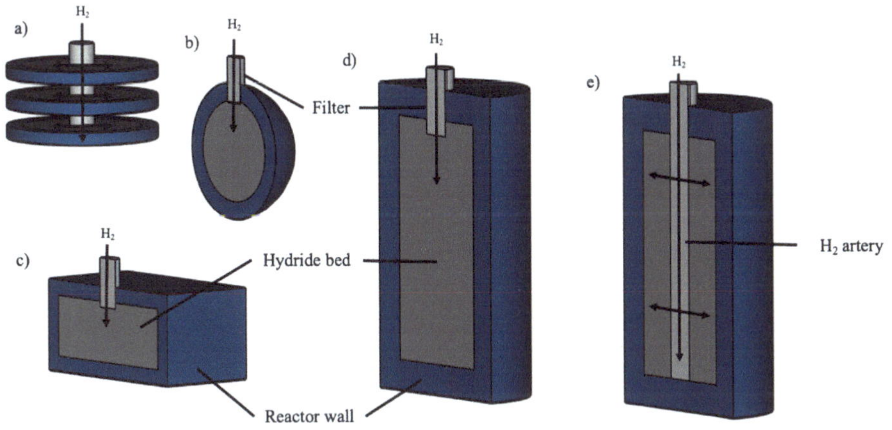

Figure 5. Common metal hydride reactor geometries: (**a**) bundle of disk reactors; (**b**) spherical reactor; (**c**) planar reactor; (**d**) cylindrical tank-type reactor; (**e**) and cylindrical tubular-type reactor.

Another aspect to consider in terms of reactor geometry is their orientation. It has been documented that horizontal reactors are less prone to stress accumulation than their vertically oriented counterparts, with particle agglomeration in the latter reaching densities so high that they were able to deform thermocouples inside the reactor [103,136]. These aspects must be taken into account when contemplating the integration of internal heat exchangers.

When designing H_2 storage systems, taking heat management strategies into consideration is extremely important. As aforementioned, if heat is not extracted/supplied at the appropriate rate from the metal hydride bed, temperature changes in the powder bed might bring the system very close to its equilibrium conditions, dramatically decreasing the hydrogen release/uptake rates. An illustrative example of the demand placed upon heat management technologies is the case of Nb_2O_5-doped MgH_2 reactors. These metal hydride beds can absorb over 5 wt.% H_2 in 30 s; if translated to a tank containing 5 kg H_2, then 500 kW of heat would need to be extracted from the reactor to successfully store said amount of hydrogen [8,137]. The most common solutions to the problem of heat accumulation, in terms of reactor design, are reducing the distance through which heat must travel to escape, increasing the heat transfer area, and improving the heat transfer coefficient [128]. Internal heat exchangers can satisfy all three requirements, while external apparatuses such as jackets are only able to contribute to the improvement of the last two. However, combining both is also a possible alternative. Developments in internal heat exchangers have mostly focused on tube shape/placement/number, and fin design. Cooling and heating tubes can be straight, U-shaped, helical or coiled, while alternatives such as embossed plate or annular heat exchangers can also be installed [124,128]. An illustration of these common types of heat-exchange tubes can be found in Figure 6.

Past studies have established the superiority of helical tubes over straight and finned tubes, even highlighting the improvements that can be made by fine tuning the pitch-to-diameter ratio [138,139]. Simulations have similarly shown that doubled U-shaped tubes allow for reduced charging times compared to straight cooling tubes [140]. In terms of the number of tubes and tube passes, a large collection of studies have confirmed that increasing both of these values within a reactor has positive effects on reaction kinetics and temperature distribution [141–146]. However, this behavior is observed only up to the point at which a certain number of tubes are added [141,142] and an optimization-based approach should also be considered, so as to keep the system weight within acceptable levels and avoid complicated manufacturing. Tube placement within the reactor is an additional factor to consider for adequate heat management. Krokos et al. studied five different configurations of nine cooling tubes in a cylindrical reactor, and found that charging and cooling times were reduced to a minimum when the tubes were equally spread out in the metal hydride bed [132]. Liu et al. studied configurations of two, three, and four pipes (arranged either in the corners of a square, or in the vertices of an equilateral triangle), and found that the least time to achieve 90% saturation of the hydride bed was obtained with the four tubes arranged as a square [141]. Additionally, they identified an overall decreasing trend in saturation time as the tubes are placed farther away from the center, but when they are placed too close to the reactor wall, the reaction rate slightly increases. Raju et al. also reported a positive correlation between H_2 uptake and tube separation; although, an increasing tube diameter for each of the cases analyzed must be considered [147].

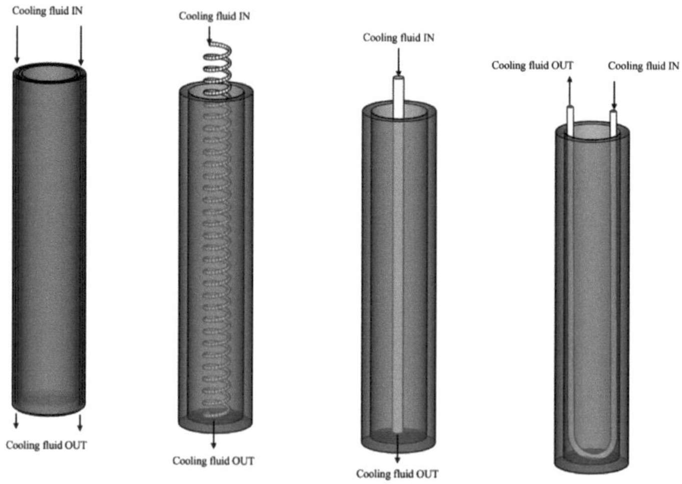

Figure 6. Various cooling tube configurations in cylindrical metal hydride tanks. From left to right: annular heat exchanger, helical/coiled tube, straight tube, and U-shaped tube.

Fin design has proved to be another aspect of great relevance for heat removal, mainly by increasing the heat-transfer area, as well as providing better performance in terms of local thermal conductivity and natural convection, for internal and external fins, respectively [148]. Typically, internal fins, which are placed inside the metal hydride bed, can be found in either transverse/radial or longitudinal form, with respect to the orientation of the heat exchange tube. It has been reported by several authors that both types of fins exhibit very similar heat-transfer performances overall, with the longitudinal variation exhibiting slightly better performance [149,150]. A wide variety of less conventional fin designs are also evaluated in the literature, such as conical fins [151], perforated disk-shaped fins [152], pin-shaped fins [153], fan or quadratically arranged fins [154], honeycomb patterned fins [155], tree-shaped fins designed by genetic algorithms [156], or more complex 3-D printed patterns [124]. Regarding other fin parameters, it has been found that an increase in fin length [157–159], and fin thickness [152,157] both improve hydrogen uptake/release rates, but none are more influential than fin number in that aspect [126,152,160]. A crucial consideration to be made when employing fins is that they can consume a considerable portion of the metal reactor volume, therefore leading to a decrease in its volumetric and gravimetric energy density. Even optimized designs can result in 29% of the reactor internal volume being occupied by fins [161].

External fins can be utilized to facilitate heat transfer between the reactor outside wall and a cooling medium. It has been found that the effectiveness of external fins is greatly dependent on the cooling fluid, whereas internal fins are not [162]. Using water jackets by themselves has been found to outperform both external fins [163] and straight cooling tubes [164], emerging as a good option. Nonetheless, they still underperform compared to helical cooling tubes [164,165], and adding fins to the jacket has in some occasions proved to be one of the best options for heat extraction from the metal hydride bed [157,164].

3. Thermal Energy Storage (TES)

3.1. Types of TES and Working Principle

TES technologies are one of the most prevalent energy storage systems (ESSs) and are, depending on the material, able to be used both as short- and long-term energy storage media [166]. The three main categories of TES systems are sensible, latent, and thermochemical energy storage, which are grouped based on the main physical phenomena that govern the ability of the TES materials to store heat [166,167].

Sensible heat storage systems are the simplest of the three types of TES technologies, and refer to the storage of heat associated with a change in the temperature of a given material, without involving a phase change [168]. Latent heat storage systems, aside from the energy stored through a change in temperature, can also store additional energy due to the presence of a phase transition, which usually results in these materials having a higher specific energy storage capacity than sensible heat storage materials [166]. Finally, thermochemical energy storage (TCS) systems make use of chemical interactions in sorption-based and reaction-based materials as a means of storing and liberating heat whenever it is required [169,170].

The materials discussed in this work are presented schematically under the appropriate classification in Figure 7. The materials were chosen based on their perceived relevance and potential for coupling with hydrogen storage systems. Also, a summary of the most important characteristics of each of these storage methods is described in Table 3. In the following sections, the selected materials are described in more detail.

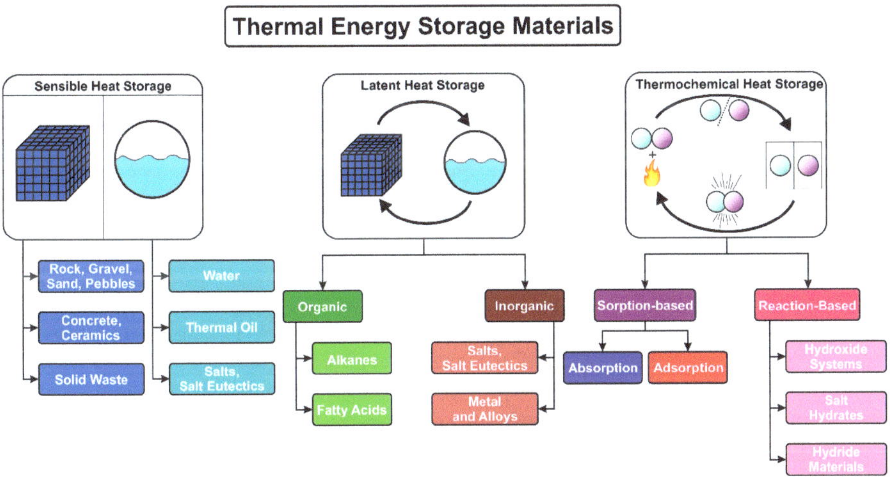

Figure 7. Classification of the thermal energy storage materials covered in this work.

Table 3. Summary of main characteristics of sensible, latent and thermochemical thermal energy storage materials.

TES Aspect	Sensible	Latent	Thermochemical
Complexity	Lowest	Intermediary	Highest
Technology Maturity	Mature	New Technology	Under development/mostly laboratory scale

Table 3. *Cont.*

TES Aspect	Sensible	Latent	Thermochemical
Type of Materials	Rock, Pebble, Sand, Gravel, Water, Thermal Oil, Salt, Salt Eutectics, etc.	Alkenes (Paraffins), Alcohols, Esters, Fatty Acids, Salt, Salt Eutectics, Metals, Metallic Alloys, etc.	Adsorption materials (e.g., Silica Gel/H_2O, Zeolite/H_2O), Absorption materials (e.g., LiBr/H_2O, LiCl/H_2O) and reaction-type materials (e.g., Hydroxides, Metal Oxides, Carbonates, Hydride Materials, Salt Hydrites)
Gravimetric Energy Storage Density [kWh/kg]	~0.02–0.03	~ 0.05–0.10	~0.5–1.0
Volumetric Energy Storage Density [kWh/m^3]	~50	~100	~500
Energy Losses Over Time	Yes	Yes	No losses associated with long-term storage (except heating and cooling)
Operation Temperature Range	From room temperature to over 1000 °C	From room temperature to around 1000 °C	From room temperature to around 1500 °C
Comments	Outlet temperature variable	Narrower outlet temperature (in relation to sensible TES)	Operation temperature is pressure-dependent (adjustable to some extent)

Some of the most important parameters for the selection of an appropriate material for TES are listed below [171]:

- ***Capacity:*** defines the energy stored in the system and varies with the size, storage technology and storage medium;
- ***Gravimetric and/or volumetric energy density:*** defines the amount of energy that the material can store per unit of mass or per unit of volume, respectively;
- ***Power:*** defines how quickly the stored energy in the system can be charged or discharged;
- ***Efficiency:*** defined as the ratio between the energy made available to use, and the energy needed to charge the system. This value quantifies the losses during the charge/discharge cycles;
- ***Storage period:*** defines how long the energy can be stored;
- ***Charge and discharge time:*** defines how long a system takes to charge or discharge;
- ***Cost:*** can be defined in terms of the cost of stored energy (€/kWh) or power (€/kW) and depends on the specific configuration of the implemented system, i.e., its CAPEX and OPEX values;
- ***Environmental impact:*** refers to considerations regarding the production of the material and the operation of the plant, or the implementation of the TES technology, that should be taken into account;
- ***Mechanical and chemical stability:*** this parameter affects not only the associated costs but also has implications on the environmental impact due to the potentially lower lifetime of the material/plant;
- ***System complexity:*** this parameter can influence both CAPEX and OPEX as well as the flexibility and reliability of a given TES technology.

3.1.1. Sensible Heat Storage

As previously mentioned, sensible heat TES refers to the storage of heat associated with a change in temperature without the occurrence of a phase change [167]. The amount of thermal energy $Q_{Sensible}$ that a given system is able to store can be determined by Equation (5):

$$Q_{sensible} = \int_{T_i}^{T_f} m \cdot c_p \, dT = m \cdot c_p \cdot (T_f - T_i) \tag{5}$$

in which m is the mass of the material, c_p is its specific heat capacity, and T_i and T_f refer to the initial and final temperatures, respectively. This technology for storing heat is the most mature and most utilized heat-storage technology [172], with applications in use and in development that include industrial heating [173], industrial heat recovery [174], district heating [175], and concentrated solar plants [171], to name a few.

While sensible heat storage materials have comparatively low gravimetric and volumetric energy densities, they present very interesting properties, such as their low complexity (compared to the two other TES technologies), their cost-effectiveness [171] and the possibility of using highly available materials, like rocks, bricks, sand, soil, and water [172]. Additionally, these materials exhibit a very wide operation temperature range, which goes from room temperature to temperatures that can exceed 1000 °C [176]. A main benefit of sensible heat storage materials over the other two types is that the materials tend to present very good reversibility, with little to no limitations in terms of their life cycle [167]. Sensible TES materials can be found in both liquid or solid form, with solid TES materials not being prone to leakages and usually having wider operation temperature ranges, which makes them eligible for higher temperature applications. Liquid materials, on the other hand, such as water, might cause corrosion and can have unfavorable properties, like having a high vapor pressure at the desired operation temperature range [166].

It is also worth mentioning that solid and liquid materials for sensible heat storage differ in the strategies available for the heat exchange between the heat storage medium and the system that is providing/extracting heat, be that during charging or discharging. In the case of solid materials, some sort of fluid needs to be used as a heat exchange medium to transfer the heat from the storage material to the process in which it is required. In the case of liquids, the heat storage material can function both as the heat storage medium and the heat transfer fluid (HTF), provided that some sort of circulation (e.g., a pump) is used to allow the heat exchange between the heat storage medium and the process requiring this heat. As such, not only does the heat storage capacity play an important role in defining a system's properties, but also the heat transfer and mass transport properties, and the design of the particular system [166]. In the next subsections, the different options for sensible heat storage materials will be discussed more in detail.

3.1.2. Rocks, Pebbles, Sand, and Gravel

Solid sensible heat storage materials include rocks [177], pebbles [178], sands, gravel, and other similar materials. These are arranged as a packed bed inside a container to be used as fillers in single tank thermocline thermal energy storage systems. Earth materials are comparatively cheap, abundant, non-toxic, non-flammable, and require the use of some heat transfer fluid for its charge and discharge cycles. This heat transfer fluid is typically put directly into contact with the heat storage medium, which in this case, doubles up as the heat transfer surface [168]. Thermal oil and water/steam are the most common heat transfer fluids [172].

Still, not every rock can be used for higher temperature operations. According to Tiskatine et al. [179], the durability of rocks depends on its chemical and mineralogical composition, the grain structure, the nature of the contacts in the rocks, as well as the crystallographic texture of the rock-forming minerals. The authors also discuss some degradation mechanisms as influencing factors, among which are the cracking of quartz crystals caused by their thermal expansion or the loss of mass in marble and limestone rocks resulting from the decomposition of $CaCO_3$ with production of CO_2 [179]. In another study [180], Tiskatine et al. conclude, based on several criteria discussed, that among 50 candidate rocks the most promising candidates for high-temperature thermal energy storage are dolerite, granodiorite, hornfels, gabbro, and quartzitic sandstone. Chemical changes associated with the thermal cycling and the selection of the HTF can induce changes that can be deleterious for the operation of TES systems, with consequences such as dust formation, micro-cracks, disintegration, mass loss, and decreases in specific heat capacity [181].

The specific heat capacity of sensible heat storage systems, as mentioned earlier, is one of the most important properties to consider during material selection. Studies have reported mean values near 2300 $kJ\cdot(m^3\cdot K)^{-1}$ for this property in rocks in general [180], which, combined with their low price, make this technology a very competitive alternative. A second important property to be considered is the thermal conductivity of the material, given that it strongly influences the charge/discharge dynamics of a system [180]. An evaluation from the data available in Zoth and Haenel [182] carried out by Tiskatine et al. [180] has found, for a selected group of rock materials, adequately high values of thermal conductivity varying from 2.4 to 4.3 $W\cdot(m\cdot K)^{-1}$ at 25 °C and decreasing to 1.4 to 2.4 $W\cdot(m\cdot K)^{-1}$ for a temperature of 500 °C.

3.1.3. Concrete and Ceramic Materials

Concrete blocks have also been considered as sensible heat storage materials [168]. These materials possess good mechanical properties, have a relatively low cost, are comparatively easy to produce, and are non-toxic and non-flammable.

Improvements to the concrete composition have been achieved by Xu et al. [183] by using an admixture of silane-coated silica fume to increase the specific gravimetric heat capacity by 50% and the thermal conductivity by 38%, to 1.05 $kJ\cdot(kg\cdot K)^{-1}$ and 0.719 $W\cdot(m\cdot K)^{-1}$, respectively. Experimental studies have shown that a high-temperature concrete with a specific gravimetric heat capacity of 0.916 $kJ\cdot(kg\cdot K)^{-1}$ and a thermal conductivity of 1.0 $W\cdot(m\cdot K)^{-1}$ at 350 °C was able to reach temperatures over 300 °C in an experimental setup designed to demonstrate the technology on a 350 kWh test unit equipped with an oil-based heat exchanger [184]. Investigations by Alonso et al. [185] have shown that a calcium aluminate cement blended with an admixture of blast furnace slag is suitable for applications with temperatures up to 550 °C. While the concrete investigated had initially a heat conductivity of 2.05 $W\cdot(m\cdot K)^{-1}$, because of the heat cycling that leads to cracking, this value decreased to 1.16 $W\cdot(m\cdot K)^{-1}$. The authors suggest the use of other materials and design techniques to improve the heat conductivity and mitigate degradation of other properties, while pointing out the importance of heat conductivity to concretes as TES materials. Kunwar et al. [186] produced concrete cylindrical blocks with drillings in different positions and different sizes to investigate the heat transfer performance of the material with air as a heat transfer fluid. In terms of heat storage capacity, the studied samples were reported to have a volumetric heat capacity of 2531 $kJ\cdot(m^3\cdot K)^{-1}$, which is close to values found for some rock materials [180].

3.1.4. Solid Waste Materials

Another alternative for solid-state heat storage systems is to use waste/inertized materials. This approach is attractive due to its low costs and sustainability. Examples of inertized products are fly ashes from municipal waste [187], industrial by-products like potash [188], by-products from the mining and metallurgy industry [189,190], post-industrial ceramic [191], recycled Nylon fiber from textile industry [192], and waste glass [193].

Investigations using material prepared from demolition waste have shown promising results as sensible heat storage materials. Using a material developed in a previous study [194], Kocak et al. [195] have made numerical and experimental investigations on a lab-scale packed-bed prototype for sensible TES that can operate at temperatures between 130 and 180 °C. The authors claim that, compared to a very common sensible heat storage medium (Therminol 66), their material could store up to 45% more heat, and that in comparison to other solid materials for sensible TES, demolition wastes can be up to tenfold cheaper. As such, the use of waste material from demolition might be an alternative to make TES more economical while simultaneously conserving natural resources.

In a recent study, Zhang et al. [196] studied the development of a sensible TES material prepared from steel slag, MgO, and refractory clay. This material presents a decent heat conductivity of 1.137 $W \cdot (m \cdot K)^{-1}$, with a maximum temperature of 1000 °C and a heat capacity of 1.29 $kJ \cdot (kg \cdot K)^{-1}$. On top of that, the authors claim that the costs per unit of energy stored are very competitive compared to other, commonly used, high-temperature sensible heat storage materials.

3.1.5. Water

Water, the most used liquid sensible heat storage material [166], has the advantage of being able to act not only as the heat storage material of the system, but also as the heat transfer fluid [168]. Alongside its non-toxicity, abundance and low cost, it has a rather high gravimetric specific heat capacity (4.184 $kJ \cdot (kg \cdot K)^{-1}$). Water can be used in a solid, liquid, or gaseous state and is an excellent candidate for home space heating, cold storage of food products, and hot water supply applications due to its non-flammable and non-toxic properties [168]. Another attractive aspect of storing heat using water is the good scalability and the low level of complexity for its operation (compared to latent and thermochemical energy storage) [175]. One of the limitations of this technology is the high vapor pressure that the material has. In cases in which higher temperatures are required, pressurization of the system is necessary in order to keep the water in the liquid state [166,175].

3.1.6. Thermal Oils

This class of materials is represented by organic fluids which possess good heat transfer capabilities. Thermal oils, much like water, can work simultaneously as thermal energy storage medium and heat transfer fluid. These materials present advantages over water because they can remain in the liquid state at much higher temperatures (usually around 250 °C under atmospheric pressure). Some materials have been reported to operate at even broader temperature ranges, like between 12 °C and 400 °C [168]. Since they have a wide temperature range, this means that they can generally have a higher sensible energy storage capacity. On top of that, their low vapor pressure allows for cost savings in the installations due to reduced pressure built up during higher temperature operations, resulting in lower container and piping costs. For example, at 374 °C, the DOWTHERM thermal oil has a vapor pressure of only 7.6 bar, while water under the same conditions would have a vapor pressure of 221 bar. On top of that, since these thermal oils are liquid at room temperature, they do not require an anti-freeze mechanism like molten salts and metallic alloys do [168].

Despite their benefits, thermal oils, in general, have lower specific heat capacities (around 2 $kJ \cdot (kg \cdot K)^{-1}$) and have very modest heat conductivity values (in the range of 0.1 $W \cdot (m \cdot K)^{-1}$) compared to water [168]. The heat conductivity, however, can be improved by introducing additives like graphene, graphite, and metal oxides. By adding up to 0.20 mg graphene per mL of heat transfer oil, an increase of up to around 25% in this property could be obtained [197]. The observed changes in the kinematic viscosity of the samples to which the additives were added were minor under the investigated temperatures, resulting in negligible changes in the liquid's properties [197]. Still, these developments should be scrutinized under the perspective of an application as a thermal energy storage material, since the time scales and operation conditions under which these materials might have to work might affect the stability of the additive nanoparticles.

While vegetable oils have the potential for cost reduction, there are concerns about their rate of degradation. According to Kenda et al. [198], exposure to oxygen, moisture, and higher temperatures affect the degradation of these materials. Furthermore, the presence of fatty acids, unsaturated components, and light transition metals with two or more valence states are some of the aspects that influence the degradation rate.

3.1.7. Salts and Salt Eutectics (as Sensible Heat Storage)

For applications of sensible TES technologies in which temperatures exceed 400 °C, molten salts are the preferred storage medium and heat transfer fluid. The use of molten salts is becoming increasingly widespread in applications like middle-sized concentrated power (CSP) plants and in generation III and III+ nuclear reactors. Properties like their high specific volumetric heat capacity, high boiling points, very high thermal stability and their very low vapor pressure even at elevated temperatures make them well-suited candidates for these types of applications. Molten salt materials are also cheap, easily available, non-toxic and non-flammable [168]. Commercially available systems can reach up to 565 °C [199].

While these materials present many interesting properties as TES materials, they have important drawbacks. The fact that their melting points are usually over 200 °C calls for the use of some anti-freezing system to avoid solidification. One way to reduce the solidification temperature of these salts is by combining one or more salts in a eutectic mixture. For example, pure KNO_3 has a melting temperature of 334 °C and pure $NaNO_3$ has a melting temperature of 307 °C. Their binary mixture with a proportion of 60% $NaNO_3$ and 40% KNO_3 (known as "Solar Salt") has a melting point of 220 °C [168]. The addition of Li-based salts like $LiNO_2$ and $LiNO_3$ is deemed as an interesting alternative to further lower the melting point of salt compositions and increase the heat storage capacity. Both these aspects also help to reduce the cost of the stored energy (by reducing costs associated with keeping the salt mixture in liquid state) [200]. Another candidate that is being used as a sensible heat storage material is the HITEC™ salt, which is composed of 53 wt.% KNO_3, 40 wt.% $NaNO_2$, and 7 wt.% $NaNO_3$ [199]. In relation to Solar Salt, this material possesses better thermophysical properties and a lower melting point.

An important drawback of molten salts as sensible heat storage materials is that they have high viscosities compared to other thermal fluids like oil and water. These poor flow properties increase costs associated with the circulation of the fluid. On top of that, molten salts have very poor heat transfer properties, with heat conductivities ranging from 0.5 to 0.6 $W \cdot (m \cdot K)^{-1}$, depending on the selected salt [168].

In the subsubsection Salts and Salt Eutectics (as latent heat storage) the application of this class of materials as latent heat storage materials is discussed.

3.1.8. Latent Heat Storage

Latent TES materials, on top of storing sensible heat energy as temperature, can additionally store energy whenever the material undergoes a phase change. For their applications as a TES medium, the transition between solid and liquid phases is most explored [166]. These materials are called phase change materials (PCMs) and can be categorized as organic and inorganic PCMs, depending on their composition.

The storage of heat Q for a latent heat storage material with a melting temperature T_m (with $T_i < T_m < T_f$), starting from solid state at some initial temperature T_i until reaching a final temperature T_f can be described in Equation (6) as follows:

$$Q_{sensible} + Q_{latent} = \int_{Ti}^{Tm} m_{sol} \cdot c_{p,\, sol}\, dT + m \cdot \Delta h_{fus} + \int_{Tm}^{Tf} m_{liq} \cdot c_{p,\, liq}\, dT \tag{6}$$

where m is the mass of the latent TES material, c_p is the specific heat capacity at constant pressure, Δh_{fus} is the enthalpy of fusion of the solid phase given in kJ/kg, and the underscripts "*sol*" and "*liq*" refer to the solid and liquid phases. On the right-hand side of Equation (7), the first term refers to the amount of sensible heat stored in the solid phase, the second term is associated with the heat involved in the phase change and the third term is the amount of sensible heat stored in the liquid state.

The use of latent TES has several advantages over the sensible TES technology. The operation approaches an isothermal regime in the vicinity of the phase change temperature, which can be beneficial for the process. On top of that, the gravimetric and volumetric heat storage capacities are significantly higher compared to the sensible heat storage [166]. However, besides the additional complexity for operation, some materials lack an acceptable long-term stability, tend to have low thermal conductivity, and present some undercooling during phase transition [201].

The difference between sensible and latent TES materials is illustrated in Figure 8. Assume two different materials, both solid at an initial temperature T_i. These materials, which have similar specific heat capacity at constant pressure c_p, experience an increase in temperature, causing both of them to store an amount of energy that is proportional to their respective c_p values. This happens until a temperature T_m is reached. While an increase in the stored energy for the sensible heat storage material is only possible with an increase in temperature, the latent heat storage material undergoes the phase change, storing energy as the transformation to liquid state progresses. The latent energy storage material keeps a nearly constant temperature until the complete conversion of the solid phase into liquid phase is completed, after which a further increase in the stored energy can only occur with an increase in the temperature until reaching the final temperature T_f.

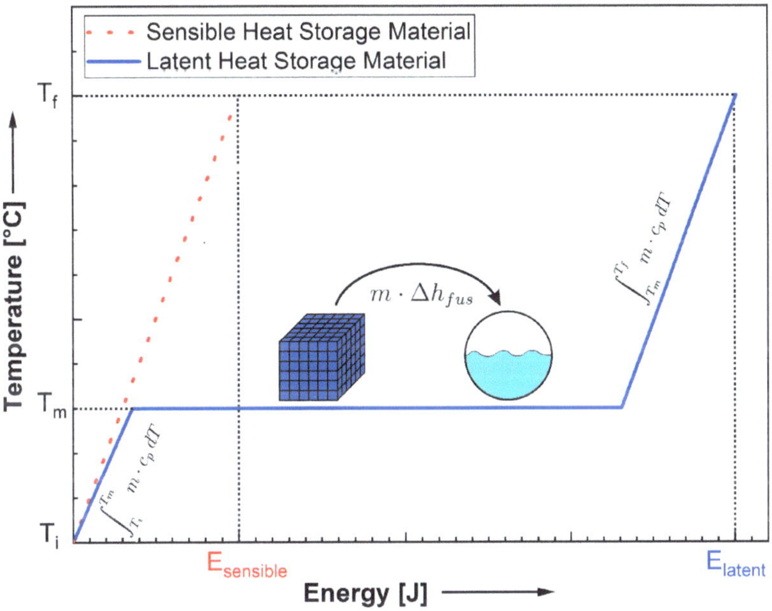

Figure 8. Schematic representation of the general behavior of sensible and latent heat storage materials for the same temperature interval.

As mentioned before, the high amount of energy that is stored in a very narrow temperature range in the vicinity of T_m is advantageous for many processes, since it gives more predictability and might simplify the control mechanisms to operate the system efficiently.

For a particular application in a system, the selection of a suitable latent TES material depends on its phase change temperature, its latent heat of fusion, the associated volume changes during phase change, and the degradation of its thermophysical characteristics with repeated cycling [166].

Latent TES systems rely on dedicated heat transfer fluids to transfer the heat from the PCMs to the relevant process under consideration. These PCMs are stored in tanks/vessels, and the system performance is very commonly limited by heat transfer between the PCMs and the heat exchange fluid [202]. Because of this fact, several techniques for the enhancement of heat exchange have been investigated and developed. Said techniques can be divided into passive and active improvement techniques.

Passive systems make use of various strategies to improve the heat transfer between the PCM and the heat transfer fluid without the need of additional energy (external power). Some of these techniques are nanoparticle dispersion, and adding heat pipes, porous matrices or conductive foams/surfaces, and extended fins [203] within the PCM storage tank/vessel [202].

When some source of external power is applied to improve the heat transfer, then it is said that an active heat transfer enhancement technique is being used. Examples for active heat exchange enhancement are mechanical aids such as pumps, vibration, jet impingement, injection, and external fields [202]. While active enhancement can improve the performance of the system in terms of power (or time to charge/discharge), it comes at the cost of considerable energy expenses and an increase in the system's complexity. These two aspects obviously impact the costs associated with the operation and construction of such a system. Because of this, passive techniques are currently more prevalent [202]. Among the passive techniques for heat conductivity enhancement, according to Al-Salami et al. [203], fins are very promising due to their minor volume occupancy, low cost, reliability, ease of manufacturing, and simplicity. In an assessment conducted by Zhang et al. [200], the latent energy storage materials considered had values of heat conductivity that ranged from 0.2 to 0.8 $W{\cdot}(m{\cdot}K)^{-1}$. These values are significantly lower than the values typically found for sensible TES materials, which further underlines the significance of heat transfer enhancement.

There are two important categories of materials used in latent heat storage systems, organic and inorganic materials. Organic materials include alkanes, fatty acids, esters, alcohols, and glycols. Due to their higher relevance and potential, only alkanes and fatty acids are covered in this review. For a brief summary of some of the properties of esters, alcohols and glycols in the context of latent heat storage materials, the reader is referred to the literature [168,203]. Inorganic materials include salts and salt eutectics, salt hydrates, metals, and alloys. A brief summary of examples and the properties of these materials is provided in the following subsections.

3.1.9. Alkanes

Alkanes are a family of organic chemical compounds with the general formula C_nH_{2n+2}. These compounds are often called paraffins, although the term "paraffin" usually refers to a mixture of alkenes [168]. Due to their interesting properties such as high enthalpy of fusion, low vapor pressure, chemical inertness, low cost, minimal tendency to separate in different phases, non-corrosivity, and non-toxicity, these materials are suitable for waste heat recovery systems and other low- to medium-temperature applications [166,204]. At room temperature, alkenes with values of n (n = number of C atoms in the molecule) between 5 and 17 are liquid, while molecules with higher n-values are waxy, solid materials [204]. The latent heat storage capacity and the melting temperature varies with the number of atoms in the molecule, with longer/heavier molecules presenting higher values for their latent heat storage capacity and higher melting points. These paraffin mixtures have melting temperatures that typically range from −5 to 100 °C [168]. In general, the values for the latent heat associated with the solid–liquid phase transition lie between 150 and 250 $kJ{\cdot}kg^{-1}$ [204].

Some of the main drawbacks of these PCMs are their low heat conductivity and high volume change during the phase transition of ca. 10% [168]. According to a summary made by Zalba et al. [205], mixtures of alkenes with different n values such as the ones designated in that work as "paraffin C_{16}-C_{28}", "paraffin C_{20}-C_{23}", and "paraffin C_{22}-C_{45}" show, respectively, melting points of ca. 43 °C, 49 °C and 59 °C. Still, all these materials exhibit the same, very low thermal conductivity of 0.21 $W{\cdot}(m{\cdot}K)^{-1}$. Some works have tried to improve the thermal conductivity of these materials, by adding, for example, 0.06 wt.%

natural graphite to an alkene material, leading to an increase in the heat conductivity to $0.38\ W\cdot(m\cdot K)^{-1}$ [204].

3.1.10. Fatty Acids

Fatty acids are organic compounds that contain a carboxylic functional group (COOH) on its aliphatic chain (i.e., do not have aromatic rings). These aliphatic chains can be saturated (i.e., only simple bonds) or unsaturated (i.e., with double or triple bonds) and have no heteroatoms in the carbon chain. Fatty acids have advantages similar to the alkenes/paraffins, as they have relatively low cost, good chemical stability, and have hardly any phase separation [168]. They also suffer from very similar drawbacks as alkanes, such as low thermal conductivities and large volume changes associated with their phase transition. Most of the fatty acids considered for latent heat storage applications are saturated fatty acids, since they present higher melting points in relation to unsaturated ones [168]. Examples of saturated fatty acids considered for heat storage applications are caprylic acid (n = 8; T_m = 16.7 °C), lauric acid (n = 12; T_m = 44.2 °C), and stearic acid (n = 18; T_m = 69.9 °C) [206]. Some of these fatty acids, like lauric acid, are naturally occurring in agricultural crops, so that they can, in principle, be extracted from low-cost, renewable sources [168] and from non-edible sources of fatty acids, like waste oils from agricultural and food processing facilities, used cooked oils or genetically modified oils that are not approved for human consumption [204].

Similarly to what has been observed for the alkanes/paraffins, attention has been given to improving the thermophysical properties of these materials, with special emphasis on thermal conductivity by means of, for example, using carbon nanotubes, as reported by Sari et al. [207].

3.1.11. Salts and Salt Eutectics (as Latent Heat Storage)

Salts have melting temperatures that make these materials attractive both as sensible and latent TES materials, depending on the operation temperatures required by the specific applications. In the subsubsection Salts and Salt Eutectics (as sensible heat storage), the application of salts in the liquid state as sensible heat storage materials was described. Here, the focus is given to salts with comparatively higher melting points, so that they can be used as latent heat storage materials in applications at higher temperatures.

There are several types of inorganic salts, which can be divided into nitrates, hydroxides, chlorides, sulfates, and fluorides. For latent heat storage applications, salts and salt eutectics with melting temperature of at least 250 °C are usually considered [168]. Nitrate salts are currently the most used in CSP plants and typically operate at temperatures of up to 565 °C [171]. Hydroxide salts have melting points between 250 °C and 600 °C, while chlorides, carbonates, sulfates, and fluorides have melting points that surpass 600 °C [168]. While the thermodynamic and thermophysical properties of pure salts are predetermined, a mixture of different salts can be used to form a eutectic mixture, in which these properties can be tuned to a degree, especially the crystallization temperature. In this mixture, the solidification and melting points are reduced and become lower than the melting points of any of the individual components. These compositions of two or more components melt and freeze congruently [200], i.e., they solidify at the same temperature. Because the phases involved solidify simultaneously, macroscopic phase separation is hindered.

According to Ong et al. [199], most commercially available salt mixtures that could work for latent TES have only a useful temperature range for applications such as hot water supply and building heating/cooling systems, although there has been some interest in finding suitable compositions and experimental efforts to determine properties relevant for high-temperature applications.

3.1.12. Metals and Alloys

Some alloys and metallic materials exhibit favorable properties for latent TES, namely, low melting points (sometimes close to or even below ambient temperatures) and very high boiling points, with negligible vapor pressure values even at elevated temperatures [168]. With the exception of cast iron/steel (which were considered as a sensible heat storage materials) [172], metallic materials and their alloys have been investigated as latent TES materials [168,208,209].

According to Costa et al. [208], metals and alloys considered for latent heat storage have a wide range of melting points, ranging from as low as −38.9 °C for pure Hg, up to 1414 °C, for pure Si. These materials present comparatively very high thermal conductivities, which is a desirable property that affects the dynamic response of the system. For instance, liquid Na exhibits a thermal conductivity of 65 $W{\cdot}(m{\cdot}K)^{-1}$ while the value for liquid Pb–Bi eutectic alloy (LBE) is 15 $W{\cdot}(m{\cdot}K)^{-1}$ [168]. Because of their high thermal conductivities, these materials do not require further additives to improve this property.

For low-temperature conditions, Ga and Ga-based alloys are the most used and investigated materials, and can be useful in applications such as thermal management. For intermediate temperature applications in the range 40 to 300 °C, Pb and Cd are the most attractive candidates based on their melting points [208]. However, the high toxicity of these elements raises concerns related to the environment and human health. For temperatures higher than 300 °C, several binary and ternary alloys have been considered, including (but not limited to) Al-, Cu-, Mg-, Si-, and Zn-based alloys. The particular properties of these alloys has been extensively covered in the work of Costa et al. [208].

These materials have been comparatively less studied due to being heavy and costly [208]. On top of that, their corrosiveness at high temperatures is a technical challenge and potential leakages can pose severe harm to the environment. A method called encapsulation can mitigate these particular issues. Different techniques have been developed for different metal alloys, including encapsulation with ceramics and other metals [208]. Furthermore, the necessary prevention of leakages or corrosion (with methods like encapsulation) reduces the storage density and can significantly increase the costs [208].

3.1.13. Thermochemical Heat Storage

Thermochemical heat storage (TCS) systems use chemical reactions to store and release thermal energy. The energy storage process of TCS materials comprises three phases, namely, charging, storage and discharging. During charging, energy in the form of heat is provided to the TCS material, which then undergoes an endothermic reaction. That means that in order to charge the material, a certain amount of energy is provided as heat to the reactant compounds and used to break the chemical bonds that hold them together. During the storage phase, the products of the first reaction are kept separate from one another until the stored energy is required. The discharge process is then carried out by combining the products of the charging reaction, and releasing heat as the reaction proceeds [169,170]. The amount of stored energy Q can be described by Equation (7):

$$Q_{sensible} + Q_{thermochemical} = \int_{Ti}^{Tr} m_{S1}{\cdot}c_{p,\,S1}\,dT + m_{S1}{\cdot}\Delta h_{reaction} + \int_{Tr}^{Tf} m_{S2}{\cdot}c_{p,\,S2}\,dT \quad (7)$$

in which m is the mass, c_p is the specific heat capacity, T_r is the temperature at which the reaction occurs at that specific pressure, $\Delta h_{reaction}$ is the amount of energy in $kJ{\cdot}kg^{-1}$ of reactant, and the underscripts S1 and S2 refer to the reactants (1) and the products (2), respectively.

There are two main classes of materials within the thermochemical heat storage group, namely, the sorption-type and reaction-type materials [169,201]. These sorption-type materials, in turn, can be divided into adsorption- and absorption-type. This schematic division is represented in Figure 9.

During adsorption, one of the reactants accumulates in, or attaches itself to, the exposed layer of the other material involved. Said material, capable of adsorbing (or 'receiving' the attaching molecules) is called an adsorbent; the material which is adsorbed (or attaches itself to the interface) is called an absorbate. Depending on the cohesive forces between the pair, a further subdivision is conducted. When the interaction is mainly caused by Van der Waals intermolecular forces, it is said that physisorption is occurring. If these forces are covalent, as the ones that are found in some chemical bonds, the phenomenon is called chemisorption. Generally, chemisorption tends to present a higher activation energy, which is related to a more significant dependence of the reaction rates on the system's temperature [201]. In adsorption, only a thin film from the solid phase takes part in the process, so that the molecules/atoms underneath the material's surface are not disturbed [170]. Examples of materials (adsorbent/adsorbate) that can store energy through sorption reactions include the silica gel/H_2O and zeolites/H_2O [201]. Most materials in this category work with H_2O as the volatile reactant and a solid material as the reactant pair.

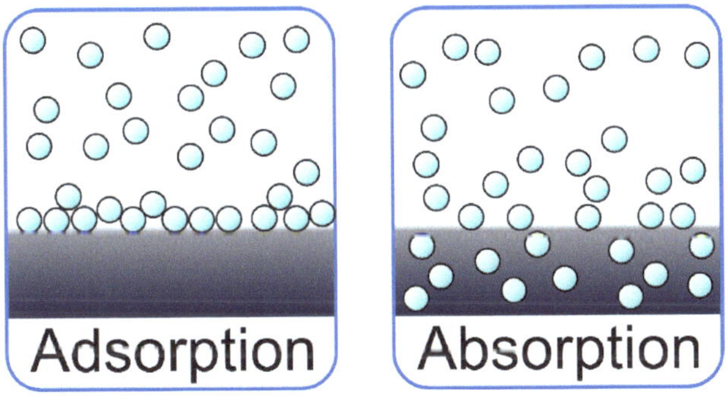

Figure 9. Schematic representation of adsorption and absorption.

While the process of adsorption is mainly related to the material's surface, absorption on the other hand, is a process in which the absorbate penetrates the bulk material, causing a change in composition, which occurs as a solution is formed. Usually, the absorbent is either solid or liquid and the absorbate is either liquid or gaseous [201].

The most relevant properties of sorption-type materials in the context of energy storage are the heat of reaction, thermal conductivity, thermal and chemical stability, toxicity, and corrosiveness [170].

Reaction-type TCS employ proper chemical reactions as a means of storing heat. A schematic representation of a reaction-type thermochemical heat storage material is presented in Figure 10. Here, some chemical A receives an amount of heat, causing its decomposition into B and C components. These components are kept separate, until the energy that is effectively stored in their chemical bonds is made available again, by combining B and C under adequate temperature and pressure conditions. Comparatively, these reactions can thermodynamically occur only at higher temperatures. While sorption-type materials tend to operate at temperatures below 200 °C [166], reaction-type materials can vary from room temperature to temperatures as high as 1400 °C [169].

A direct comparison of the gravimetric storage capacities from sorption-type and reaction-type materials reveal that most of the sorption-type systems have gravimetric energy storage densities around 2880 $kJ \cdot kg^{-1}$, with some candidates reaching as high as 4320 $kJ \cdot kg^{-1}$, like the $LiCl/H_2O$ pair [201]. Reaction-type TCS has some candidates with very high gravimetric energy storage densities, such as MgH_2/H_2 (2880 $kJ \cdot kg^{-1}$) and CaH_2/H_2 (<4430 $kJ \cdot kg^{-1}$) to name a few examples [169].

Some advantages of TCS are the long storage duration with low thermal losses (if the reaction products are stored separately), and the high gravimetric and volumetric heat storage densities. Furthermore, minimal thermal losses are experienced if the heat storage process is performed at room temperature [169,170]. This is, of course, of particular interest for long-term/seasonal TES. Also, unlike latent TES materials, the temperature of the heat storage/release process can be adjusted (within a material-specific operating window) by changing the system state (typically, a change in the partial pressure of some reactant in gaseous form) [210]. This allows highly flexible operation of a TCS system.

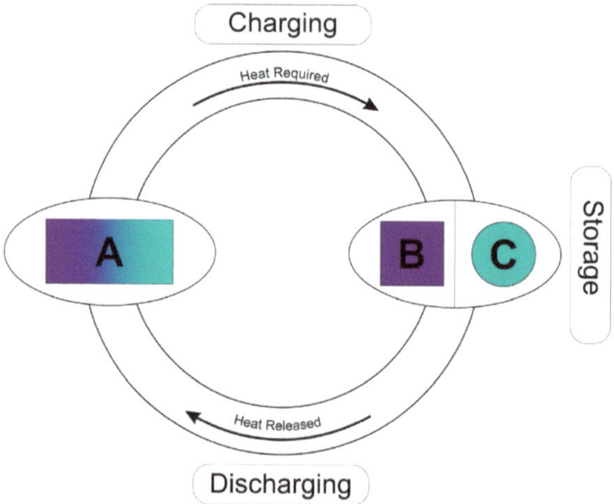

Figure 10. Schematic representation of TCS charging–storage–discharging cycle.

To define the operation windows and screen for useful materials for each application, an initial approach is to consider the equilibrium conditions for each TCS system. The equilibrium conditions of these systems can be better understood by making use of the Clausius–Clapeyron or the Van't Hoff equations (which, although coming from different formulations, yield the same expressions for the case of TCS) [211–213]. The Van't Hoff equation was previously defined as Equation (2) in chapter "Metal Hydrides". Specifically for reaction-based TCS materials, the reaction kinetics can be described in a very similar way as described in the Metal Hydrides chapter under the Section 2.3. Examples of such an approach can be seen for $MgO/Mg(OH)_2$ [214,215] and $CaO/Ca(OH)_2$ [216–219].

3.1.14. Hydroxide Systems

The two main hydroxide systems which have been investigated for heat storage applications are the $MgO/Mg(OH)_2$ and $CaO/Ca(OH)_2$ systems. The $CaO/Ca(OH)_2$ system is able to store/release heat at temperatures between 400 and 600 °C, depending on the pressure conditions, making it an interesting option for applications such as CSPs [216,218,220]. It has a reaction enthalpy of -104 kJ/mol, good cyclability and presents a high material volumetric energy of 195 kWh/m^3 [216–219,221].

Working at a lower temperature range, around 300 and 400 °C, the $MgO/Mg(OH)_2$ system also shows promise as a TES material. This material has a high gravimetric storage density of 1389 $kJ{\cdot}kg^{-1}$; however, it suffers from very low conversion rates between $Mg(OH)_2$ to MgO. Consequently, in its pristine state, storage densities are only limited to around 270 $kJ{\cdot}kg^{-1}$ [222]. This phenomena originates in the high activation energy associated with the hydration/dehydration cycles, as well as in the low thermal conductivity, and a tendency to agglomerate larger particles with the number of cycles, which leads to a decrease in porosity [222]. Strategies to counteract this low conversion include changing the material's microstructure and improving thermal conductivity. This has been achieved mainly through the use of additives such as C-based [223,224], Fe-based [222], Li-based [225] additives, among others. In particular, Tian et al. [222] investigated the cycling properties of pristine MgO and Fe-doped MgO particles. The addition of Fe led to the formation of a layered structure and enabled the material to reach a TES capacity of approximately 50 (i.e., around 700 $kJ{\cdot}kg^{-1}$) of its theoretical heat storage capacity even after 10 cycles. While promising, it remains to be seen whether the materials being investigated can maintain their layered structure after hundreds or even thousands of cycles.

Another strategy to improve the performance is through compression to produce pellets. According to Myagmarjav et al. [226], the addition of LiBr and expanded graphite to $Mg(OH)_2$ and a combination of compressing and drying of the obtained pellets, resulted in an increase in the thermal conductivity (measured in the perpendicular direction) from 0.28 $W\cdot(m\cdot K)^{-1}$ for pure $Mg(OH)_2$ pellets to 1.91 $W\cdot(m\cdot K)^{-1}$ for the obtained composite material.

3.1.15. Salt Hydrates

This class of materials is represented by mixtures between inorganic salts and water, which form a crystalline solid with general formula $M_pX_q\cdot m(H_2O)$ [227]. During the thermal charging phase, these materials undergo dehydration reactions, in which one or more molecules of water are separated for every molecule of salt, along with some compound with general formula $M_pX_q\cdot n(H_2O)$, in which $m > n$. When the stored energy is required, the material can react with water molecules once again (hydration). While these compounds are already used as latent TES materials [168], several authors have investigated them as TCS materials as well [170,227]. These materials have high energy densities and, on top of that, water is abundant, non-toxic, and inexpensive [227]. However, problems with reversibility are one of the major issues these materials face.

During dehydration, if the salt product is completely soluble in the water used for crystallization, then it is said that there is a congruent melting. If there is only partial solubility of the product salt, then it is said that there is an incongruent melting. Due to the difference in density between the salt product and water, a phase separation can occur, making subsequent reactions impossible. This phenomena can progressively decrease the heat storage capacity of these materials [168].

Due to their potential as materials for energy storage, investigations related to the development of new compositions/materials have been undertaken. As an example of these efforts, Kiyabu et al. [227] have used high-throughput density functional theory (DFT) calculations to look for suitable candidates for TCS applications. The authors claim to have found several hydrates which surpass the U.S. Department of Energy's system energy density target (at least 95 kWh/m^3 or 342 $MJ\cdot m^{-3}$).

Salt hydrates are generally divided as inorganic, nitrate, carbonate, chloride, sulfate, and fluoride salts. Several systems have been investigated so far. For applications in the range between 50 °C and 100 °C, salt hydrates like $CaCl_2$, $MgSO_4$, $SrBr_2$, Na_2S, and $MgCl_2$, are some of the most relevant [168,170].

3.1.16. Hydride Materials

In the Metal Hydrides chapter, the basic concepts and the development of these materials is reviewed as a means of storing hydrogen. In this section, hydride materials are considered for their ability to serve as TES materials. According to Sunku et al. [228], MHs have been investigated primarily for high-temperature applications, and the main candidates for those have been MgH_2, TiH_2, and CaH_2. Other interesting systems are complex metal hydrides, due to their thermodynamic properties. Carrillo et al. [229] notes that MHs have an outstanding energy storage density with an operation temperature that ranges from 250 °C to > 1000 °C. To put it into perspective, a sensible TES system based on molten salt exhibits a storage density of ca. 150 $kJ\cdot kg^{-1}$ per 100 °C of temperature change, while MgH_2 exhibits approximately 2811 $kJ\cdot kg^{-1}$ of heat storage capacity [229,230].

Among the materials investigated, MgH_2 is certainly one of the most studied. The abundance of Mg sources, its comparatively low cost, high H_2-gravimetric capacity, and high heat storage capacity make it a very good candidate for this kind of application [228]. Mg-based materials also have the advantage of containing no critical and/or hazardous components [229]. Compared to some other hydrides, pristine MgH_2 possesses a decent cyclability, but the material will experience a decrease in capacity over repeated cycling. This, however, can be improved by the use of additives [231,232]. Other Mg-based materials have been investigated for this application, such as Mg_2Ni/Mg_2NiH_6 [233,234] and Mg_2Fe/Mg_2FeH_6 [33] systems.

Metal hydrides have demonstrated their cyclic stability up of at least 2000 cycles; however, cost concerns [235–237] and the slow reaction kinetics remain the biggest hurdles for the deployment of these materials as a heat storage technology [228]. Their low thermal conductivity, problems related to sintering, and low tolerance towards gas phase impurities further impair their performance, particularly for scaled-up applications [230,238].

4. H_2-Storage Systems Coupled with TES

Up to this day, various strategies have been employed to enhance the performance of metal hydride hydrogen storage systems. As was previously mentioned in the Section 2, these include improving the intrinsic properties of the MH materials, optimizing the manufacturing processes, and the design of advanced tank configurations, which stand among some of the latest efforts found in the recent literature. Nonetheless, in most cases, these developments have not sufficiently improved the performance of MH reactors to make their scalability possible, for a couple of reasons. The most prominent example is associated with the comparatively low effective heat conductivity of the powder beds, which still demands more effort in the integration of different systems in innovative designs and configurations, as improvements in kinetic behavior are typically outweighed by heat management limitations [66]. Another frequently observed constraint is the desorption temperature of some MH candidates which present high H_2-storage capacities, like MgH_2, Mg_2Ni, $NaAlH_4$, and Li-RHC, but are not able to operate at near-room temperature and pressure conditions [10,42,72]. As has been previously mentioned in this review, a common solution to said problem is the addition of substitutional metal additives to modify the hydride's properties (like ΔH and ΔS), but this comes at the expense of cost. Some good candidates for H_2-storage also incorporate these expensive metals as part of their basic composition, or use them as a catalysts/additives to enhance kinetic behavior. Examples of these components and catalysts include precious metals such as Pd, Ru, and Pt [239], and more common metals like Co and V, which can still cost around three times the price of widely available materials like Mn [240]. However, all of these different hydrides share the same limitation: alone, they would require external energy to maintain their desorption process, making their overall efficiency very limited. It is in this context that the first advantage of coupling MH with TES systems comes in. Temporarily storing the heat of absorption in TES materials opens the door for the use of inexpensive high-temperature metal hydrides such as MgH_2 in stationary applications, as little to no additional heat has to be supplied [241]. Similarly, the coupling of MH storage tanks with TES solutions allows these systems to be installed independently of the existing heat-related operations that occur in their surroundings. Hence, H_2 could now be stored in locations that are physically distant from engines, turbines, fuel cells, or any other equipment that needed to be previously considered as a supply of waste heat for desorption. This allows for a more versatile system design and increases the number of potential applications for the use of H_2 as an energy carrier. Furthermore, by using TES materials as heat sinks/reservoirs, heat transfer in and out of the hydride bed is facilitated, leading to faster hydrogenation/dehydrogenation times [242], and thus potentially reducing the need to incorporate expensive metal catalysts to improve kinetics.

Having stated these advantages, the integration of metal hydride (MH) hydrogen storage systems with thermal energy storage (TES) is reviewed in the following section. Although MH–TES systems are still in the development stage, some applications scenarios highlight the potential of these systems, especially in industries where both hydrogen and heat are produced or consumed. Some fields that would benefit from MH–TES are the following:

- Power Generation: Surplus electricity produced from both renewable or non-renewable sources during low-demand periods could be converted to H_2 via an electrolyzer and stored in MH–TES systems for posterior grid load balancing in periods of peak demand. Waste heat from combined heat and power (CHP) plants could also be stored, using systems based on metal hydride pairs [243]. The incorporation of a TES system not only minimizes the need of additional energy inputs, but also allows for the subsequent re-conversion of hydrogen into electricity in times of the day where the processes that would otherwise provide heat for H_2 desorption are not operating. The application of MH–TES solutions is particularly attractive nowadays due to the ongoing switch to H_2 gas turbines.

- Industrial Processes and Waste Heat Recovery: Industries that produce or consume substantial amounts of either heat or hydrogen, such as the cement, petrochemical, glass, or steel industries, could integrate MH–TES systems and benefit from them in several ways. Waste heat from industrial unit operations could be captured by installed MH–TES systems and stored for later use in different processes. Alternatively, the heat generated during hydrogen absorption reactions can be leveraged through the use of TES materials and repurposed in other industrial operations. Processes that require molecular hydrogen could also directly obtain it from nearby MH tanks. On top of that, the incorporation of TES material might simultaneously guarantee a steady supply of H_2 for industrial operations, and benefit the system's overall energy efficiency, since TES materials can be used to control the temperature of the MH vessel, and therefore the rate of release of H_2 can be adjusted to meet the required inputs.
- District Heating and Cooling: MH–TES systems could be employed in district heating networks. Waste thermal energy generated by district power plants or by day-to-day residential activities can be used to operate MH heat pumps based on coupled MH pairs, and subsequently, either cool or provide heating for homes. This heat-pump concept has already been studied by Kumar et al. [244]. The benefit of including TES materials in these devices, similarly to other applications, would be to provide a more careful control of the H_2 release and uptake by each of the MH beds, and thus allow for the fine tuning of at home heating/cooling.

Table 4 offers an overview of the developments on MH hydrogen storage coupled with TES. The results of some key publications are then presented in more detail. Table 4 and the subsequent discussions are subdivided according to the TES method (sensible, latent, and thermochemical). Significantly more numerical studies were performed than experimental studies. Therefore, all experimental studies will be discussed in detail, while only selected numerical studies will be discussed.

Table 4. Overview on numerical and experimental studies on metal hydrides coupled with thermal energy storage.

TES Method	Metal Hydride	TES Material	Heat Exchange Strategies	Operating Temperature [°C]	Operation Pressure in Metal Hydride [bar]	H_2-Capacity [kg]	Key Findings	Study Type	Year	Reference
Sensible	$LaNi_5$	Concrete	Embedded cooling tubes with water as HTF	20–100	30 (abs)/0.1–0.9 (des)	0.5 kg H_2	The system stores 37% of the heat generated during hydrogen absorption. Optimized system with eight cooling tubes. Sensible TES reduces the need for external heating.	Numerical	2019	[245]
Sensible	$LaNi_5$	Water	Helical tube embedded in both reactors	20–97	4–20 (abs)/0.01–1.00 (des)	2.5 kg powder/0.0375 kg H_2	Integration with thermocline storage achieved 81.7% storage efficiency during absorption and delivered 60% of required heat during desorption with over 70% discharging efficiency.	Numerical	2023	[246]
Latent	MgH_2+4 at.-% Ti-V-Cr	Mg-Zn-Al Alloy	Cylindrical tank with internal heat exchanger	300–360	10 (abs)/1 (des)	10 kg material, 0.63 g H_2	The PCM effectively stored the heat of reaction during hydrogen absorption and provided it during desorption, resulting in a daily storage efficiency of 69%.	Experimental	2013	[241]
Latent	MgH_2	$Mg_{69}Zn_{28}Al_3$ (at.-%)	Cylindrical, annular reactor with metal hydride bed in the core and the PCM surrounding the metal hydride in the annular gap	300–370	8 (abs)	10 kg material, 0.63 g H_2	A numerical model was proposed which agrees well with the experimental results of a MgH_2 tank. It was shown that neglecting the natural convection in the fully melted PCM provides good results while simplifying the calculations.	Numerical	2013	[247]
Latent	MgH_2	$Mg_{69}Zn_{28}Al_3$ (at.-%)	Metal hydride and PCM were separated by steel wall.	300–360	10 (abs)/1 (des)	-	Two technical tank configurations were analyzed numerically. The spherical tank was found to have a higher system performance than a cylindrical tank.	Numerical	2015	[248]
Latent	Mg_2Ni	$NaNO_3$	Cylindrical tubes filled with PCM concentrically inside the MH bed	344 (abs)/270 (des)	15 (abs)/2 (des)	-	Integration of PCM reduced filling time by 58.1%. Aluminum foam enhances thermal conductivity, reducing charging/discharging times.	Numerical	2016	[249]
Latent	$LaNi_5$	$LiNO_3$–3 H_2O	Heat exchange through steel wall between metal hydride tank and PCM. The PCM is located in four cylindrical tubes within the $LaNi_5$ hydride bed.	20–70	10 (abs)/1 (des)	0.309 kg of $LaNi_5$ alloy~0.004 kg H_2 (est.)	Up to 80% of the hydrogen can be released utilizing the heat from the PCM.	Numerical	2016	[250]
Latent	$LaNi_5$	Paraffin RT35	Heat exchange through steel wall between metal hydride tank and PCM.	20–70	8 (abs)	0.270 kg powder/~0.004 kg H_2 (est.)	The influence of thermal conductivity, the mass and latent heat of the PCM on the system performance was studied. The high thermal conductivity of the PCM is beneficial to achieve a high capacity of the metal hydride storage system. The incorporation of metal foams further improves the system performance, with copper foam showing better performance than aluminum foam.	Numerical	2019	[251]

Table 4. *Cont.*

TES Method	Metal Hydride	TES Material	Heat Exchange Strategies	Operating Temperature [°C]	Operation Pressure in Metal Hydride [bar]	H_2-Capacity [kg]	Key Findings	Study Type	Year	Reference
Latent	Mg_2Ni	$NaNO_3$	Cylindrical sandwich bed packed with PCM	307	12 (abs)/3 (des)	-	A new MH-PCM configuration reduced the time for hydrogenation by 81.5% and dehydrogenation by 73% compared to conventional systems.	Numerical	2020	[242]
Latent	MgH_2	$NaNO_3$	Two reactor configurations: 1. Annular reactor with MgH_2 in the central part and the PCM located in the annular gap. Heat exchange through the steel wall. 2. Sandwiched MH-PCM units with steel brackets separating the units.	280–372	10 (abs)/1 bar (des)	-	The configuration of the PCM affects the heat transfer. The sandwiched configuration exhibits improved performances. Absorption and desorption times are 78% and 59% shorter, respectively, for the sandwiched configuration.	Numerical	2021	[252]
Latent	$LaNi_5$	$LiNO_3$·3 H_2O	The application of metal foam was investigated. PCM channels within the metal hydride bed were investigated.	20–70	10 (abs)/1 (des)	0.4813 kg of alloy; ~0.006 kg H_2 (est.)	PCM channels within hydride bed show superior performance compared to alternative structures of MH-PCM contacting.	Numerical	2021	[253]
Latent	MmNiMnCo (AB_5 alloy)	RT35HC (Organic PCM)	Integrated PCM unit for MH thermal management	35–40	10 (abs)/3 (des)	-	PCM-based thermal storage system captured heat released during hydrogen absorption and utilized it during hydrogen desorption, improving overall system efficiency.	Numerical	2022	[254]
Latent	MmNiMnCo (AB_5 alloy)	RT35HC (Organic PCM)	Integrated PCM unit for MH thermal management	35–40	10 (abs)/3 (des)	0.072 kg	With the introduction of the PCM, hydrogen flows could be maintained significantly longer than without the PCM. Mixing Cu foam significantly enhanced the H_2 charge and discharge performance.	Experimental	2022	[255]
Latent	$LaNi_5$	Paraffin + Metal Foams (Cu, Al, Ni, Ti)	Tube-and-shell heat exchanger	22–56	50 (abs)	-	Metal foams enhance PCM thermal conductivity and melting rate. The Cu-metal foam showed the best thermal performance. Economic analysis pointed out as necessary to choose best metal foam.	Numerical	2022	[256]
Latent	$LaNi_5$	Paraffin C13-C24 pure and with 5 wt.% metal oxide (Al_2O_3, MgO, SiO_2, SnO_2) nanoparticle additives	Heat exchange through steel wall between metal hydride tank and PCM. The PCM is located in four cylindrical tubes within the $LaNi_5$ hydride bed.	20–60	50 (abs)	-	Mixing metal oxide nanoparticles with PCM was studied on a large-scale metal hydride storage reactor system. No effect was found of including 5 wt.% metal oxide additives on the overall thermal and hydrogen storage performance of the reactor.	Numerical	2022	[257]

Table 4. *Cont.*

TES Method	Metal Hydride	TES Material	Heat Exchange Strategies	Operating Temperature [°C]	Operation Pressure in Metal Hydride [bar]	H_2-Capacity [kg]	Key Findings	Study Type	Year	Reference
Latent	$LaNi_5$	Paraffin RT35	Heat exchange through steel wall between metal hydride tank and PCM. The PCM is located in four cylindrical tubes within the $LaNi_5$ hydride bed.	20–70	6–12 (abs)	-	Natural convection in the PCM increases the average hydrogen storage by 13%. Optimizing the thermal conductivity of the PCM is important to increase the system thermal performance. Higher inlet pressure improves storage rate but increases compression costs. Optimal inlet pressure for the system found to be approximately 10 bar.	Numerical	2024	[258]
Latent	Mg_2FeH_6/$LaNi_5$	$C_2H_3O_2Na{\cdot}3H_2O$	Heat exchange between the $LaNi_5$ LTMH bed and the coupled PCM outer jacket occurs through the LTMH tank wall.	46.85–61.85	5.6–9.7 (abs and des)	0.0499 kg H_2	The proposed system couples Mg_2FeH_6 as a heat storage medium receiving waste heat and providing H_2 to the low-temperature MH ($LaNi_5$), which is coupled with a latent TES for thermal management. The temperature rise in the $LaNi_5$ powder bed during H_2 absorption is more pronounced at the center of the tank. Due to the melting of the PCM, its temperature rise is smaller than the one experienced by the MH powder bed itself. Desorption of H_2 occurs faster at the MH/PCM interface. Heat is absorbed faster by the PCM in the form of latent heat, than sensible heat. Authors reported an efficiency of around 87% for the proposed system.	Numerical	2024	[259]
Thermochemical	MgH_2 + 5 wt.% expanded natural graphite (ENG)	$Mg(OH)_2$ + ENG (Pellet, block and powder + Al foam)	Integrated thermochemical storage	305–370	10 (abs)/2 (des)	-	The system reduces the mass of the heat storage media by a factor of 4 compared to PCM-based systems. Offers flexibility in operating pressure conditions, enabling shorter hydrogen absorption times.	Numerical	2016	Bhouri et al. [260]
Thermochemical	MgH_2	$Mg(OH)_2$/MgO	Adiabatic storage reactor	280–330	1 (des)	-	Mg/MgH_2–MgO/$Mg(OH)_2$ system investigated numerically. Feasibility of H_2-desorption with MgO hydration in the system demonstrated. Compatible reaction times were identified.	Numerical	2019	Lutz et al. [210]
Thermochemical	$Mg_{90}Ni_{10}$	$Mg(OH)_2$/MgO	Adiabatic storage reactor	370 (abs)/300 (des)	9 (abs)/1 (des)	0.012 kg H_2	Absorption under 9 bar H_2 and 60 mbar H_2O; H_2 release and CaO hydration at 10 bar H_2O and 300 °C demonstrated. High volumetric capacity at low pressures.	Numerical and Experimental	2020	Lutz et al. [261]

Table 4. *Cont.*

TES Method	Metal Hydride	TES Material	Heat Exchange Strategies	Operating Temperature [°C]	Operation Pressure in Metal Hydride [bar]	H_2-Capacity [kg]	Key Findings	Study Type	Year	Reference
Thermochemical	$LaNi_5$	$Na_2HPO_3 \cdot 7 H_2O$	Metal hydride contained in a steel tank surrounded by PCM directly.	20–70	6–10 (abs)	0.078 kg powder; ~0.001 kg H_2 (est.)	A 3D model of a metal hydride reactor with PCM was presented. The effects of operating pressure and PCM properties were investigated. A higher operating pressure is beneficial in achieving a higher reaction fraction, e.g., the reacted fraction at 2400 s increases from 53% at 6 bar to 77% at 10 bar. Reported metal hydride reactor gravimetric capacity of 0.58 wt.%. Electrical power of FC coupled system 8.2 kJ.	Numerical and Experimental	2020	Yao et al. [262]
Thermochemical	MgH_2	$Mg(OH)_2/MgO$	A heat exchanger with fins was modeled. The heat exchanger fins were optimized as part of the study.	300–370	10 (abs)	-	The heat of reaction of the Mg/MgH_2 system is stored using the thermochemical dehydrogenation of $Mg(OH)_2$. Through topology optimization of the heat exchanger and the metal hydride and TES material, the charging performance can be improved.	Numerical	2022	Shi et al. [263]
Thermochemical	MgH_2	$Mg(OH)_2/MgO$	No HT fluid, TCS material around MH	300–360	10 (abs)	-	Absorption takes place at 10 bar and desorption is studied at 5 bar backpressures. The thermal management can be optimized by adjusting the water vapor pressure. An increase in water pressure decreases the dehydrogenation time of the MgH_2 material.	Numerical	2022	Shi et al. [264]
Thermochemical	$LaNi_5$	$LiNO_3 \cdot 3 H_2O$	Double-walled cylinder with MH/PCM (no thermal losses)	20–120	6 (abs)/0.068 (des)	0.0324 kg H_2	Development of a non-dimensional methodology to identify MH-PCM/TCS materials pairings based on charge/discharge and equivalent power.	Numerical	2024	Maggini et al. [265]

4.1. Metal Hydride: Sensible Thermal Energy Storage

While sensible TES systems are the most used in current heat storage applications [168], they are not frequently considered as candidates for coupling with MH systems. There are two important reasons for this. First, the comparatively low storage density of sensible TES significantly increases the size of the whole hydrogen storage system, and thus reduces the volumetric and gravimetric H_2 densities. Second, the non-isothermal behavior during the thermal charging/discharging process is not optimal for MH-based H_2 storage systems, which ideally require a constant temperature of the heat sink and source during the hydrogenation and dehydrogenation reactions, respectively. Latent and thermochemical TES systems exhibit higher storage densities and store/release heat more isothermally than sensible TES systems.

Two numerical studies where sensible heat storage materials and hydride systems were coupled are presented in more detail in this section. Tiwari et al. [245] studied the heat exchange between a $LaNi_5$-containing H_2 storage tank and a concrete-based TES system, using a 3D numerical model developed in COMSOL Multiphysics 5.2 software. The proposed system is shown in Figure 11 and consists of two cylindrical tanks containing the metal hydride and the TES material. The heat exchange was facilitated by water as a heat transfer fluid, flowing through embedded cooling tubes.

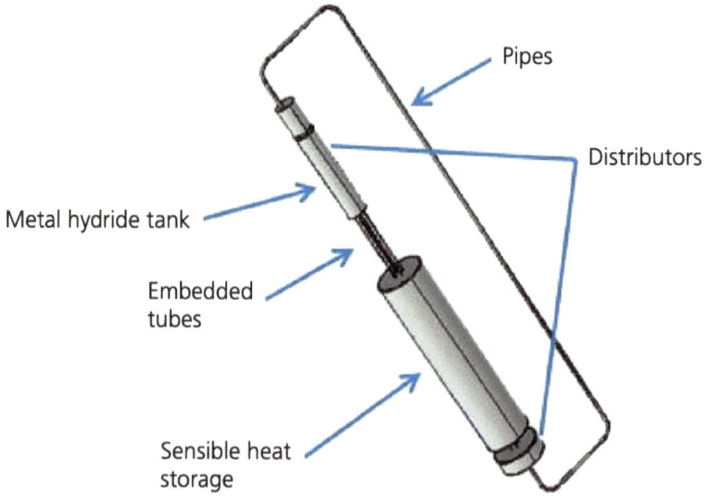

Figure 11. Integrated hydrogen storage system with a sensible heat-storage system. "Reprinted with permission from Ref. [245]. 2019, Elsevier".

The influence of the dimensions of the hydride tank on the reaction rates was studied. The highest reaction rates and lowest H_2 charging times were found for tank dimensions of $D{\cdot}L = (0.15\text{ m}){\cdot}(0.7\text{ m})$. The number and arrangement of the cooling tubes was studied, and it was found that eight tubes homogeneously distributed in the hydride bed were the optimal number (the maximum number of tubes studied was twelve). The velocity of the heat transfer fluid was also evaluated. Lower velocities allowed for higher heat transfer rates to the sensible storage system, but reduced cooling efficiency in the metal hydride tank. A velocity of 0.1 m/s was considered optimal for balancing these effects. The study found that, with the optimal hydride tank design, a maximum of 37% of the heat generated during hydrogen absorption could be stored in the concrete-based system. This limit of 37% was attributed to the low thermal conductivity of concrete [245].

In a later study [246], the same authors numerically investigated a $LaNi_5$ hydride storage tank in combination with a water-based sensible TES system. The system design is shown in Figure 12.

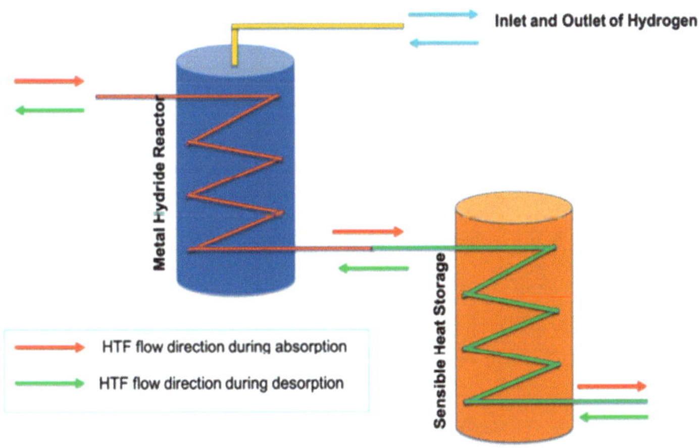

Figure 12. Schematic of sensible heat storage system coupled with metal hydride reactor. "Reprinted with permission from Ref. [246]. 2023, Elsevier".

Here, the metal hydride reactor was equipped with helical heat exchanger tubes, through which the heat exchange fluid (water) could flow. During the H_2 absorption process, the water temperature increased as it flowed through the metal hydride bed and reached a maximum increment of 32 K. A total of 81.7% of the heat emitted during absorption was successfully stored by the TES system (at an H_2 supply pressure of 16 bar). During the subsequent desorption process, 60% of the heat required to discharge the metal hydride reactor could be supplied by the TES system. The authors mention that the system's efficiency can be further improved by optimizing the helical tube design and operating parameters inside the metal hydride storage reactor [246].

4.2. Metal Hydride: Latent Thermal Energy Storage

The earliest studies of coupling latent TES systems with metal hydrides were published in 2013 by Garrier et al. [241] and Marty et al. [247]. In these studies, the possibility of reducing the amount of external heat required for H_2 desorption from MgH_2 was explored. MgH_2 requires high-temperature heat (>300 °C) to release its hydrogen. Such heat is classified as high-grade heat, which is thermodynamically and economically valuable. Thus, there is strong motivation to reduce the amount of externally supplied heat for H_2 desorption from MgH_2 by storing and re-using the heat from absorption.

Garrier et al. [241] built a 10 kg MgH_2/0.63 kg H_2 prototype reactor, which employed a metallic Mg–Zn alloy ($Mg_{69}Zn_{28}Al_3$) as PCMs for latent TES. Compacted disks of MgH_2 mixed with a 4 at.% Ti–V–Cr alloy were inserted into a cylindrical reactor. This reactor was surrounded by an external cylindrical tube containing pure Mg–Zn alloy. The heat released during Mg hydrogenation flowed through the tank wall (which separated the two compartments) and melted the Mg–Zn alloy. The heat released during the solidification of the alloy was then used for hydrogen desorption.

Figure 13 shows the temperature and pressure conditions of the metal hydride and Mg Zn alloy PCM during the absorption and desorption processes. The red curve represents the equilibrium conditions of the Mg/MgH_2 system. At any point below that curve, the equilibrium shifts towards the formation Mg and H_2. Above that curve, the equilibrium favors the formation of the hydride, MgH_2. In order to maintain a temperature gradient between the metal hydride and the PCM, the temperature and pressure conditions in the MgH_2 reactor were very different during absorption and desorption. The PCM melted and solidified at ca. 340 °C. Absorption took place at around 375 °C and 10 bar, and desorption took place at around 310 °C and 2 bar.

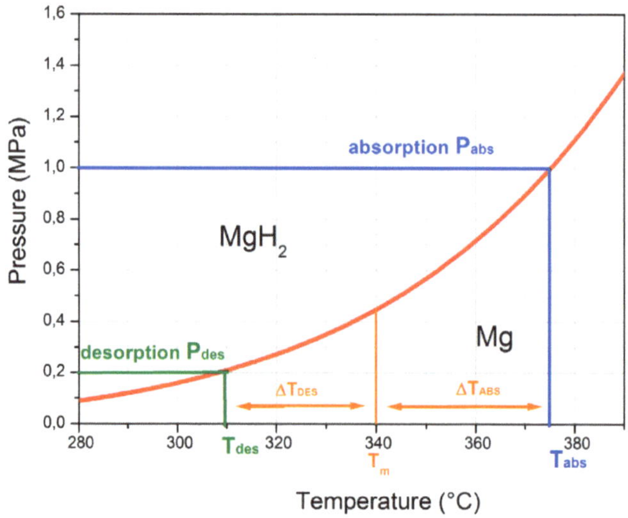

Figure 13. Temperature and pressure conditions of the Mg/MgH_2 system coupled with the Mg–Zn alloy phase change material. "Reprinted with permission from Ref. [241]. 2013, Elsevier".

Heat was exchanged between hydride and PCM through the steel wall, which separated the two materials. No additional heat transfer enhancement methods were employed, such as fins or a heat transfer fluid.

Considering a 24 h charge/discharge cycle, the daily efficiency of the reported tank was about 69%. The authors propose better insulation and a scale-up of the system to further improve this efficiency. They further note that the efficiency of the first prototype was already much better than the efficiency of a conventional MgH_2 tank, which had a daily efficiency of 37% [241].

To investigate the influence of the tank shape on the system performance, Mellouli et al. [248] numerically studied different geometrical configurations of MgH_2 tanks coupled with $Mg_{69}Zn_{28}Al_3$ PCM to compare their heat exchange efficiency. Cylindrical and spherical tank shapes were studied and compared with each other in regard to the time required to absorb and desorb H_2. It was found that a spherical tank shape is superior in terms of the hydrogen charging time. A 22% improvement was achieved by spherical tank, in terms of the time required to reach 80% of its hydrogen storage capacity, compared to the cylindrical tank.

Ye et al. [252] numerically investigated two different reactor types for the coupling of MgH_2 with a $NaNO_3$ TES material. The first reactor configuration had an annular reactor-type design similar to the reactor reported by Garrier et al. [241]. The second reactor had a cylindrical, sandwich-type design, in which circular metal hydride and PCM disks were alternately layered on top each other as shown in Figure 14.

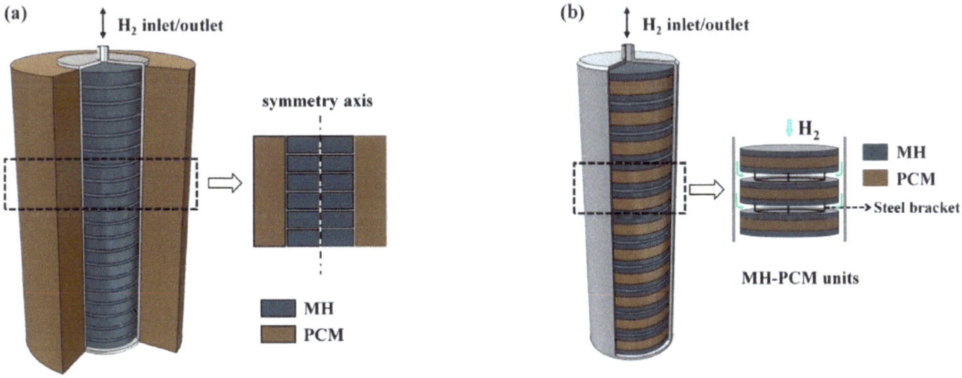

Figure 14. Two reactor designs studied by Ye et al. [252]: (**a**) the PCM surrounding layout (annular reactor type) and (**b**) sandwiched metal hydride—PCM units. "Reprinted with permission from Ref. [252]. 2021, Elsevier".

It was found that the hydrogen absorption and desorption time of the sandwiched MH–PCM unis were 78% and 59% shorter, respectively, than in the case of annular reactor design. This behavior was attributed to the enhanced heat transfer between the hydride and PCM disks in the case of the sandwich reactor. For the same hydrogen storage capacity, the heat transfer area of the sandwich reactor was 2.5 times larger than in the annular reactor. Thermal resistances between the hydride and PCM were also considerably lower [252].

The studies by Garrier et al. [241], Mellouli et al. [248] and Ye et al. [252] on MgH_2 coupled with latent TES materials highlight the importance of the tank design for its performance.

The Mg_2Ni/Mg_2NiH_4 material is another high-temperature hydride system, which operates at similar temperature conditions as Mg/MgH_2 [266]. No experimental studies have been reported for the coupling of this system with a TES material yet. However, in a numerical study by Mellouli et al. [249], Mg_2Ni/Mg_2NiH_4 was investigated with sodium nitrate ($NaNO_3$) as latent TES material. Mg_2Ni/Mg_2NiH_4 is able to absorb and desorb hydrogen at temperatures around 344 °C and 272 °C and pressures of 15 bar and 2 bar for absorption and desorption, respectively. These temperature ranges make the selection of $NaNO_3$ adequate, since it melts at a temperature of 307 °C [267]. An adequate temperature gradient is provided for both processes.

In their design, aluminum foam was incorporated to improve the heat exchange in both the metal hydride reactor as well as the PCM compartment. Four configurations were investigated, as shown in Figure 15.

- Design 1: A basic configuration with a disk of Mg_2Ni alloy surrounded by PCM stacked on a central filter tube.
- Design 2: Similarly to Design 1 but with 48 encapsulated spherical shells filled with PCM, arranged concentrically inside the MH bed.
- Design 3: Each spherical shell in Design 2 is replaced by a hexagonal tube filled with PCM.
- Design 4: Each hexagonal tube in Design 3 is replaced by a cylindrical tube filled with PCM.

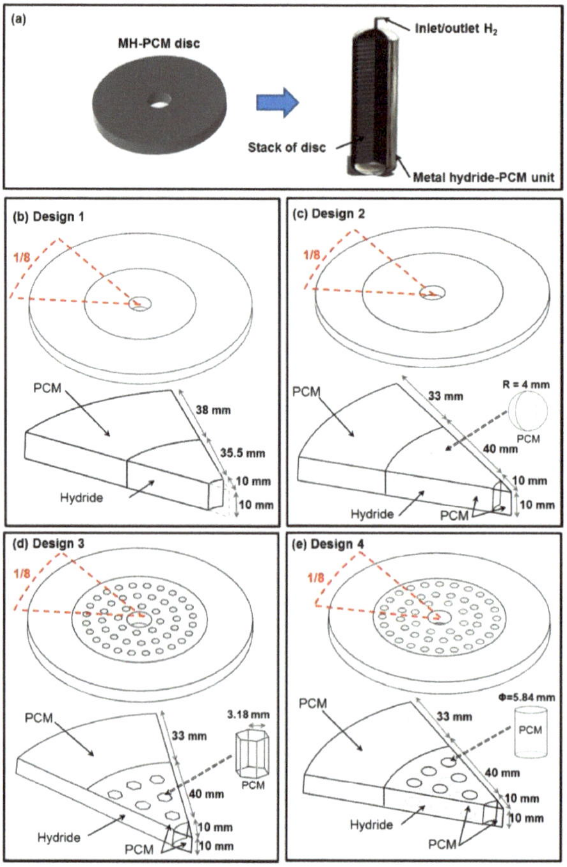

Figure 15. Different geometrical configurations of metal hydride—PCM reactors as studied numerically by Mellouli et al. [249]. (**a**) Base design for MH-TES reactor; (**b**) Design 1: Disks with no PCM embedded in MH; (**c**) Design 2: Disks with embedded PCM spheres; (**d**) Design 3: Disks with embedded PCM hexagonal tubes; (**e**) Design 4: Disks with embedded PCM cylindrical tubes "Reprinted with permission from Ref. [249]. 2016, Elsevier".

Mellouli et al. found that Design 4 was the most efficient option. This conclusion was rationalized based on the observation that cylindrical tubes filled with PCM inside the hydride bed favored a more effective heat exchange than the spherical shells or the hexagonal tubes. The geometry of the cylindrical tubes provides a larger surface area for heat exchange, resulting in more efficient thermal management. Design 4 showed a 58.1% decrease in filling time, relative to the basic configuration (Design 1) [249]. It was also found that the aluminum foam enhances the effective thermal conductivity of the hydride bed as well as that of the PCM medium. This improvement is critical, as it results in reduced heat accumulation in the hydride bed, which accelerates the heat transfer between metal hydride and TES material. It was also mentioned that a balance must be found between the enhancement of the heat transfer characteristics of the hydride and PCM beds and an acceptable hydrogen storage capacity, which is reduced by the addition of the foam [249].

In a later numerical study conducted by Alqahtani et al. [242], the heat transfer efficiency of two different cylindrical reactor configurations that also employed Mg_2Ni/Mg_2NiH_4, with $NaNO_3$ as a PCM, was analyzed. The following designs were investigated, as shown in Figure 16:

- Design 1: A metal hydride reactor was surrounded by a single cylindrical jacket packed with the PCM. The heat exchange took place at the interface between the hydride reactor and the PCM jacket.
- Design 2: A metal hydride reactor was enclosed by PCM in a cylindrical sandwich bed, which increased the heat transfer area by adding more interfaces between the MH reactor and the PCM.

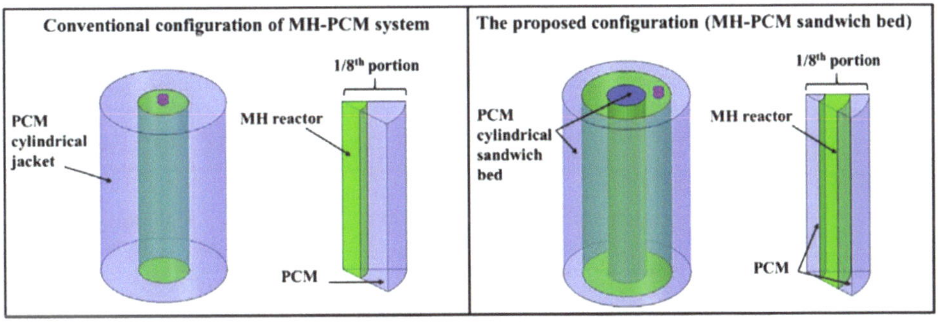

Figure 16. Different geometrical configurations of MH–PCM cylindrical reactors as studied by Alqahtani et al. [242]. "Reprinted with permission from Ref. [242]. 202, Elsevier".

It was found that Design 2 showed a dramatic improvement in terms of the hydrogenation and dehydrogenation rates compared to Design 1. The hydrogenation rate improved by 81.5%, and the dehydrogenation by 73%. This was attributed to a greater heat transfer rate in Design 2, which was achieved as a result of the increase in heat transfer area [242].

In addition to high-temperature hydrides such as MgH_2 and Mg_2NiH_4, the coupling of low-temperature hydrides such as $LaNi_5$ and other AB_5-type alloys with latent TES materials has been investigated (see Table 4 for a comprehensive list). For this class of hydrides, the number of numerical studies also greatly exceeds the number of experimental studies.

Nguyen et al. [255] experimentally studied the application of metal foam as a means to improve heat transfer between a $LaNi_5$ hydrogen storage canister and paraffin wax (RT28HC) latent TES material, as shown in Figure 17.

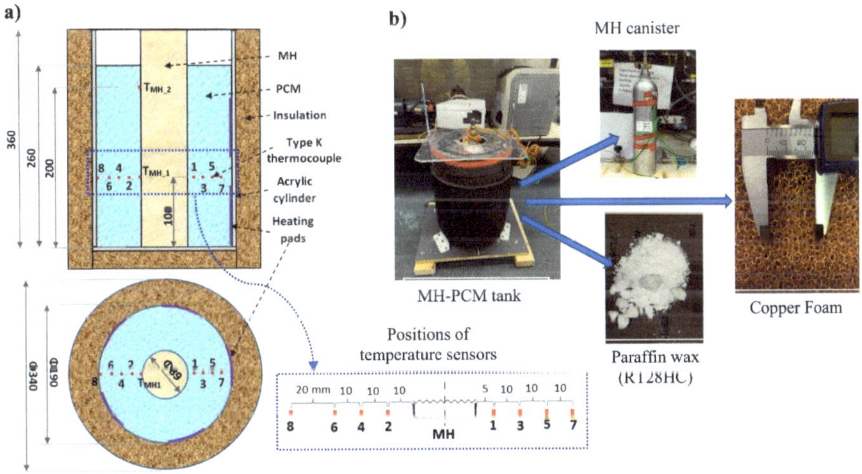

Figure 17. (**a**) Schematic of the $LaNi_5$ hydrogen storage (MH) with PCM thermal management system and the positions of thermocouples; (**b**) Photograph of MH–PCM hydrogen storage setup (with and without metal foam (MF)). "Reprinted with permission from Ref. [255]. 2022, Elsevier".

An 800-NL hydride canister was tested for its ability to charge and discharge hydrogen at flow rates of 1.1 slpm, 1.5 slpm, and 2 slpm. Without the PCM, the canister could only absorb or release hydrogen for a short duration, until a maximum of 12% of the total storage capacity was reached. Introducing RT28HC paraffin wax PCM allowed the system to maintain hydrogen flows for longer periods, achieving up to 26% and 29% of the total storage capacity during charging and discharging at 1.1 slpm. The study also found that the PCM's thermal conductivity significantly affects the MH system's performance. Pure PCM with low thermal conductivity (0.2 W/(m·K)) limited the effectiveness of the thermal management solution. Enhancing the PCM with copper foam increased its effective thermal conductivity to about 4.3 W/(m·K), improving its thermal and charging/discharging performance. The copper foam-embedded PCM maintained the hydride canister's temperature around 28 °C, which is ideal for the PCM's functionality. The enhanced system achieved much higher storage utilization, reaching 62.5% and 84% of the canister's theoretical capacity at 1.1 slpm and 1.5 slpm over 450 min [255].

In a numerical study by Chibani et al. [256], different metal foams were investigated to enhance heat transfer between $LaNi_5$ hydrogen storage material and a paraffin PCM. The setup design is shown in Figure 18. It consists of a cylindrical tank filled with $LaNi_5$ metal hydride and a central hydrogen feed tube surrounded by four PCM tubes which were arranged in a rhombus pattern. These tubes were embedded with metal foams (aluminum, copper, nickel, and titanium) to enhance heat transfer.

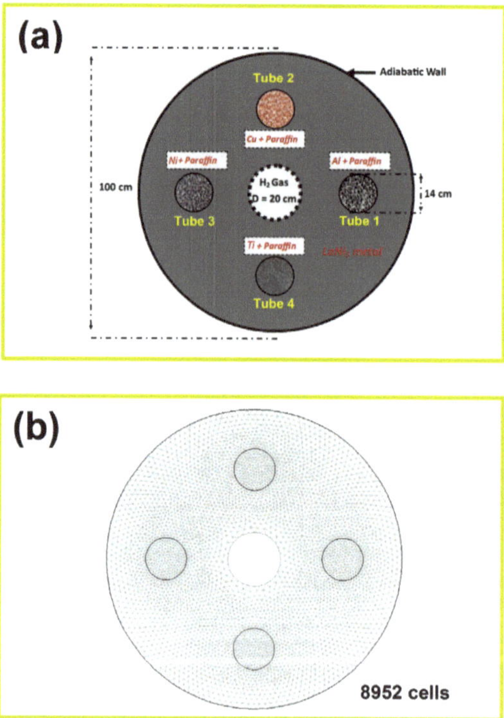

Figure 18. Dimensional schematic of the metal hydride—PCM reactor (**a**) and mesh configuration after grid independence analysis (**b**) as shown by Chibani et al. [256]. "Reprinted with permission from Ref. [256]. 2022, Elsevier". The incorporation of metal foams significantly improved heat transfer. Copper showed the best results, followed by aluminum, nickel, and titanium. The porosity of the foam also played a crucial role, particularly, on the melting time of the paraffin. Lower porosities (<80%) resulted in higher thermal conductivities within the foam-embedded PCM phase, and led to a considerable reduction in the melting time of paraffin (from over 16,000 s to less than 500 s) [256].

4.3. Metal Hydride: Thermochemical Thermal Energy Storage

This section deals with the reports on the coupling of metal hydrides with thermochemical TES systems. Thermochemical TES systems exhibit the highest theoretical thermal storage capacities of the three TES methods. This attribute is attractive for their coupling with hydrogen storage systems, where the system gravimetric and volumetric hydrogen densities are important parameters.

In this section, a sequence of three consecutive studies is presented, which all originated from the same laboratory at the German Aerospace Center (DLR) in Germany. These studies properly illustrate the sequence of steps needed to transition from a feasibility study to a numerical study, and finally to a working prototype. In the first study, performed by Bhouri et al. [260] in 2016, a feasibility analysis of a solid-state hydrogen storage reactor is presented. MgH_2 was used as the metal hydride, and $Mg(OH)_2$ was used as the thermochemical TES material. An advantage of thermochemical TES materials is their ability to store and release heat energy at different temperature and pressure conditions. Besides their high storage density, this behavior makes them attractive for coupling with metal hydrides. By adjusting the pressure of the heat storage and release reactions, it is possible to perfectly match the operating conditions of the thermochemical TES system to the optimal operating conditions of the metal hydride.

Bhouri et al. analytically investigated a novel reactor concept, in which the heat released during hydrogen absorption in Mg was stored in the $Mg(OH)_2$ bed by driving its dehydration reaction, and vice versa. Figure 19 shows the operating principle.

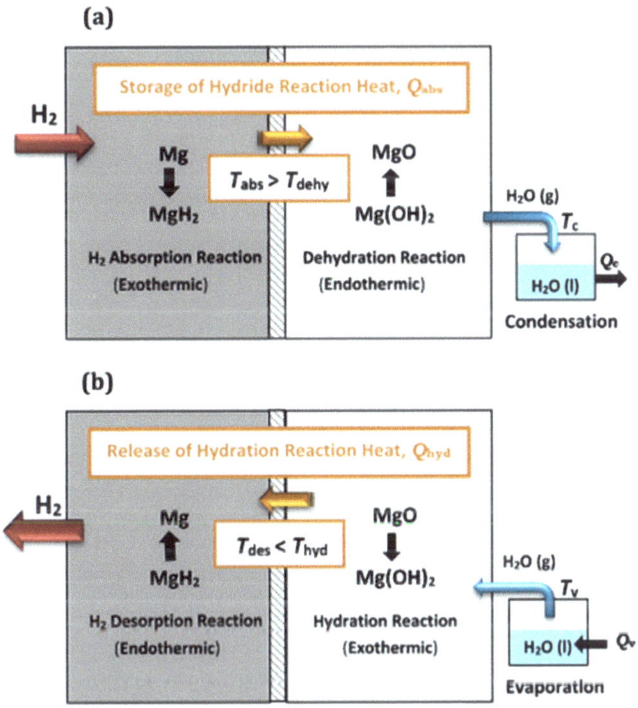

Figure 19. Operating principle of Mg/MgH_2 reactor using $Mg(OH)_2$ as heat storage media: (**a**) absorption of H_2/dehydration of $Mg(OH)_2$/condensation of H_2O and (**b**) desorption of H_2/hydration of MgO/evaporation of H_2O. "Reprinted with permission from Ref. [260]. 2016, Elsevier".

The study concluded that the use of $Mg(OH)_2$ could reduce the overall mass of the heat storage system by a factor of four compared to traditional PCMs. This, in turn, resulted in an increase in the system's gravimetric H_2 storage capacity compared to the utilization of PCMs as TES solutions. Garrier et al. reported a system gravimetric H_2 capacity as 0.315 wt.% when utilizing a Mg–Zn–Al PCM [241]. The concept proposed by Bhouri et al. exhibited a H_2 capacity of 1.5 wt.% [260]. Additionally, the TES material costs of the thermochemical TES system were also reported to be considerably cheaper compared to the PCM costs (by a factor of around 12), thus demonstrating the high potential of this concept. During the analysis, the thickness of the metal hydride and thermochemical material layers was optimized, and additives were used to enhance the thermal conductivity of $Mg(OH)_2$, emphasizing their importance [260].

The first analytical feasibility study by Bhouri et al. [260] was followed up by two numerical studies performed by Bhouri et al. [268] and by Lutz et al. [210], in which the H_2 absorption and desorption processes were investigated, respectively. The same material combination was used as in the initial feasibility study [260]. In this review, the results of the H_2 desorption study [210] are presented in detail. In said study, the results were obtained by running a 2D numerical simulation. The reactor was designed with a double-walled cylindrical geometry, separating the metal hydride bed from the thermochemical bed. The basic design of the reactor is shown in Figure 20.

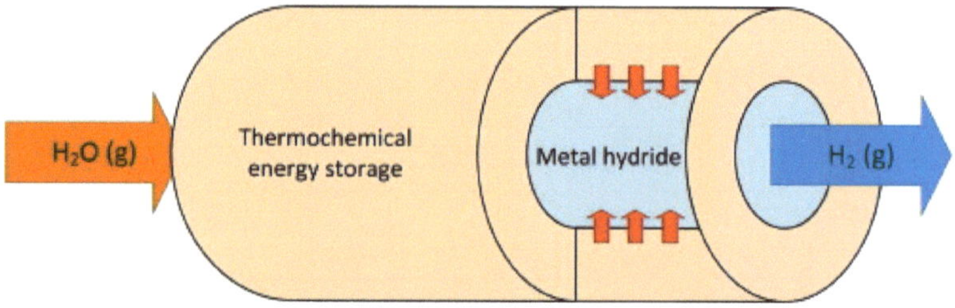

Figure 20. Design proposed by Lutz et al. [210] as shown in the original article as a graphical abstract. "Reprinted with permission from Ref. [210]. 2019, Elsevier".

It was found that the hydrogen release process was completed in 132 min, which the authors of the study suggest would be suitable for stationary applications. It was also found that the efficiency of the system is highly dependent on the thermal conductivity of the reactive beds, with the MgO bed being particularly sensitive. A concept was proposed to utilize the waste heat of a high-temperature PEM fuel cell in combination with a superheater to generate water vapor at the desired temperature and pressure level of 350 °C and 10 bar, which was found to be adequate for the hydration reaction. It was also found that adjusting the water vapor pressure for higher temperatures during the MgO hydration could enhance heat transfer between the TES and metal hydride compartments. This resulted in a faster dehydrogenation [210].

On the basis of the three previous studies [210,260,268], Lutz et al. [261] performed an experimental study on a H_2 storage reactor prototype including peripheral components, combining MgH_2 in the form of a 90% Mg/10% Ni alloy ($Mg_{90}Ni_{10}$) as the metal hydride and $Mg(OH)_2$ as the thermochemical TES material. The authors did not provide an explanation for their switch from pure Mg, which was utilized in the previous studies, to the Mg–Ni alloy. The experimental setup (Figure 21) consisted of three main components:

- Water vapor infrastructure, where water vapor was supplied or withdrawn using a tube bundle that acted as an evaporator/condenser, regulated by thermal oils. This allowed the control of the water vapor pressure in the reactor, and the measurement of the reacted fraction of the thermochemical system via liquid level changes and pressure sensors.
- The hydrogen infrastructure, where hydrogen was supplied or withdrawn from the reactor using volume flow controllers.

- The reactor, which consisted of a double-walled tube with the metal hydride in the form of 300 mm diameter pellets located in the inner tube, and $Mg(OH)_2$ powder in the outer tube for heat absorption. The reactor included thermocouples for temperature monitoring and was designed to exclude mass transfer limitations.

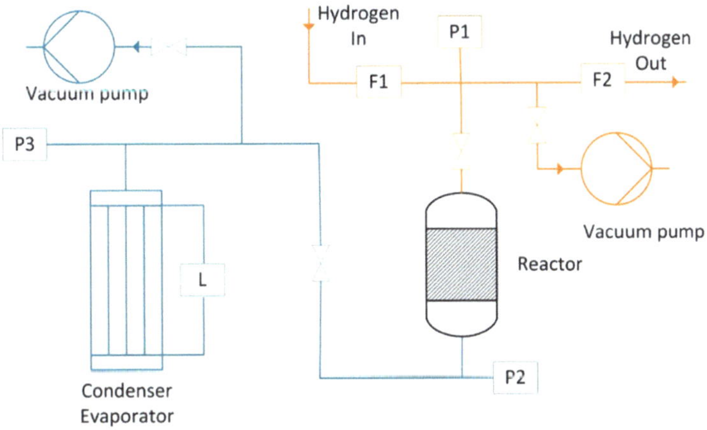

Figure 21. Design of the test bench schematic proposed by Lutz et al. [261]; H_2 and H_2O vapor infrastructure are represented in orange and blue, respectively; F: Flow meter; P: Pressure sensor; L: Level meter. "Reprinted with permission from Ref. [261]. 2020, Elsevier".

The study demonstrated that hydrogen storage was achieved at pressures as low as 9 bar, storing hydrogen at 20.8 g H_2/L based on the reactive materials alone. It was found that the optimal operating temperature for the magnesium oxide hydration was around 300 °C, achieved at a water vapor pressure of 9.75 bar. This temperature was sufficient for the dehydrogenation of the magnesium hydride. Thermal losses and poor kinetics of the $Mg(OH)_2$/MgO system prevented full conversion during simultaneous thermochemical cooling and heating. However, the proposed reactor concept allows for high-capacity hydrogen storage at low pressures, making it a promising option for stationary storage applications.

5. Future Developments

Throughout this review, metal hydrides have been presented as a promising means to store hydrogen, and consequently contribute to the integration of hydrogen-based technologies into the global energy infrastructure. However, further research is still necessary to successfully optimize the properties of metal hydride beds themselves and make H_2 storage in metal hydrides industrially viable. Addressing the thermal management challenges inherent in metal hydrides remains especially critical for achieving efficient, cost-effective, and reliable hydrogen storage. For this reason, coupling MH systems with thermal management solutions such as TES emerges as a favorable approach, offering a synergistic solution which maximizes energy efficiency by reusing stored thermal energy. In the following sections, key topics related to the improvement of MH–TES hydrogen storage systems, such as material optimization, heat transfer improvements, and cost reduction strategies are discussed, highlighting the importance of heat integration alongside other metal hydride improving strategies.

5.1. Material and System Development

Despite the progress made in coupling metal hydride systems with thermal energy storage solutions (MH–TES), these technologies still have a low degree of maturity. Research and development are key to the success of hydrogen storage in metal hydrides, with material optimization and the design of efficient systems that can adequately manage heat transfer being crucial aspects to improve upon. The latter is particularly important, as the efficiency of MH systems depends on the ability to store and reuse heat during the hydrogen absorption/desorption processes.

Regarding developments in the properties of hydride materials, some widely known compounds such as MgH_2 have been the subject of ample investigation, with existing works focusing on its various alloys, system configurations, and enhancement additives. This, however, is not the case of complex metal hydrides (for example), whose performance still needs to be improved, despite presenting some favorable properties. Complex metal hydrides show great promise as hydrogen storage materials for stationary applications, mainly due to their high H_2 capacity and potentially low raw material cost, but their relatively high operating temperatures of over 100 °C makes effective thermal management imperative. Here, the application of TES materials could play an important role in the future; however, there have been no studies on complex hydrides coupled with TES systems to date.

As has been mentioned throughout this review, a significant challenge in improving these systems rests in the poor thermal conductivity of both metal hydrides and TES storage materials. In metal hydrides, this deficiency leads to uneven temperature distributions during hydrogen absorption/desorption, which complicates the heat exchange process and results in operational inefficiencies. Meanwhile, poor thermal conductivity in TES materials has been shown to be a limiting factor, and reduce system performance [261]. Adding metal foams has been suggested as a solution to improve thermal conductivity of PCMs at low temperatures. However, for high-temperature applications, the potential reactivity of metals with TES materials could become an issue. In this regard, encapsulating PCMs with ceramic materials for operation at high temperatures has been explored as a potential solution [269].

Cyclic stability has also presented itself as an issue in metal/hydrogen systems which must be solved, particularly in the case of complex hydrides and Mg-based systems [118]. It has been long known that many metal hydrides tend to deactivate or passivate in the presence of impurities in the gas phase [270] or when brought in contact with ambient air [19]. These unfortunate behaviors greatly reduce the potential of metal hydrides for practical applications, as significant efforts to purify H_2 gas would be required before exposing it to the metal hydride, and the implementation of costly safety measures during the operation and maintenance of metal hydride systems becomes necessary. More research is required to reduce the sensitivity of metal hydrides when they are in contact with gasses other than H_2.

While all the above-mentioned developments might greatly contribute to the competitiveness of MH and MH–TES technologies, significant challenges lie in how these coupled systems are designed. Heat management plays a crucial role in their effectiveness. It is thus important not only to take into consideration strategies that might minimize heat losses to the environment, but also ways to improve the heat exchange between MH and TES during operation. In these systems, having heat losses means that additional energy inputs will need to be factored in to compensate for the lost heat, which directly translates into higher operational costs and, somewhat indirectly, into higher capital expenditures, because of the additional measures that need to be devised to mitigate these losses. Due to the importance of these two aspects, more research efforts should be dedicated to investigating optimum design strategies/geometries for coupling MH–TES systems in different use-case scenarios, while minimizing environmental heat losses whenever possible.

5.2. Economic Feasibility

Building a commercial-scale MH–TES system requires numerous materials and technical components. The costs of hydride materials remain high, especially for complex metal hydrides. Thermochemical materials can be expensive as they often contain rare metals. Catalysts and thermal additives are also costly, but necessary, as many MHs depend on them to guarantee reversible storage behavior, and for property enhancements. Besides elevated investment costs, the operating costs of MH systems can also be considerable. Inflated operating costs could potentially be offset by integrating TES materials, as they allow for the recovery and reuse of heat during the absorption/desorption cycles. This reduces or could even eliminate the need for additional energy inputs, improving the overall energtic and economic efficiency of the system. Durability is also crucial for economic viability, with systems needing to cycle around 11,000 times over 30 years with minimal degradation [11,271].

A practical aspect to consider is the integration of metal hydride hydrogen storage systems into existing infrastructures with established heat exchange systems. By leveraging preexisting heat exchange systems, the additional costs of TES implementation may be reduced. This approach could be especially relevant to ensure viability in industrial applications, where heat exchange networks are already in place.

Efficient synthetic routes are critical to process large volumes of materials quickly and cost-effectively, and only through research will overall production prices be eventually reduced. For instance, Wang et al. proposed a new direct synthesis of complex metal hydrides ($NaAlH_4$ and $LiAlH_4$) through hydrogenation of their respective metal hydrides with aluminum in the presence of a catalyst and a liquid complexing agent. This process produces $NaAlH_4$ and $LiAlH_4$ in high yield at ambient temperature and near ambient pressure conditions, a breakthrough that will lead to reductions in production costs [30]. Low-cost synthesis routes like this one will encourage researchers and engineers to integrate complex hydrides in large-scale developments.

Another important aspect rests in the sustainability of these systems. Recycling metal hydrides after usage and incorporating recycled materials into metal hydride synthesis would not only be beneficial environmentally, but also economically. As an example, Guerrero-Ortiz et al. [272] have developed a viable synthetic route from Al-cans into $NaAlH_4$ metal hydride using NaH and TiF_3 as a catalyst. This approach could create a much needed synergy between the recycling and energy industries, potentially leading to reduced costs with further development [272]. Passing et al. [273] studied the use of scrap Mg/Al metal waste for hydrogen storage. Al served as an enhancement additive to MgH_2, increasing its thermal conductivity. The incorporation of recycling processes in the production of metal hydride hydrogen storage systems is still in the early stages of development but holds significant promise in driving the overall economic viability of hydrogen storage technologies. Finally, it is worth noting that coupled MH–TES hydrogen storage systems are made-up of numerous components which are usually expensive. Unless the production costs of MH materials can be reduced, MH–TES systems will present themselves as financially unattractive solutions. Therefore, maintaining a balance between material and system costs remains a critical factor.

6. Conclusions

Despite the growing interest in the development of MH–TES systems, different aspects impact their economic and technical viability. Regarding MH materials themselves, much effort has been dedicated to not only develop new materials, but also to improve the kinetic behavior and propose efficient ways to improve the effective heat conductivity of powders and compacts. Other important developments in this field are related to the production methods of these materials, the utilization of low-cost and earth-abundant raw materials, and the use of recycled materials.

Developments related to the systems themselves have focused on the optimization of geometry, with numerous works numerically investigating the behavior of these systems. Special emphasis has been given to the heat management of these systems. The positioning of heat exchange surfaces and the use of fins and baffles are some of the most prominent examples of investigated aspects in the field.

The diversity of TES materials is overwhelming and the properties of each class of materials affect their use. Further developments of TES materials being currently used in renewable energy sites might also produce new opportunities for MH–TES systems.

While no systematic bibliographical review was carried out, some conclusions with the sampling presented here can be pointed out. The surveyed works suggest that sensible heat storage materials have been studied the least in relation to the other two TES types, which might suggest that further consideration and a careful analysis of the merits and shortcomings of this pairing should be made to avoid overlooking good potential candidates.

A significant amount of the studies surveyed sought to couple MH systems with latent TES materials. This trend suggests that researchers are looking for materials with higher thermal capacity. For the thermochemical TES materials coupled with MH systems, few studies were found. This might be related either to the complexity of these systems or the amount of knowledge, models and data required to enable the development of such systems. Still, due to their outstanding heat storage capacities, the further development of new pairings might be one of the alternatives to obtain systems with promising characteristics for hydrogen and energy storage. It is possible that one of the bottlenecks for the development of thermochemical TES materials with MH systems is the inherent difficulty and complexity of finding candidates that match well in operation temperature as well as thermodynamic and kinetic behavior. The development of a framework for the selection of material pairing in this regard would be highly desirable.

Most of the studies available in the literature for MH–TES systems use different numerical methods to evaluate the system. However, the vast majority of the experimental studies also involved numerical investigations or were based on previous work, in which the designed system was first investigated numerically. This shows how fundamental these numerical studies have become for the development of laboratory-scale prototypes and proof-of-concept designs.

In spite of the current achievements, an accelerating demand for zero-emission energy technologies is putting more pressure on the race to develop economically competitive solutions. Most of the possible MH and TES combinations are yet to be explored, and the leap from theory to practice is only just beginning. Therefore, this work emerges as a starting point for future research efforts on the matter, be it in the form of optimizing already studied MH–TES systems or conceiving new ones.

Author Contributions: Conceptualization: U.U.; Resources: J.J., L.H. and U.U.; Writing–Original Draft Preparation, M.D.C., M.R.d.T.A., A.M.N. and U.U.; Writing–Review and Editing, M.D.C., A.M.N., L.H., J.J. and U.U.; Supervision: U.U.; Project Administration: U.U.; Funding Acquisition: J.J. and U.U. All authors have read and agreed to the published version of the manuscript.

Funding: Funding by the Energie Campus Nürnberg—Nachwuchsprogramm and Technische Hochschule Nürnberg Georg Simon Ohm is gratefully acknowledged. This publication is partly funded by dtec.bw—Digitalization and Technology Research Center of the Bundeswehr which we gratefully acknowledge. dtec.bw is funded by the European Union—NextGenerationEU.

Conflicts of Interest: The authors declare no conflicts of interest.

References

1. IEA Report Electricity. 2024. Available online: https://www.iea.org/reports/electricity-2024 (accessed on 13 September 2024).
2. IEA Report World Energy Outlook. 2023. Available online: https://www.iea.org/reports/world-energy-outlook-2023 (accessed on 13 September 2024).
3. Møller, K.T.; Jensen, T.R.; Akiba, E.; Li, H. Hydrogen—A Sustainable Energy Carrier. *Prog. Nat. Sci. Mater. Int.* **2017**, *27*, 34–40. [CrossRef]
4. Scita, R.; Raimondi, P.P.; Noussan, M. *Green Hydrogen: The Holy Grail of Decarbonisation? An Analysis of the Technical and Geopolitical Implications of the Future Hydrogen Economy*; Fondazione Eni Enrico Mattei (FEEM): Milan, Italy, 2020. [CrossRef]
5. AlZohbi, G.; Almoaikel, A.; AlShuhail, L. An Overview on the Technologies Used to Store Hydrogen. *Energy Rep.* **2023**, *9*, 28–34. [CrossRef]
6. Drawer, C.; Lange, J.; Kaltschmitt, M. Metal Hydrides for Hydrogen Storage—Identification and Evaluation of Stationary and Transportation Applications. *J. Energy Storage* **2024**, *77*, 109988. [CrossRef]
7. Chatterjee, S.; Parsapur, R.K.; Huang, K.-W. Limitations of Ammonia as a Hydrogen Energy Carrier for the Transportation Sector. *ACS Energy Lett.* **2021**, *6*, 4390–4394. [CrossRef]

8. Hirscher, M. (Ed.) *Handbook of Hydrogen Storage: New Materials for Future Energy Storage*; Wiley-VCH: Weinheim, Germany, 2010; ISBN 978-3-527-32273-2.
9. Kukkapalli, V.K.; Kim, S.; Thomas, S.A. Thermal Management Techniques in Metal Hydrides for Hydrogen Storage Applications: A Review. *Energies* **2023**, *16*, 3444. [CrossRef]
10. Klopčič, N.; Grimmer, I.; Winkler, F.; Sartory, M.; Trattner, A. A Review on Metal Hydride Materials for Hydrogen Storage. *J. Energy Storage* **2023**, *72*, 108456. [CrossRef]
11. Sandrock, G. A Panoramic Overview of Hydrogen Storage Alloys from a Gas Reaction Point of View. *J. Alloys Compd.* **1999**, *293–295*, 877–888. [CrossRef]
12. Liu, J.; Yang, F.; Wu, Z.; Zhang, Z. A Review of Thermal Coupling System of Fuel Cell-Metal Hydride Tank: Classification, Control Strategies, and Prospect in Distributed Energy System. *Int. J. Hydrogen Energy* **2024**, *51*, 274–289. [CrossRef]
13. Cetinkaya, S.A.; Disli, T.; Soyturk, G.; Kizilkan, O.; Colpan, C.O. A Review on Thermal Coupling of Metal Hydride Storage Tanks with Fuel Cells and Electrolyzers. *Energies* **2022**, *16*, 341. [CrossRef]
14. Nguyen, H.Q.; Shabani, B. Proton Exchange Membrane Fuel Cells Heat Recovery Opportunities for Combined Heating/Cooling and Power Applications. *Energy Convers. Manag.* **2020**, *204*, 112328. [CrossRef]
15. Somo, T.R.; Maponya, T.C.; Davids, M.W.; Hato, M.J.; Lototskyy, M.V.; Modibane, K.D. A Comprehensive Review on Hydrogen Absorption Behaviour of Metal Alloys Prepared through Mechanical Alloying. *Metals* **2020**, *10*, 562. [CrossRef]
16. Alapati, S.V.; Karl Johnson, J.; Sholl, D.S. Using First Principles Calculations to Identify New Destabilized Metal Hydride Reactions for Reversible Hydrogen Storage. *Phys. Chem. Chem. Phys.* **2007**, *9*, 1438. [CrossRef]
17. Salman, M.S.; Pratthana, C.; Lai, Q.; Wang, T.; Rambhujun, N.; Srivastava, K.; Aguey-Zinsou, K.-F. Catalysis in Solid Hydrogen Storage: Recent Advances, Challenges, and Perspectives. *Energy Technol.* **2022**, *10*, 2200433. [CrossRef]
18. Meyers, R.A. (Ed.) *Encyclopedia of Physical Science and Technology*, 3rd ed.; Academic Press: San Diego, CA, USA, 2002; ISBN 978-0-12-227410-7.
19. Huston, E.L.; Sandrock, G.D. Engineering Properties of Metal Hydrides. *J. Common. Met.* **1980**, *74*, 435–443. [CrossRef]
20. Graetz, J. Metastable Metal Hydrides for Hydrogen Storage. *ISRN Mater. Sci.* **2012**, *2012*, 863025. [CrossRef]
21. Sakintuna, B.; Lamaridarkrim, F.; Hirscher, M. Metal Hydride Materials for Solid Hydrogen Storage: A Review. *Int. J. Hydrogen Energy* **2007**, *32*, 1121–1140. [CrossRef]
22. Poeppelmeier, K.R.; Reedijk, J. *Comprehensive Inorganic Chemistry II: From Elements to Applications*; Elsevier: Amsterdam, The Netherlands, 2013; ISBN 978-0-08-096529-1.
23. Züttel, A. FUELS—HYDROGEN STORAGE | Hydrides. In *Encyclopedia of Electrochemical Power Sources*; Elsevier: Amsterdam, The Netherlands, 2009; pp. 440–458. ISBN 978-0-444-52745-5.
24. Huheey, J.E. *Inorganic Chemistry: Principles of Structure and Reactivity*, 3rd ed.; Harper & Row: New York, NY, USA, 1983; ISBN 978-0-06-042987-4.
25. Lys, A.; Fadonougbo, J.O.; Faisal, M.; Suh, J.-Y.; Lee, Y.-S.; Shim, J.-H.; Park, J.; Cho, Y.W. Enhancing the Hydrogen Storage Properties of AxBy Intermetallic Compounds by Partial Substitution: A Short Review. *Hydrogen* **2020**, *1*, 38–63. [CrossRef]
26. Liu, W.; Webb, C.J.; Gray, E.M. Review of Hydrogen Storage in AB 3 Alloys Targeting Stationary Fuel Cell Applications. *Int. J. Hydrogen Energy* **2016**, *41*, 3485–3507. [CrossRef]
27. Marinelli, M.; Santarelli, M. Hydrogen Storage Alloys for Stationary Applications. *J. Energy Storage* **2020**, *32*, 101864. [CrossRef]
28. Abdin, Z.; Tang, C.; Liu, Y.; Catchpole, K. Current State and Challenges for Hydrogen Storage Technologies. In *Towards Hydrogen Infrastructure*; Elsevier: Amsterdam, The Netherlands, 2024; pp. 101–132. ISBN 978-0-323-95553-9.
29. Andersson, J.; Grönkvist, S. Large-Scale Storage of Hydrogen. *Int. J. Hydrogen Energy* **2019**, *44*, 11901–11919. [CrossRef]
30. Wang, L.; Aguey-Zinsou, K.-F. Synthesis of LiAlH4 Nanoparticles Leading to a Single Hydrogen Release Step upon Ti Coating. *Inorganics* **2017**, *5*, 38. [CrossRef]
31. Chen, Z.; Ma, Z.; Zheng, J.; Li, X.; Akiba, E.; Li, H.-W. Perspectives and Challenges of Hydrogen Storage in Solid-State Hydrides. *Chin. J. Chem. Eng.* **2021**, *29*, 1–12. [CrossRef]
32. Puszkiel, J.; Garroni, S.; Milanese, C.; Gennari, F.; Klassen, T.; Dornheim, M.; Pistidda, C. Tetrahydroborates: Development and Potential as Hydrogen Storage Medium. *Inorganics* **2017**, *5*, 74. [CrossRef]
33. Bogdanović, B.; Reiser, A.; Schlichte, K.; Spliethoff, B.; Tesche, B. Thermodynamics and Dynamics of the Mg–Fe–H System and Its Potential for Thermochemical Thermal Energy Storage. *J. Alloys Compd.* **2002**, *345*, 77–89. [CrossRef]
34. Thiangviriya, S.; Plerdsranoy, P.; Hagenah, A.; Le, T.T.; Kidkhunthod, P.; Utke, O.; Dornheim, M.; Klassen, T.; Pistidda, C.; Utke, R. Effects of Ni-Loading Contents on Dehydrogenation Kinetics and Reversibility of Mg2FeH6. *Int. J. Hydrogen Energy* **2021**, *46*, 32099–32109. [CrossRef]
35. Züttel, A.; Borgschulte, A.; Orimo, S.-I. Tetrahydroborates as New Hydrogen Storage Materials. *Scr. Mater.* **2007**, *56*, 823–828. [CrossRef]
36. Puszkiel, J.; Gasnier, A.; Amica, G.; Gennari, F. Tuning $LiBH_4$ for Hydrogen Storage: Destabilization, Additive, and Nanoconfinement Approaches. *Molecules* **2019**, *25*, 163. [CrossRef]
37. Mauron, P.; Buchter, F.; Friedrichs, O.; Remhof, A.; Bielmann, M.; Zwicky, C.N.; Züttel, A. Stability and Reversibility of $LiBH_4$. *J. Phys. Chem. B* **2008**, *112*, 906–910. [CrossRef]
38. Ali, N.A.; Sazelee, N.A.; Ismail, M. An Overview of Reactive Hydride Composite (RHC) for Solid-State Hydrogen Storage Materials. *Int. J. Hydrogen Energy* **2021**, *46*, 31674–31698. [CrossRef]

39. Vajo, J.J.; Skeith, S.L.; Mertens, F. Reversible Storage of Hydrogen in Destabilized $LiBH_4$. *J. Phys. Chem. B* **2005**, *109*, 3719–3722. [CrossRef]
40. Ding, Z.; Li, S.; Zhou, Y.; Chen, Z.; Yang, W.; Ma, W.; Shaw, L. LiBH4 for Hydrogen Storage—New Perspectives. *Nano Mater. Sci.* **2020**, *2*, 109–119. [CrossRef]
41. Neves, A.M.; Puszkiel, J.; Capurso, G.; Bellosta Von Colbe, J.M.; Milanese, C.; Dornheim, M.; Klassen, T.; Jepsen, J. Modeling the Kinetic Behavior of the Li-RHC System for Energy-Hydrogen Storage: (I) Absorption. *Int. J. Hydrogen Energy* **2021**, *46*, 32110–32125. [CrossRef]
42. Neves, A.M.; Puszkiel, J.; Capurso, G.; Bellosta Von Colbe, J.M.; Klassen, T.; Jepsen, J. Development of a New Approach for the Kinetic Modeling of the Lithium Reactive Hydride Composite (Li-RHC) for Hydrogen Storage under Desorption Conditions. *Chem. Eng. J.* **2023**, *464*, 142274. [CrossRef]
43. Laughlin, D.E.; Hono, K. *Physical Metallurgy*, 5th ed.; Elsevier: Amsterdam, The Netherlands, 2014; ISBN 978-0-444-53770-6.
44. Aguey-Zinsou, K.-F.; Ares-Fernández, J.-R. Hydrogen in Magnesium: New Perspectives toward Functional Stores. *Energy Environ. Sci.* **2010**, *3*, 526. [CrossRef]
45. Li, Q.; Lu, Y.; Luo, Q.; Yang, X.; Yang, Y.; Tan, J.; Dong, Z.; Dang, J.; Li, J.; Chen, Y.; et al. Thermodynamics and Kinetics of Hydriding and Dehydriding Reactions in Mg-Based Hydrogen Storage Materials. *J. Magnes. Alloys* **2021**, *9*, 1922–1941. [CrossRef]
46. Siegmann, H.C.; Schlapbach, L.; Brundle, C.R. Self-Restoring of the Active Surface in the Hydrogen Sponge La Ni_5. *Phys. Rev. Lett.* **1978**, *40*, 972–975. [CrossRef]
47. Kirchheim, R.; Pundt, A. Hydrogen in Metals. In *Physical Metallurgy*; Elsevier: Amsterdam, The Netherlands, 2014; pp. 2597–2705. ISBN 978-0-444-53770-6.
48. Lototskyy, M.V.; Tarasov, B.P.; Yartys, V.A. Gas-Phase Applications of Metal Hydrides. *J. Energy Storage* **2023**, *72*, 108165. [CrossRef]
49. Bloch, J.; Mintz, M.H. Kinetics and Mechanisms of Metal Hydrides Formation—A Review. *J. Alloys Compd.* **1997**, *253–254*, 529–541. [CrossRef]
50. Atilio Puszkiel, J. Tailoring the Kinetic Behavior of Hydride Forming Materials for Hydrogen Storage. In *Gold Nanoparticles—Reaching New Heights*; Rahman, M., Mohammed Asiri, A., Eds.; IntechOpen: London, UK, 2019; ISBN 978-1-78984-998-1.
51. Campo, M.; Tanaka, A.; Mendes, A.; Sousa, J.M. Characterization of Membranes for Energy and Environmental Applications. In *Advanced Membrane Science and Technology for Sustainable Energy and Environmental Applications*; Elsevier: Amsterdam, The Netherlands, 2011; pp. 56–89. ISBN 978-1-84569-969-7.
52. Wilde, G. (Ed.) Nanostructured Materials. In *Frontiers of Nanoscience*, 1st ed.; Elsevier: Oxford, UK; Boston, MA, USA, 2009; ISBN 978-0-08-044965-4.
53. Balasubramaniam, R. Hysteresis in Metal–Hydrogen Systems. *J. Alloys Compd.* **1997**, *253–254*, 203–206. [CrossRef]
54. Qian, S.; Northwood, D. Hysteresis in Metal-Hydrogen Systems: A Critical Review of the Experimental Observations and Theoretical Models. *Int. J. Hydrogen Energy* **1988**, *13*, 25–35. [CrossRef]
55. Flanagan, T.B.; Clewley, J.D. Hysteresis in Metal Hydrides. *J. Common. Met.* **1982**, *83*, 127–141. [CrossRef]
56. Schwarz, R.B.; Khachaturyan, A.G. Thermodynamics of Open Two-Phase Systems with Coherent Interfaces: Application to Metal–Hydrogen Systems. *Acta Mater.* **2006**, *54*, 313–323. [CrossRef]
57. Wang, X.; Suda, S. Reaction Kinetics of Hydrogen-Metal Hydride Systems. *Int. J. Hydrogen Energy* **1990**, *15*, 569–577. [CrossRef]
58. Pang, Y.; Li, Q. A Review on Kinetic Models and Corresponding Analysis Methods for Hydrogen Storage Materials. *Int. J. Hydrogen Energy* **2016**, *41*, 18072–18087. [CrossRef]
59. Murdoch, J.R. What Is the Rate-Limiting Step of a Multistep Reaction? *J. Chem. Educ.* **1981**, *58*, 32. [CrossRef]
60. Wu, H. Strategies for the Improvement of the Hydrogen Storage Properties of Metal Hydride Materials. *ChemPhysChem* **2008**, *9*, 2157–2162. [CrossRef]
61. Dornheim, M. Thermodynamics of Metal Hydrides: Tailoring Reaction Enthalpies of Hydrogen Storage Materials. In *Thermodynamics—Interaction Studies—Solids, Liquids and Gases*; Moreno Pirajn, J.C., Ed.; InTech: Tokyo, Japan, 2011; ISBN 978-953-307-563-1.
62. Ulmer, U.; Oertel, D.; Diemant, T.; Bonatto Minella, C.; Bergfeldt, T.; Dittmeyer, R.; Behm, R.J.; Fichtner, M. Performance Improvement of V–Fe–Cr–Ti Solid State Hydrogen Storage Materials in Impure Hydrogen Gas. *ACS Appl. Mater. Interfaces* **2018**, *10*, 1662–1671. [CrossRef]
63. Zhang, L.; Sun, Z.; Yao, Z.; Yang, L.; Yan, N.; Lu, X.; Xiao, B.; Zhu, X.; Chen, L. Excellent Catalysis of Mn_3O_4 Nanoparticles on the Hydrogen Storage Properties of MgH_2: An Experimental and Theoretical Study. *Nanoscale Adv.* **2020**, *2*, 1666–1675. [CrossRef]
64. Ali, N.A.; Idris, N.H.; Sazelee, N.A.; Yahya, M.S.; Yap, F.A.H.; Ismail, M. Catalytic Effects of $MgFe_2O_4$ Addition on the Dehydrogenation Properties of LiAlH4. *Int. J. Hydrogen Energy* **2019**, *44*, 28227–28234. [CrossRef]
65. Dematteis, E.M.; Berti, N.; Cuevas, F.; Latroche, M.; Baricco, M. Substitutional Effects in TiFe for Hydrogen Storage: A Comprehensive Review. *Mater. Adv.* **2021**, *2*, 2524–2560. [CrossRef]
66. Lozano, G.A.; Bellosta Von Colbe, J.M.; Klassen, T.; Dornheim, M. Transport Phenomena versus Intrinsic Kinetics: Hydrogen Sorption Limiting Sub-Process in Metal Hydride Beds. *Int. J. Hydrogen Energy* **2014**, *39*, 18952–18957. [CrossRef]
67. Stampfer, J.F.; Holley, C.E.; Suttle, J.F. The Magnesium-Hydrogen System[1–3]. *J. Am. Chem. Soc.* **1960**, *82*, 3504–3508. [CrossRef]
68. Bogdanović, B.; Bohmhammel, K.; Christ, B.; Reiser, A.; Schlichte, K.; Vehlen, R.; Wolf, U. Thermodynamic Investigation of the Magnesium–Hydrogen System. *J. Alloys Compd.* **1999**, *282*, 84–92. [CrossRef]

69. Dematteis, E.M.; Dreistadt, D.M.; Capurso, G.; Jepsen, J.; Cuevas, F.; Latroche, M. Fundamental Hydrogen Storage Properties of TiFe-Alloy with Partial Substitution of Fe by Ti and Mn. *J. Alloys Compd.* **2021**, *874*, 159925. [CrossRef]
70. Luo, W.; Craft, A.; Kuji, T.; Chung, H.S.; Flanagan, T.B. Thermodynamic Characterization of the ZrNi H System by Reaction Calorimetry and P-c-t Measurements. *J. Common. Met.* **1990**, *162*, 251–266. [CrossRef]
71. Kölbig, M.; Bürger, I.; Linder, M. Characterization of Metal Hydrides for Thermal Applications in Vehicles below 0 °C. *Int. J. Hydrogen Energy* **2019**, *44*, 4878–4888. [CrossRef]
72. Khan, D.; Zou, J.; Zeng, X.; Ding, W. Hydrogen Storage Properties of Nanocrystalline Mg_2Ni Prepared from Compressed $2MgH_2$-Ni Powder. *Int. J. Hydrogen Energy* **2018**, *43*, 22391–22400. [CrossRef]
73. Zhu, Z.; Zhu, S.; Lu, H.; Wu, J.; Yan, K.; Cheng, H.; Liu, J. Stability of LaNi5-Co Alloys Cycled in Hydrogen—Part 1 Evolution in Gaseous Hydrogen Storage Performance. *Int. J. Hydrogen Energy* **2019**, *44*, 15159–15172. [CrossRef]
74. Klyamkin, S.N.; Zakharkina, N.S. Hysteresis and Related Irreversible Phenomena in $CeNi_5$-Based Intermetallic Hydrides. *J. Alloys Compd.* **2003**, *361*, 200–205. [CrossRef]
75. Bellosta Von Colbe, J.; Ares, J.-R.; Barale, J.; Baricco, M.; Buckley, C.; Capurso, G.; Gallandat, N.; Grant, D.M.; Guzik, M.N.; Jacob, I.; et al. Application of Hydrides in Hydrogen Storage and Compression: Achievements, Outlook and Perspectives. *Int. J. Hydrogen Energy* **2019**, *44*, 7780–7808. [CrossRef]
76. Bogdanović, B.; Brand, R.A.; Marjanović, A.; Schwickardi, M.; Tölle, J. Metal-Doped Sodium Aluminium Hydrides as Potential New Hydrogen Storage Materials. *J. Alloys Compd.* **2000**, *302*, 36–58. [CrossRef]
77. Martelli, P.; Caputo, R.; Remhof, A.; Mauron, P.; Borgschulte, A.; Züttel, A. Stability and Decomposition of $NaBH_4$. *J. Phys. Chem. C* **2010**, *114*, 7173–7177. [CrossRef]
78. Puszkiel, J.A.; Larochette, P.A.; Gennari, F.C. Thermodynamic and Kinetic Studies of Mg–Fe–H after Mechanical Milling Followed by Sintering. *J. Alloys Compd.* **2008**, *463*, 134–142. [CrossRef]
79. Chen, P.; Xiong, Z.; Luo, J.; Lin, J.; Tan, K.L. Interaction of Hydrogen with Metal Nitrides and Imides. *Nature* **2002**, *420*, 302–304. [CrossRef] [PubMed]
80. Kojima, Y.; Kawai, Y. IR Characterizations of Lithium Imide and Amide. *J. Alloys Compd.* **2005**, *395*, 236–239. [CrossRef]
81. Puszkiel, J.A.; Castro Riglos, M.V.; Ramallo-López, J.M.; Mizrahi, M.; Karimi, F.; Santoru, A.; Hoell, A.; Gennari, F.C.; Larochette, P.A.; Pistidda, C.; et al. A Novel Catalytic Route for Hydrogenation–Dehydrogenation of 2LiH + MgB_2 via In Situ Formed Core–Shell Li_x TiO_2 Nanoparticles. *J. Mater. Chem. A* **2017**, *5*, 12922–12933. [CrossRef]
82. Ye, J.; Li, Z.; Zhang, L.; Wang, S.; Jiang, L. Measurement and the Improvement of Effective Thermal Conductivity for a Metal Hydride Bed—A Review. *RSC Adv.* **2022**, *12*, 25722–25743. [CrossRef]
83. Broom, D.P. *Hydrogen Storage Materials: The Characterisation of Their Storage Properties*; Green Energy and Technology; Springer: London, UK; Heidelberg, Germany, 2011; ISBN 978-0-85729-220-9.
84. Jepsen, J.; Milanese, C.; Girella, A.; Lozano, G.A.; Pistidda, C.; Bellosta Von Colbe, J.M.; Marini, A.; Klassen, T.; Dornheim, M. Compaction Pressure Influence on Material Properties and Sorption Behaviour of $LiBH_4$–MgH_2 Composite. *Int. J. Hydrogen Energy* **2013**, *38*, 8357–8366. [CrossRef]
85. Pohlmann, C.; Röntzsch, L.; Weißgärber, T.; Kieback, B. Heat and Gas Transport Properties in Pelletized Hydride–Graphite-Composites for Hydrogen Storage Applications. *Int. J. Hydrogen Energy* **2013**, *38*, 1685–1691. [CrossRef]
86. Atalmis, G.; Sattarkhanov, K.; Kaplan, R.N.; Demiralp, M.; Kaplan, Y. The Effect of Powder and Pellet Forms of Added Metal Hydride Materials on Reaction Kinetics and Storage. *Int. J. Hydrogen Energy* **2024**, *75*, 98–105. [CrossRef]
87. Lozano, G.A.; Bellosta Von Colbe, J.M.; Bormann, R.; Klassen, T.; Dornheim, M. Enhanced Volumetric Hydrogen Density in Sodium Alanate by Compaction. *J. Power Sources* **2011**, *196*, 9254–9259. [CrossRef]
88. Safyari, M.; Gneiger, S.; Simson, C.; Moshtaghi, M. A New Methodology for Extra Enhancement of Hydrogen Storage Capacity of Mg–Ni Based Alloys: The Role of Gaseous O_2/H_2 Mixture. *Int. J. Hydrogen Energy* **2024**, *73*, 761–767. [CrossRef]
89. Modi, P.; Aguey-Zinsou, K.-F. Room Temperature Metal Hydrides for Stationary and Heat Storage Applications: A Review. *Front. Energy Res.* **2021**, *9*, 616115. [CrossRef]
90. Kim, D.; Kim, J.B.; Lee, J.; Lee, B.J. Measurement of Effective Thermal Conductivity of $LaNi_5$ Powder Packed Bed. *Int. J. Heat Mass Transf.* **2021**, *165*, 120735. [CrossRef]
91. Bird, J.E.; Humphries, T.D.; Paskevicius, M.; Poupin, L.; Buckley, C.E. Thermal Properties of Thermochemical Heat Storage Materials. *Phys. Chem. Chem. Phys.* **2020**, *22*, 4617–4625. [CrossRef]
92. Albert, R.; Urbanczyk, R.; Felderhoff, M. Thermal Conductivity Measurements of Magnesium Hydride Powder Beds under Operating Conditions for Heat Storage Applications. *Int. J. Hydrogen Energy* **2019**, *44*, 29273–29281. [CrossRef]
93. Ishido, Y.; Kawamura, M.; Ono, S. Thermal Conductivity of Magnesium-Nickel Hydride Powder Beds in a Hydrogen Atmosphere. *Int. J. Hydrogen Energy* **1982**, *7*, 173–182. [CrossRef]
94. Hahne, E.; Kallweit, J.J. Thermal Conductivity of Metal Hydride Materials for Storage of Hydrogen: Experimental Investigation. *Int. J. Hydrogen Energy* **1998**, *23*, 107–114. [CrossRef]
95. Suda, S.; Kobayashi, N.; Yoshida, K. Thermal Conductivity in Metal Hydride Beds. *Int. J. Hydrogen Energy* **1981**, *6*, 521–528. [CrossRef]
96. Dedrick, D.E.; Kanouff, M.P.; Replogle, B.C.; Gross, K.J. Thermal Properties Characterization of Sodium Alanates. *J. Alloys Compd.* **2005**, *389*, 299–305. [CrossRef]

97. Ghafir, M.F.A.; Batcha, M.F.M.; Raghavan, V.R. Prediction of the Thermal Conductivity of Metal Hydrides—The Inverse Problem. *Int. J. Hydrogen Energy* **2009**, *34*, 7125–7130. [CrossRef]
98. Cheng, G.; Gan, J.; Xu, D.; Yu, A. Evaluation of Effective Thermal Conductivity in Random Packed Bed: Heat Transfer through Fluid Voids and Effect of Packing Structure. *Powder Technol.* **2020**, *361*, 326–336. [CrossRef]
99. Wang, H.; Prasad, A.K.; Advani, S.G. Hydrogen Storage Systems Based on Hydride Materials with Enhanced Thermal Conductivity. *Int. J. Hydrogen Energy* **2012**, *37*, 290–298. [CrossRef]
100. Huang, L.J.; Lin, H.J.; Wang, H.; Ouyang, L.Z.; Zhu, M. Amorphous Alloys for Hydrogen Storage. *J. Alloys Compd.* **2023**, *941*, 168945. [CrossRef]
101. Yadav, T.; Mukhopadhyay, N. Quasicrystal: A Low-Frictional Novel Material. *Curr. Opin. Chem. Eng.* **2018**, *19*, 163–169. [CrossRef]
102. Takasaki, A.; Kelton, K.F. Hydrogen Storage in Ti-Based Quasicrystal Powders Produced by Mechanical Alloying. *Int. J. Hydrogen Energy* **2006**, *31*, 183–190. [CrossRef]
103. Borzenko, V.I.; Romanov, I.A.; Dunikov, D.O.; Kazakov, A.N. Hydrogen Sorption Properties of Metal Hydride Beds: Effect of Internal Stresses Caused by Reactor Geometry. *Int. J. Hydrogen Energy* **2019**, *44*, 6086–6092. [CrossRef]
104. Au, M.; Wu, J.; Wang, Q.-D. Some Engineering Methods for Eliminating Deformation and Expansion Damage of Hydride Storage Containers. *J. Common. Met.* **1991**, *172–174*, 1168–1174. [CrossRef]
105. Zohra, F.T.; Webb, C.J.; Lamb, K.E.; Gray, E.M. Degradation of Metal Hydrides in Hydrogen-Based Thermodynamic Machines: A Review. *Int. J. Hydrogen Energy* **2024**, *64*, 417–438. [CrossRef]
106. Rusman, N.A.A.; Dahari, M. A Review on the Current Progress of Metal Hydrides Material for Solid-State Hydrogen Storage Applications. *Int. J. Hydrogen Energy* **2016**, *41*, 12108–12126. [CrossRef]
107. Yu, X.; Tang, Z.; Sun, D.; Ouyang, L.; Zhu, M. Recent Advances and Remaining Challenges of Nanostructured Materials for Hydrogen Storage Applications. *Prog. Mater. Sci.* **2017**, *88*, 1–48. [CrossRef]
108. Arsad, A.Z.; Hannan, M.A.; Al-Shetwi, A.Q.; Begum, R.A.; Hossain, M.J.; Ker, P.J.; Mahlia, T.I. Hydrogen Electrolyser Technologies and Their Modelling for Sustainable Energy Production: A Comprehensive Review and Suggestions. *Int. J. Hydrogen Energy* **2023**, *48*, 27841–27871. [CrossRef]
109. Abdin, Z.; Tang, C.; Liu, Y.; Catchpole, K. Large-Scale Stationary Hydrogen Storage via Liquid Organic Hydrogen Carriers. *iScience* **2021**, *24*, 102966. [CrossRef] [PubMed]
110. Lee, S.; Kim, T.; Han, G.; Kang, S.; Yoo, Y.-S.; Jeon, S.-Y.; Bae, J. Comparative Energetic Studies on Liquid Organic Hydrogen Carrier: A Net Energy Analysis. *Renew. Sustain. Energy Rev.* **2021**, *150*, 111447. [CrossRef]
111. Morales-Ospino, R.; Celzard, A.; Fierro, V. Strategies to Recover and Minimize Boil-off Losses during Liquid Hydrogen Storage. *Renew. Sustain. Energy Rev.* **2023**, *182*, 113360. [CrossRef]
112. Wang, X.; Li, B.; Han, B.; Jin, X.; Zhang, D.; Bi, M. Explosion of High Pressure Hydrogen Tank in Fire: Mechanism, Criterion, and Consequence Assessment. *J. Energy Storage* **2023**, *72*, 108455. [CrossRef]
113. Hasegawa, H.; Ohki, Y. Development of a Model of On-Board Pemfc Powered Locomotive with a Metal Hydride Cylinder. *MRS Proc.* **1995**, *393*, 145. [CrossRef]
114. Hsiao, D.-R.; Huang, B.-W.; Shih, N.-C. Development and Dynamic Characteristics of Hybrid Fuel Cell-Powered Mini-Train System. *Int. J. Hydrogen Energy* **2012**, *37*, 1058–1066. [CrossRef]
115. Lototskyy, M.V.; Tolj, I.; Parsons, A.; Smith, F.; Sita, C.; Linkov, V. Performance of Electric Forklift with Low-Temperature Polymer Exchange Membrane Fuel Cell Power Module and Metal Hydride Hydrogen Storage Extension Tank. *J. Power Sources* **2016**, *316*, 239–250. [CrossRef]
116. Bevan, A.I.; Züttel, A.; Book, D.; Harris, I.R. Performance of a Metal Hydride Store on the "Ross Barlow" Hydrogen Powered Canal Boat. *Faraday Discuss.* **2011**, *151*, 353. [CrossRef]
117. Target Explanation Document: Onboard Hydrogen Storage for Light-Duty Fuel Cell Vehicles; US DRIVE Partnership. 2017. Available online: https://www.energy.gov/eere/fuelcells/articles/target-explanation-document-onboard-hydrogen-storage-light-duty-fuel-cell (accessed on 12 July 2024).
118. Afzal, M. Metal Hydride Hydrogen Storage: A Systems Perspective. In *Towards Hydrogen Infrastructure*; Elsevier: Amsterdam, The Netherlands, 2024; pp. 257–272. ISBN 978-0-323-95553-9.
119. Baricco, M.; Bang, M.; Fichtner, M.; Hauback, B.; Linder, M.; Luetto, C.; Moretto, P.; Sgroi, M. SSH2S: Hydrogen Storage in Complex Hydrides for an Auxiliary Power Unit Based on High Temperature Proton Exchange Membrane Fuel Cells. *J. Power Sources* **2017**, *342*, 853–860. [CrossRef]
120. Gray, E.M.; Webb, C.J.; Andrews, J.; Shabani, B.; Tsai, P.J.; Chan, S.L.I. Hydrogen Storage for Off-Grid Power Supply. *Int. J. Hydrogen Energy* **2011**, *36*, 654–663. [CrossRef]
121. Malleswararao, K.; Dutta, P.; Murthy, S.S. Applications of Metal Hydride Based Thermal Systems: A Review. *Appl. Therm. Eng.* **2022**, *215*, 118816. [CrossRef]
122. Tarasov, B.P.; Fursikov, P.V.; Volodin, A.A.; Bocharnikov, M.S.; Shimkus, Y.Y.; Kashin, A.M.; Yartys, V.A.; Chidziva, S.; Pasupathi, S.; Lototskyy, M.V. Metal Hydride Hydrogen Storage and Compression Systems for Energy Storage Technologies. *Int. J. Hydrogen Energy* **2021**, *46*, 13647–13657. [CrossRef]
123. Muthukumar, P.; Kumar, A.; Raju, N.N.; Malleswararao, K.; Rahman, M.M. A Critical Review on Design Aspects and Developmental Status of Metal Hydride Based Thermal Machines. *Int. J. Hydrogen Energy* **2018**, *43*, 17753–17779. [CrossRef]

124. Kudiiarov, V.; Elman, R.; Pushilina, N.; Kurdyumov, N. State of the Art in Development of Heat Exchanger Geometry Optimization and Different Storage Bed Designs of a Metal Hydride Reactor. *Materials* **2023**, *16*, 4891. [CrossRef]
125. Weckerle, C.; Bürger, I.; Linder, M. Numerical Optimization of a Plate Reactor for a Metal Hydride Open Cooling System. *Int. J. Hydrogen Energy* **2019**, *44*, 16862–16876. [CrossRef]
126. Gkanas, E.I.; Statheros, T.; Khzouz, M. Heat Management on Rectangular Metal Hydride Tanks for Green Building Applications. *Int. J. Hydrogen Energy* **2019**, *44*, 19267–19274. [CrossRef]
127. Mazzucco, A.; Dornheim, M.; Sloth, M.; Jensen, T.R.; Jensen, J.O.; Rokni, M. Bed Geometries, Fueling Strategies and Optimization of Heat Exchanger Designs in Metal Hydride Storage Systems for Automotive Applications: A Review. *Int. J. Hydrogen Energy* **2014**, *39*, 17054–17074. [CrossRef]
128. Sreeraj, R.; Aadhithiyan, A.K.; Anbarasu, S. Integration of Thermal Augmentation Methods in Hydride Beds for Metal Hydride Based Hydrogen Storage Systems: Review and Recommendation. *J. Energy Storage* **2022**, *52*, 105039. [CrossRef]
129. Yang, F.S.; Wang, G.X.; Zhang, Z.X.; Meng, X.Y.; Rudolph, V. Design of the Metal Hydride Reactors—A Review on the Key Technical Issues. *Int. J. Hydrogen Energy* **2010**, *35*, 3832–3840. [CrossRef]
130. Veerraju, C.; Gopal, M.R. Heat and Mass Transfer Studies on Elliptical Metal Hydride Tubes and Tube Banks. *Int. J. Hydrogen Energy* **2009**, *34*, 4340–4350. [CrossRef]
131. Bhouri, M.; Goyette, J.; Hardy, B.J.; Anton, D.L. Numerical Modeling and Performance Evaluation of Multi-Tubular Sodium Alanate Hydride Finned Reactor. *Int. J. Hydrogen Energy* **2012**, *37*, 1551–1567. [CrossRef]
132. Krokos, C.A.; Nikolic, D.; Kikkinides, E.S.; Georgiadis, M.C.; Stubos, A.K. Modeling and Optimization of Multi-Tubular Metal Hydride Beds for Efficient Hydrogen Storage. *Int. J. Hydrogen Energy* **2009**, *34*, 9128–9140. [CrossRef]
133. Sunku Prasad, J.; Muthukumar, P. Design and Performance Analysis of an Annular Metal Hydride Reactor for Large-Scale Hydrogen Storage Applications. *Renew. Energy* **2022**, *181*, 1155–1166. [CrossRef]
134. Sunku Prasad, J.; Muthukumar, P. Design of Metal Hydride Reactor for Medium Temperature Thermochemical Energy Storage Applications. *Therm. Sci. Eng. Prog.* **2023**, *37*, 101570. [CrossRef]
135. Wang, D.; Wang, Y.; Wang, F.; Zheng, S.; Guan, S.; Zheng, L.; Wu, L.; Yang, X.; Lv, M.; Zhang, Z. Optimal Design of Disc Mini-Channel Metal Hydride Reactor with High Hydrogen Storage Efficiency. *Appl. Energy* **2022**, *308*, 118389. [CrossRef]
136. Qin, F. Pulverization, Expansion of $La_{0.6}Y_{0.4}Ni_{4.8}Mn_{0.2}$ during Hydrogen Absorption–Desorption Cycles and Their Influences in Thin-Wall Reactors. *Int. J. Hydrogen Energy* **2008**, *33*, 709–717. [CrossRef]
137. Hanada, N.; Ichikawa, T.; Hino, S.; Fujii, H. Remarkable Improvement of Hydrogen Sorption Kinetics in Magnesium Catalyzed with Nb_2O_5. *J. Alloys Compd.* **2006**, *420*, 46–49. [CrossRef]
138. Wu, Z.; Yang, F.; Zhang, Z.; Bao, Z. Magnesium Based Metal Hydride Reactor Incorporating Helical Coil Heat Exchanger: Simulation Study and Optimal Design. *Appl. Energy* **2014**, *130*, 712–722. [CrossRef]
139. Tong, L.; Xiao, J.; Yang, T.; Bénard, P.; Chahine, R. Complete and Reduced Models for Metal Hydride Reactor with Coiled-Tube Heat Exchanger. *Int. J. Hydrogen Energy* **2019**, *44*, 15907–15916. [CrossRef]
140. Bai, X.-S.; Yang, W.-W.; Zhang, W.-Y.; Yang, F.-S.; Tang, X.-Y. Hydrogen Absorption Performance of a Novel Cylindrical MH Reactor with Combined Loop-Type Finned Tube and Cooling Jacket Heat Exchanger. *Int. J. Hydrogen Energy* **2020**, *45*, 28100–28115. [CrossRef]
141. Liu, Y.; Wang, H.; Prasad, A.K.; Advani, S.G. Role of Heat Pipes in Improving the Hydrogen Charging Rate in a Metal Hydride Storage Tank. *Int. J. Hydrogen Energy* **2014**, *39*, 10552–10563. [CrossRef]
142. Anbarasu, S.; Muthukumar, P.; Mishra, S.C. Thermal Modeling of $LmNi_{4.91}Sn_{0.15}$ Based Solid State Hydrogen Storage Device with Embedded Cooling Tubes. *Int. J. Hydrogen Energy* **2014**, *39*, 15549–15562. [CrossRef]
143. Keith, M.D.; Kukkapalli, V.K.; Kim, S. Phase Change Cooling of a Metal Hydride Reactor for Rapid Hydrogen Absorption. *Energies* **2022**, *15*, 2490. [CrossRef]
144. Bao, Z.; Yang, F.; Wu, Z.; Nyallang Nyamsi, S.; Zhang, Z. Optimal Design of Metal Hydride Reactors Based on CFD–Taguchi Combined Method. *Energy Convers. Manag.* **2013**, *65*, 322–330. [CrossRef]
145. Boukhari, A.; Bessaïh, R. Numerical Heat and Mass Transfer Investigation of Hydrogen Absorption in an Annulus-Disc Reactor. *Int. J. Hydrogen Energy* **2015**, *40*, 13708–13717. [CrossRef]
146. Muthukumar, P.; Singhal, A.; Bansal, G.K. Thermal Modeling and Performance Analysis of Industrial-Scale Metal Hydride Based Hydrogen Storage Container. *Int. J. Hydrogen Energy* **2012**, *37*, 14351–14364. [CrossRef]
147. Raju, M.; Kumar, S. Optimization of Heat Exchanger Designs in Metal Hydride Based Hydrogen Storage Systems. *Int. J. Hydrogen Energy* **2012**, *37*, 2767–2778. [CrossRef]
148. Cui, Y.; Zeng, X.; Xiao, J.; Kou, H. The Comprehensive Review for Development of Heat Exchanger Configuration Design in Metal Hydride Bed. *Int. J. Hydrogen Energy* **2022**, *47*, 2461–2490. [CrossRef]
149. Garrison, S.L.; Hardy, B.J.; Gorbounov, M.B.; Tamburello, D.A.; Corgnale, C.; vanHassel, B.A.; Mosher, D.A.; Anton, D.L. Optimization of Internal Heat Exchangers for Hydrogen Storage Tanks Utilizing Metal Hydrides. *Int. J. Hydrogen Energy* **2012**, *37*, 2850–2861. [CrossRef]
150. Parida, A.; Muthukumar, P. Reactor Design and Numerical Study on Metal Hydride Based Finned Reactor Configurations for Hydrogen Compression Application. *Int. J. Hydrogen Energy* **2023**, *48*, 37930–37943. [CrossRef]
151. Chandra, S.; Sharma, P.; Muthukumar, P.; Tatiparti, S.S.V. Modeling and Numerical Simulation of a 5 Kg $LaNi_5$-Based Hydrogen Storage Reactor with Internal Conical Fins. *Int. J. Hydrogen Energy* **2020**, *45*, 8794–8809. [CrossRef]

152. Singh, A.; Prakash Maiya, M.; Srinivasa Murthy, S. Performance of a Solid State Hydrogen Storage Device with Finned Tube Heat Exchanger. *Int. J. Hydrogen Energy* **2017**, *42*, 26855–26871. [CrossRef]
153. Keshari, V.; Maiya, M.P. Design and Investigation of Hydriding Alloy Based Hydrogen Storage Reactor Integrated with a Pin Fin Tube Heat Exchanger. *Int. J. Hydrogen Energy* **2018**, *43*, 7081–7095. [CrossRef]
154. Zhang, S.; Yang, F.; Zhou, L.; Zhang, Y.; Wu, Z.; Zhang, Z.; Wang, Y. A Novel Multilayer Fin Structure for Heat Transfer Enhancement in Hydride-Based Hydrogen Storage Reactor. *Int. J. Energy Res.* **2018**, *42*, 3837–3850. [CrossRef]
155. George, M.; Mohan, G. Constructal Design of Weight Optimized Metal Hydride Storage Device Embedded with Ribbed Honeycomb. *Appl. Therm. Eng.* **2023**, *219*, 119368. [CrossRef]
156. Bai, X.-S.; Yang, W.-W.; Tang, X.-Y.; Yang, F.-S.; Jiao, Y.-H.; Yang, Y. Optimization of Tree-Shaped Fin Structures towards Enhanced Absorption Performance of Metal Hydride Hydrogen Storage Device: A Numerical Study. *Energy* **2021**, *220*, 119738. [CrossRef]
157. Gupta, S.; Sharma, V.K. Design and Analysis of Metal Hydride Reactor Embedded with Internal Copper Fins and External Water Cooling. *Int. J. Energy Res.* **2021**, *45*, 1836–1856. [CrossRef]
158. Nyamsi, S.N.; Tolj, I.; Pasupathi, S. Multi-Objective Optimization of a Metal Hydride Reactor Coupled with Phase Change Materials for Fast Hydrogen Sorption Time. *J. Energy Storage* **2023**, *71*, 108089. [CrossRef]
159. Liu, M.; Zhao, B.; Li, Y.; Wang, Z.; Zhang, X.; Tong, L.; Yang, T.; Li, X.; Xiao, J. Parametric Study on Fin Structure and Injection Tube in Metal Hydride Tank Packed with LaNi5 Alloy for Efficient and Safe Hydrogen Storage. *Sustainability* **2023**, *15*, 9735. [CrossRef]
160. Ma, J.; Wang, Y.; Shi, S.; Yang, F.; Bao, Z.; Zhang, Z. Optimization of Heat Transfer Device and Analysis of Heat & Mass Transfer on the Finned Multi-Tubular Metal Hydride Tank. *Int. J. Hydrogen Energy* **2014**, *39*, 13583–13595. [CrossRef]
161. Visaria, M.; Mudawar, I.; Pourpoint, T. Enhanced Heat Exchanger Design for Hydrogen Storage Using High-Pressure Metal Hydride: Part 1. Design Methodology and Computational Results. *Int. J. Heat Mass Transf.* **2011**, *54*, 413–423. [CrossRef]
162. Andreasen, G.; Melnichuk, M.; Ramos, S.; Corso, H.L.; Visintin, A.; Triaca, W.E.; Peretti, H.A. Hydrogen Desorption from a Hydride Container under Different Heat Exchange Conditions. *Int. J. Hydrogen Energy* **2013**, *38*, 13352–13359. [CrossRef]
163. Kaplan, Y. Effect of Design Parameters on Enhancement of Hydrogen Charging in Metal Hydride Reactors. *Int. J. Hydrogen Energy* **2009**, *34*, 2288–2294. [CrossRef]
164. Satya Sekhar, B.; Lototskyy, M.; Kolesnikov, A.; Moropeng, M.L.; Tarasov, B.P.; Pollet, B.G. Performance Analysis of Cylindrical Metal Hydride Beds with Various Heat Exchange Options. *J. Alloys Compd.* **2015**, *645*, S89–S95. [CrossRef]
165. Mellouli, S.; Askri, F.; Dhaou, H.; Jemni, A.; Ben Nasrallah, S. Numerical Study of Heat Exchanger Effects on Charge/Discharge Times of Metal–Hydrogen Storage Vessel. *Int. J. Hydrogen Energy* **2009**, *34*, 3005–3017. [CrossRef]
166. Tawalbeh, M.; Khan, H.A.; Al-Othman, A.; Almomani, F.; Ajith, S. A Comprehensive Review on the Recent Advances in Materials for Thermal Energy Storage Applications. *Int. J. Thermofluids* **2023**, *18*, 100326. [CrossRef]
167. Mitali, J.; Dhinakaran, S.; Mohamad, A.A. Energy Storage Systems: A Review. *Energy Storage Sav.* **2022**, *1*, 166–216. [CrossRef]
168. Alva, G.; Lin, Y.; Fang, G. An Overview of Thermal Energy Storage Systems. *Energy* **2018**, *144*, 341–378. [CrossRef]
169. Zhao, J.; Korba, D.; Mishra, A.; Klausner, J.; Randhir, K.; AuYeung, N.; Li, L. Particle-Based High-Temperature Thermochemical Energy Storage Reactors. *Prog. Energy Combust. Sci.* **2024**, *102*, 101143. [CrossRef]
170. Desai, F.; Sunku Prasad, J.; Muthukumar, P.; Rahman, M.M. Thermochemical Energy Storage System for Cooling and Process Heating Applications: A Review. *Energy Convers. Manag.* **2021**, *229*, 113617. [CrossRef]
171. Khan, M.I.; Asfand, F.; Al-Ghamdi, S.G. Progress in Research and Technological Advancements of Thermal Energy Storage Systems for Concentrated Solar Power. *J. Energy Storage* **2022**, *55*, 105860. [CrossRef]
172. Koçak, B.; Fernandez, A.I.; Paksoy, H. Review on Sensible Thermal Energy Storage for Industrial Solar Applications and Sustainability Aspects. *Sol. Energy* **2020**, *209*, 135–169. [CrossRef]
173. Kant, K.; Shukla, A.; Sharma, A.; Kumar, A.; Jain, A. Thermal Energy Storage Based Solar Drying Systems: A Review. *Innov. Food Sci. Emerg. Technol.* **2016**, *34*, 86–99. [CrossRef]
174. Miró, L.; Gasia, J.; Cabeza, L.F. Thermal Energy Storage (TES) for Industrial Waste Heat (IWH) Recovery: A Review. *Appl. Energy* **2016**, *179*, 284–301. [CrossRef]
175. Guelpa, E.; Verda, V. Thermal Energy Storage in District Heating and Cooling Systems: A Review. *Appl. Energy* **2019**, *252*, 113474. [CrossRef]
176. Paul, A.; Holy, F.; Textor, M.; Lechner, S. High Temperature Sensible Thermal Energy Storage as a Crucial Element of Carnot Batteries: Overall Classification and Technical Review Based on Parameters and Key Figures. *J. Energy Storage* **2022**, *56*, 106015. [CrossRef]
177. Bruch, A.; Fourmigué, J.F.; Couturier, R. Experimental and Numerical Investigation of a Pilot-Scale Thermal Oil Packed Bed Thermal Storage System for CSP Power Plant. *Sol. Energy* **2014**, *105*, 116–125. [CrossRef]
178. Zanganeh, G.; Pedretti, A.; Zavattoni, S.; Barbato, M.; Steinfeld, A. Packed-Bed Thermal Storage for Concentrated Solar Power—Pilot-Scale Demonstration and Industrial-Scale Design. *Sol. Energy* **2012**, *86*, 3084–3098. [CrossRef]
179. Tiskatine, R.; Eddemani, A.; Gourdo, L.; Abnay, B.; Ihlal, A.; Aharoune, A.; Bouirden, L. Experimental Evaluation of Thermo-Mechanical Performances of Candidate Rocks for Use in High Temperature Thermal Storage. *Appl. Energy* **2016**, *171*, 243–255. [CrossRef]
180. Tiskatine, R.; Oaddi, R.; Ait El Cadi, R.; Bazgaou, A.; Bouirden, L.; Aharoune, A.; Ihlal, A. Suitability and Characteristics of Rocks for Sensible Heat Storage in CSP Plants. *Sol. Energy Mater. Sol. Cells* **2017**, *169*, 245–257. [CrossRef]

181. Knobloch, K.; Ulrich, T.; Bahl, C.; Engelbrecht, K. Degradation of a Rock Bed Thermal Energy Storage System. *Appl. Therm. Eng.* **2022**, *214*, 118823. [CrossRef]
182. Zoth, G.; Haenel, R. Appendix. In *Handbook of Terrestrial Heat-Flow Density Determination*; Haenel, R., Rybach, L., Stegena, L., Eds.; Springer: Dordrecht, The Netherlands, 1988; pp. 449–468. ISBN 978-94-010-7780-4.
183. Xu, Y.; Chung, D.D.L. Cement of High Specific Heat and High Thermal Conductivity, Obtained by Using Silane and Silica Fume as Admixtures. *Cem. Concr. Res.* **2000**, *30*, 1175–1178. [CrossRef]
184. Laing, D.; Steinmann, W.-D.; Tamme, R.; Richter, C. Solid Media Thermal Storage for Parabolic Trough Power Plants. *Sol. Energy* **2006**, *80*, 1283–1289. [CrossRef]
185. Alonso, M.C.; Vera-Agullo, J.; Guerreiro, L.; Flor-Laguna, V.; Sanchez, M.; Collares-Pereira, M. Calcium Aluminate Based Cement for Concrete to Be Used as Thermal Energy Storage in Solar Thermal Electricity Plants. *Cem. Concr. Res.* **2016**, *82*, 74–86. [CrossRef]
186. Kunwar, A.; Kumar, M.; Gupta, A.; Mangrulkar, C.K.; Chamoli, S. Experimental Investigation of a Packed-Bed Thermal Energy Storage System Fitted with Perforated Cylindrical Elements. *Heat Mass Transf.* **2019**, *55*, 2723–2737. [CrossRef]
187. Faik, A.; Guillot, S.; Lambert, J.; Véron, E.; Ory, S.; Bessada, C.; Echegut, P.; Py, X. Thermal Storage Material from Inertized Wastes: Evolution of Structural and Radiative Properties with Temperature. *Sol. Energy* **2012**, *86*, 139–146. [CrossRef]
188. Miró, L.; Navarro, M.E.; Suresh, P.; Gil, A.; Fernández, A.I.; Cabeza, L.F. Experimental Characterization of a Solid Industrial By-Product as Material for High Temperature Sensible Thermal Energy Storage (TES). *Appl. Energy* **2014**, *113*, 1261–1268. [CrossRef]
189. Agalit, H.; Zari, N.; Maaroufi, M. Thermophysical and Chemical Characterization of Induction Furnace Slags for High Temperature Thermal Energy Storage in Solar Tower Plants. *Sol. Energy Mater. Sol. Cells* **2017**, *172*, 168–176. [CrossRef]
190. Navarro, M.E.; Martínez, M.; Gil, A.; Fernández, A.I.; Cabeza, L.F.; Olives, R.; Py, X. Selection and Characterization of Recycled Materials for Sensible Thermal Energy Storage. *Sol. Energy Mater. Sol. Cells* **2012**, *107*, 131–135. [CrossRef]
191. Motte, F.; Falcoz, Q.; Veron, E.; Py, X. Compatibility Tests between Solar Salt and Thermal Storage Ceramics from Inorganic Industrial Wastes. *Appl. Energy* **2015**, *155*, 14–22. [CrossRef]
192. Ozger, O.B.; Girardi, F.; Giannuzzi, G.M.; Salomoni, V.A.; Majorana, C.E.; Fambri, L.; Baldassino, N.; Di Maggio, R. Effect of Nylon Fibres on Mechanical and Thermal Properties of Hardened Concrete for Energy Storage Systems. *Mater. Des.* **2013**, *51*, 989–997. [CrossRef]
193. Gutierrez, A.; Miró, L.; Gil, A.; Rodríguez-Aseguinolaza, J.; Barreneche, C.; Calvet, N.; Py, X.; Inés Fernández, A.; Grágeda, M.; Ushak, S.; et al. Advances in the Valorization of Waste and By-Product Materials as Thermal Energy Storage (TES) Materials. *Renew. Sustain. Energy Rev.* **2016**, *59*, 763–783. [CrossRef]
194. Koçak, B.; Paksoy, H. Using Demolition Wastes from Urban Regeneration as Sensible Thermal Energy Storage Material. *Int. J. Energy Res.* **2019**, *43*, 6454–6460. [CrossRef]
195. Kocak, B.; Paksoy, H. Performance of Laboratory Scale Packed-Bed Thermal Energy Storage Using New Demolition Waste Based Sensible Heat Materials for Industrial Solar Applications. *Sol. Energy* **2020**, *211*, 1335–1346. [CrossRef]
196. Zhang, J.; Guo, Z.; Zhu, Y.; Zhang, H.; Yan, M.; Liu, D.; Hao, J. Preparation and Characterization of Novel Low-Cost Sensible Heat Storage Materials with Steel Slag. *J. Energy Storage* **2024**, *76*, 109643. [CrossRef]
197. Wang, N.; Xu, G.; Li, S.; Zhang, X. Thermal Properties and Solar Collection Characteristics of Oil-Based Nanofluids with Low Graphene Concentration. *Energy Procedia* **2017**, *105*, 194–199. [CrossRef]
198. Kenda, E.S.; N'Tsoukpoe, K.E.; Ouédraogo, I.W.K.; Coulibaly, Y.; Py, X.; Ouédraogo, F.M.A.W. Jatropha Curcas Crude Oil as Heat Transfer Fluid or Thermal Energy Storage Material for Concentrating Solar Power Plants. *Energy Sustain. Dev.* **2017**, *40*, 59–67. [CrossRef]
199. Ong, T.-C.; Sarvghad, M.; Bell, S.; Will, G.; Steinberg, T.A.; Yin, Y.; Andersson, G.; Lewis, D. Review on the Challenges of Salt Phase Change Materials for Energy Storage in Concentrated Solar Power Facilities. *Appl. Therm. Eng.* **2024**, *238*, 122034. [CrossRef]
200. Zhang, H.; Baeyens, J.; Cáceres, G.; Degrève, J.; Lv, Y. Thermal Energy Storage: Recent Developments and Practical Aspects. *Prog. Energy Combust. Sci.* **2016**, *53*, 1–40. [CrossRef]
201. Yu, N.; Wang, R.Z.; Wang, L.W. Sorption Thermal Storage for Solar Energy. *Prog. Energy Combust. Sci.* **2013**, *39*, 489–514. [CrossRef]
202. Shank, K.; Tiari, S. A Review on Active Heat Transfer Enhancement Techniques within Latent Heat Thermal Energy Storage Systems. *Energies* **2023**, *16*, 4165. [CrossRef]
203. Al-Salami, H.A.; Dhaidan, N.S.; Abbas, H.H.; Al-Mousawi, F.N.; Homod, R.Z. Review of PCM Charging in Latent Heat Thermal Energy Storage Systems with Fins. *Therm. Sci. Eng. Prog.* **2024**, *51*, 102640. [CrossRef]
204. Yang, Z.L.; Walvekar, R.; Wong, W.P.; Sharma, R.K.; Dharaskar, S.; Khalid, M. Advances in Phase Change Materials, Heat Transfer Enhancement Techniques, and Their Applications in Thermal Energy Storage: A Comprehensive Review. *J. Energy Storage* **2024**, *87*, 111329. [CrossRef]
205. Zalba, B.; Marín, J.M.; Cabeza, L.F.; Mehling, H. Review on Thermal Energy Storage with Phase Change: Materials, Heat Transfer Analysis and Applications. *Appl. Therm. Eng.* **2003**, *23*, 251–283. [CrossRef]
206. Beare-Rogers, J.L.; Dieffenbacher, A.; Holm, J.V. Lexicon of Lipid Nutrition (IUPAC Technical Report). *Pure Appl. Chem.* **2001**, *73*, 685–744. [CrossRef]

207. Sarı, A.; Bicer, A.; Al-Ahmed, A.; Al-Sulaiman, F.A.; Zahir, M.H.; Mohamed, S.A. Silica Fume/Capric Acid-Palmitic Acid Composite Phase Change Material Doped with CNTs for Thermal Energy Storage. *Sol. Energy Mater. Sol. Cells* **2018**, *179*, 353–361. [CrossRef]
208. Costa, S.C.; Kenisarin, M. A Review of Metallic Materials for Latent Heat Thermal Energy Storage: Thermophysical Properties, Applications, and Challenges. *Renew. Sustain. Energy Rev.* **2022**, *154*, 111812. [CrossRef]
209. Lin, Y.; Alva, G.; Fang, G. Review on Thermal Performances and Applications of Thermal Energy Storage Systems with Inorganic Phase Change Materials. *Energy* **2018**, *165*, 685–708. [CrossRef]
210. Lutz, M.; Bhouri, M.; Linder, M.; Bürger, I. Adiabatic Magnesium Hydride System for Hydrogen Storage Based on Thermochemical Heat Storage: Numerical Analysis of the Dehydrogenation. *Appl. Energy* **2019**, *236*, 1034–1048. [CrossRef]
211. Carrero, J.I. Application of the van't Hoff Equation to Phase Equilibria. *ChemTexts* **2024**, *10*, 4. [CrossRef]
212. Neveu, P.; Castaing, J. Solid-Gas Chemical Heat Pumps: Field of Application and Performance of the Internal Heat of Reaction Recovery Process. *Heat Recovery Syst. CHP* **1993**, *13*, 233–251. [CrossRef]
213. Goetz, V.; Elie, F.; Spinner, B. The Structure and Performance of Single Effect Solid-Gas Chemical Heat Pumps. *Heat Recovery Syst. CHP* **1993**, *13*, 79–96. [CrossRef]
214. Kato, Y.; Yamashita, N.; Kobayashi, K.; Yoshizawa, Y. Kinetic Study of the Hydration of Magnesium Oxide for a Chemical Heat Pump. *Appl. Therm. Eng.* **1996**, *16*, 853–862. [CrossRef]
215. Bratton, R.J.; Brindley, G.W. Kinetics of Vapour Phase Hydration of Magnesium Oxide. Part 2.—Dependence on Temperature and Water Vapour Pressure. *Trans. Faraday Soc.* **1965**, *61*, 1017–1025. [CrossRef]
216. Criado, Y.A.; Alonso, M.; Abanades, J.C. Kinetics of the $CaO/Ca(OH)_2$ Hydration/Dehydration Reaction for Thermochemical Energy Storage Applications. *Ind. Eng. Chem. Res.* **2014**, *53*, 12594–12601. [CrossRef]
217. Schaube, F.; Koch, L.; Wörner, A.; Müller-Steinhagen, H. A Thermodynamic and Kinetic Study of the De- and Rehydration of $Ca(OH)_2$ at High H2O Partial Pressures for Thermo-Chemical Heat Storage. *Thermochim. Acta* **2012**, *538*, 9–20. [CrossRef]
218. Yuan, Y.; Li, Y.; Zhao, J. Development on Thermochemical Energy Storage Based on CaO-Based Materials: A Review. *Sustainability* **2018**, *10*, 2660. [CrossRef]
219. Risthaus, K.; Bürger, I.; Lutz, M.; Funayama, S.; Kato, Y.; Linder, M.; Schmidt, M. Experimental and Numerical Investigation of the Dehydration of $Ca(OH)_2$ at Low Steam Pressures. *Processes* **2022**, *10*, 325. [CrossRef]
220. Schmidt, M.; Szczukowski, C.; Roßkopf, C.; Linder, M.; Wörner, A. Experimental Results of a 10 kW High Temperature Thermochemical Storage Reactor Based on Calcium Hydroxide. *Appl. Therm. Eng.* **2014**, *62*, 553–559. [CrossRef]
221. Pardo, P.; Anxionnaz-Minvielle, Z.; Rougé, S.; Cognet, P.; Cabassud, M. $Ca(OH)_2/CaO$ Reversible Reaction in a Fluidized Bed Reactor for Thermochemical Heat Storage. *Sol. Energy* **2014**, *107*, 605–616. [CrossRef]
222. Tian, Z.; Zhang, J.; Zhang, Y.; Fang, Y.; Han, K.; Li, Y. Thermochemical Heat Storage Performance of Fe-Doped MgO/Mg(OH)2: Experimental and DFT Investigation. *J. Energy Storage* **2024**, *86*, 111388. [CrossRef]
223. Li, S.; Yang, X.; Li, X.; Qu, W.; Zhou, T.; Dong, T.; Deng, L.; Zhang, J.; Zhao, J. A High Energy Density 3D Nano-Carbon Based Magnesium Hydroxide Reversible Chemical Reaction Heat Storage Material Synthesis and Heat Transfer Performance Investigation. *J. Energy Storage* **2022**, *50*, 104260. [CrossRef]
224. Zamengo, M.; Ryu, J.; Kato, Y. Thermochemical Performance of Magnesium Hydroxide–Expanded Graphite Pellets for Chemical Heat Pump. *Appl. Therm. Eng.* **2014**, *64*, 339–347. [CrossRef]
225. Shkatulov, A.I.; Aristov, Y. Thermochemical Energy Storage Using $LiNO_3$-Doped $Mg(OH)_2$: A Dehydration Study. *Energy Technol.* **2018**, *6*, 1844–1851. [CrossRef]
226. Myagmarjav, O.; Zamengo, M.; Ryu, J.; Kato, Y. Energy Density Enhancement of Chemical Heat Storage Material for Magnesium Oxide/Water Chemical Heat Pump. *Appl. Therm. Eng.* **2015**, *91*, 377–386. [CrossRef]
227. Kiyabu, S.; Girard, P.; Siegel, D.J. Discovery of Salt Hydrates for Thermal Energy Storage. *J. Am. Chem. Soc.* **2022**, *144*, 21617–21627. [CrossRef]
228. Sunku Prasad, J.; Muthukumar, P.; Desai, F.; Basu, D.N.; Rahman, M.M. A Critical Review of High-Temperature Reversible Thermochemical Energy Storage Systems. *Appl. Energy* **2019**, *254*, 113733. [CrossRef]
229. Carrillo, A.J.; González-Aguilar, J.; Romero, M.; Coronado, J.M. Solar Energy on Demand: A Review on High Temperature Thermochemical Heat Storage Systems and Materials. *Chem. Rev.* **2019**, *119*, 4777–4816. [CrossRef]
230. Song, M.; Zhang, L.; Wu, F.; Zhang, H.; Zhao, H.; Chen, L.; Li, H. Recent Advances of Magnesium Hydride as an Energy Storage Material. *J. Mater. Sci. Technol.* **2023**, *149*, 99–111. [CrossRef]
231. Felderhoff, M.; Bogdanović, B. High Temperature Metal Hydrides as Heat Storage Materials for Solar and Related Applications. *Int. J. Mol. Sci.* **2009**, *10*, 325–344. [CrossRef] [PubMed]
232. Paskevicius, M.; Sheppard, D.A.; Williamson, K.; Buckley, C.E. Metal Hydride Thermal Heat Storage Prototype for Concentrating Solar Thermal Power. *Energy* **2015**, *88*, 469–477. [CrossRef]
233. Kawamura, M.; Ono, S.; Mizuno, Y. Dynamic Characteristics of a Hydride Heat Storage System. *J. Common. Met.* **1983**, *89*, 365–372. [CrossRef]
234. Kawamura, M.; Ono, S.; Higano, S. Experimental Studies on the Behaviours of Hydride Heat Storage System. *Energy Convers. Manag.* **1982**, *22*, 95–102. [CrossRef]

235. Ulmer, U.; Dieterich, M.; Pohl, A.; Dittmeyer, R.; Linder, M.; Fichtner, M. Study of the Structural, Thermodynamic and Cyclic Effects of Vanadium and Titanium Substitution in Laves-Phase AB2 Hydrogen Storage Alloys. *Int. J. Hydrogen Energy* **2017**, *42*, 20103–20110. [CrossRef]
236. Ulmer, U.; Asano, K.; Patyk, A.; Enoki, H.; Nakamura, Y.; Pohl, A.; Dittmeyer, R.; Fichtner, M. Cost Reduction Possibilities of Vanadium-Based Solid Solutions—Microstructural, Thermodynamic, Cyclic and Environmental Effects of Ferrovanadium Substitution. *J. Alloys Compd.* **2015**, *648*, 1024–1030. [CrossRef]
237. Ulmer, U.; Asano, K.; Bergfeldt, T.; Chakravadhanula, V.S.K.; Dittmeyer, R.; Enoki, H.; Kübel, C.; Nakamura, Y.; Pohl, A.; Fichtner, M. Effect of Oxygen on the Microstructure and Hydrogen Storage Properties of V–Ti–Cr–Fe Quaternary Solid Solutions. *Int. J. Hydrogen Energy* **2014**, *39*, 20000–20008. [CrossRef]
238. Ulmer, U.; Cholewa, M.; Diemant, T.; Bonatto Minella, C.; Dittmeyer, R.; Behm, R.J.; Fichtner, M. Thermochemical Energy Storage through De/Hydrogenation of Organic Liquids: Reactions of Organic Liquids on Metal Hydrides. *ACS Appl. Mater. Interfaces* **2016**, *8*, 13993–14003. [CrossRef]
239. Luo, Y.; Wang, Q.; Li, J.; Xu, F.; Sun, L.; Zou, Y.; Chu, H.; Li, B.; Zhang, K. Enhanced Hydrogen Storage/Sensing of Metal Hydrides by Nanomodification. *Mater. Today Nano* **2020**, *9*, 100071. [CrossRef]
240. National Minerals Information Center. *Mineral Commodity Summaries 2024: Mineral Commodity Summaries*; U.S. Geological Survey, National Minerals Information Center: Reston, VA, USA, 2024. [CrossRef]
241. Garrier, S.; Delhomme, B.; De Rango, P.; Marty, P.; Fruchart, D.; Miraglia, S. A New MgH_2 Tank Concept Using a Phase-Change Material to Store the Heat of Reaction. *Int. J. Hydrogen Energy* **2013**, *38*, 9766–9771. [CrossRef]
242. Alqahtani, T.; Mellouli, S.; Bamasag, A.; Askri, F.; Phelan, P.E. Thermal Performance Analysis of a Metal Hydride Reactor Encircled by a Phase Change Material Sandwich Bed. *Int. J. Hydrogen Energy* **2020**, *45*, 23076–23092. [CrossRef]
243. Mellouli, S.; Alqahtani, T.; Askri, F.; Bashiri, A.H.; Algarni, S. Method for Screening Paired Metal Hydrides with Appropriate Phase Change Material for a Thermochemical Energy Storage System. *Appl. Therm. Eng.* **2024**, *236*, 121781. [CrossRef]
244. Kumar, S.; Dutta, P.; Srinivasa Murthy, S.; Aristov, Y.I.; Gordeeva, L.; Li, T.X.; Wang, R.Z. Studies on a Metal Hydride Based Year-Round Comfort Heating and Cooling System for Extreme Climates. *Energy Build.* **2021**, *244*, 111042. [CrossRef]
245. Tiwari, S.; Sharma, P. Simulations of Hydrogen-Storage System Integrated with Sensible Storage System. *Nanomater. Energy* **2019**, *8*, 33–41. [CrossRef]
246. Tiwari, S.; Sharma, P. Integration of Metal Hydride Reactor with Thermocline Based Heat Storage System. *J. Energy Storage* **2023**, *59*, 106506. [CrossRef]
247. Marty, P.; De Rango, P.; Delhomme, B.; Garrier, S. Various Tools for Optimizing Large Scale Magnesium Hydride Storage. *J. Alloys Compd.* **2013**, *580*, S324–S328. [CrossRef]
248. Mellouli, S.; Ben Khedher, N.; Askri, F.; Jemni, A.; Ben Nasrallah, S. Numerical Analysis of Metal Hydride Tank with Phase Change Material. *Appl. Therm. Eng.* **2015**, *90*, 674–682. [CrossRef]
249. Mellouli, S.; Abhilash, E.; Askri, F.; Ben Nasrallah, S. Integration of Thermal Energy Storage Unit in a Metal Hydride Hydrogen Storage Tank. *Appl. Therm. Eng.* **2016**, *102*, 1185–1196. [CrossRef]
250. Ben Mâad, H.; Miled, A.; Askri, F.; Ben Nasrallah, S. Numerical Simulation of Absorption-Desorption Cyclic Processes for Metal-Hydrogen Reactor with Heat Recovery Using Phase-Change Material. *Appl. Therm. Eng.* **2016**, *96*, 267–276. [CrossRef]
251. Tong, L.; Xiao, J.; Bénard, P.; Chahine, R. Thermal Management of Metal Hydride Hydrogen Storage Reservoir Using Phase Change Materials. *Int. J. Hydrogen Energy* **2019**, *44*, 21055–21066. [CrossRef]
252. Ye, Y.; Ding, J.; Wang, W.; Yan, J. The Storage Performance of Metal Hydride Hydrogen Storage Tanks with Reaction Heat Recovery by Phase Change Materials. *Appl. Energy* **2021**, *299*, 117255. [CrossRef]
253. Lewis, S.D.; Chippar, P. Analysis of Heat and Mass Transfer During Charging and Discharging in a Metal Hydride—Phase Change Material Reactor. *J. Energy Storage* **2021**, *33*, 102108. [CrossRef]
254. Nguyen, H.Q.; Shabani, B. Thermal Management of Metal Hydride Hydrogen Storage Using Phase Change Materials for Standalone Solar Hydrogen Systems: An Energy/Exergy Investigation. *Int. J. Hydrogen Energy* **2022**, *47*, 1735–1751. [CrossRef]
255. Nguyen, H.Q.; Mourshed, M.; Paul, B.; Shabani, B. An Experimental Study of Employing Organic Phase Change Material for Thermal Management of Metal Hydride Hydrogen Storage. *J. Energy Storage* **2022**, *55*, 105457. [CrossRef]
256. Chibani, A.; Merouani, S.; Gherraf, N.; Benguerba, Y. Thermodynamics and Kinetics Analysis of Hydrogen Absorption in Large-Scale Metal Hydride Reactor Coupled to Phase Change Material-Metal Foam-Based Latent Heat Storage System. *Int. J. Hydrogen Energy* **2022**, *47*, 27617–27632. [CrossRef]
257. Chibani, A.; Merouani, S.; Bougriou, C.; Dehane, A. Heat and Mass Transfer Characteristics of Charging in a Metal Hydride-Phase Change Material Reactor with Nano Oxide Additives: The Large Scale-Approach. *Appl. Therm. Eng.* **2022**, *213*, 118622. [CrossRef]
258. Wang, Y.; Dai, H.; Chen, Z.; He, S.; Wang, W.; Gao, M. Simulation Study on a Novel Solid–Gas Coupling Hydrogen Storage Method for Photovoltaic Hydrogen Production Systems. *Energy Convers. Manag.* **2024**, *299*, 117866. [CrossRef]
259. Mellouli, S.; Alqahtani, T.; Algarni, S.; Askri, F.; Naimi, S. Numerical Simulation of a Combined Thermochemical-Latent Energy Storage System Based on Paired Metal Hydrides and Phase Change Material. *J. Energy Storage* **2024**, *86*, 111216. [CrossRef]
260. Bhouri, M.; Bürger, I.; Linder, M. Feasibility Analysis of a Novel Solid-State H2 Storage Reactor Concept Based on Thermochemical Heat Storage: MgH2 and Mg(OH)2 as Reference Materials. *Int. J. Hydrogen Energy* **2016**, *41*, 20549–20561. [CrossRef]
261. Lutz, M.; Linder, M.; Bürger, I. High Capacity, Low Pressure Hydrogen Storage Based on Magnesium Hydride and Thermochemical Heat Storage: Experimental Proof of Concept. *Appl. Energy* **2020**, *271*, 115226. [CrossRef]

262. Yao, J.; Zhu, P.; Qian, C.; Hamidullah, U.; Kurko, S.; Yang, F.; Zhang, Z.; Wu, Z. Study of an Autothermal-Equilibrium Metal Hydride Reactor by Reaction Heat Recovery as Hydrogen Source for the Application of Fuel Cell Power System. *Energy Convers. Manag.* **2020**, *213*, 112864. [CrossRef]
263. Shi, T.; Xu, H. Integration of Hydrogen Storage and Heat Storage in Thermochemical Reactors Enhanced with Optimized Topological Structures: Charging Process. *Appl. Energy* **2022**, *327*, 120138. [CrossRef]
264. Shi, T.; Xu, H.J.; Ke, H.B.; Zhao, C.Y. Thermal Transport of Charging/Discharging for Hydrogen Storage in a Metal Hydride Reactor Coupled with Thermochemical Heat Storage Materials. *Energy Convers. Manag.* **2022**, *273*, 116421. [CrossRef]
265. Maggini, M.; Falcucci, G.; Rosati, A.; Ubertini, S.; Facci, A.L. Non-Dimensional Numerical Analysis of Coupled Metal Hydride-Phase Change Material Hydrogen Storage System. *J. Energy Storage* **2024**, *93*, 112230. [CrossRef]
266. Zaluski, L.; Zaluska, A.; Ström-Olsen, J.O. Hydrogen Absorption in Nanocrystalline Mg2Ni Formed by Mechanical Alloying. *J. Alloys Compd.* **1995**, *217*, 245–249. [CrossRef]
267. Sharma, S.D.; Sagara, K. Latent Heat Storage Materials and Systems: A Review. *Int. J. Green Energy* **2005**, *2*, 1–56. [CrossRef]
268. Bhouri, M.; Bürger, I. Numerical Investigation of H2 Absorption in an Adiabatic High-Temperature Metal Hydride Reactor Based on Thermochemical Heat Storage: MgH2 and $Mg(OH)_2$ as Reference Materials. *Int. J. Hydrogen Energy* **2017**, *42*, 16632–16644. [CrossRef]
269. Chang, Y.; Yao, X.; Chen, Y.; Huang, L.; Zou, D. Review on Ceramic-Based Composite Phase Change Materials: Preparation, Characterization and Application. *Compos. Part B Eng.* **2023**, *254*, 110584. [CrossRef]
270. Sandrock, G.D.; Goodell, P.D. Cyclic Life of Metal Hydrides with Impure Hydrogen: Overview and Engineering Considerations. *J. Common. Met.* **1984**, *104*, 159–173. [CrossRef]
271. Ward, P.A.; Corgnale, C.; Teprovich, J.A.; Motyka, T.; Hardy, B.; Sheppard, D.; Buckley, C.; Zidan, R. Technical Challenges and Future Direction for High-Efficiency Metal Hydride Thermal Energy Storage Systems. *Appl. Phys. A* **2016**, *122*, 462. [CrossRef]
272. Guerrero-Ortiz, R.; Tena-García, J.R.; Flores-Jacobo, A.; Suárez-Alcántara, K. From the Can to the Tank: NaAlH4 from Recycled Aluminum. *Int. J. Hydrogen Energy* **2019**, *44*, 20183–20190. [CrossRef]
273. Passing, M.; Pistidda, C.; Capurso, G.; Jepsen, J.; Metz, O.; Dornheim, M.; Klassen, T. Development and Experimental Validation of Kinetic Models for the Hydrogenation/Dehydrogenation of Mg/Al Based Metal Waste for Energy Storage. *J. Magnes. Alloys* **2022**, *10*, 2761–2774. [CrossRef]

inorganics

MDPI

Article

Nickel Foam-Supported FeP Encapsulated in N, P Co-Doped Carbon Matrix for Efficient Electrocatalytic Hydrogen Evolution

Jianguo Zhong [1], Ting Zhang [1,*], Jianqiang Tian [1], Wei Gao [2] and Yuxin Wang [1,*]

[1] School of Materials Science and Engineering, Jiangsu University of Science and Technology, Zhenjiang 212114, China; 231210601424@stu.just.edu.cn (J.Z.); jianqiangtian@outlook.com (J.T.)

[2] Department of Chemical and Materials Engineering, Faculty of Engineering, The University of Auckland, Auckland 1142, New Zealand; w.gao@auckland.ac.nz

* Correspondence: tingzhang@just.edu.cn (T.Z.); ywan943@163.com (Y.W.)

Abstract: Transition metal phosphides (TMPs) show great potential as catalysts for the hydrogen evolution reaction (HER). FeP stands out as an efficient and cost-effective non-noble metal-based HER catalyst. However, FeP tends to aggregate and suffer from instability during the reaction. To tackle these challenges, we developed an efficient and straightforward approach to load metal-organic framework-derived N/P co-doped carbon encapsulated FeP nanoparticles onto a nickel foam substrate (FeP@NPC/NF-450). This catalyst exhibits exceptional HER activity in 0.5 M H_2SO_4 and 1.0 M KOH solutions, with overpotentials of 68.3 mV and 106.1 mV at a current density of 10 mA cm^{-2}, respectively. Furthermore, it demonstrates excellent stability with negligible decay over 48 h in both acidic and alkaline solutions. The outstanding hydrogen evolution catalytic performance of FeP@NPC/NF-450 is mainly due to the N, P co-doped carbon matrix, which safeguards the FeP nanoparticles from aggregation and surface oxidation. Consequently, this enhances the availability of active sites during the hydrogen evolution reaction (HER), leading to improved stability. Moreover, introducing nickel foam offers a larger specific surface area and enhances charge transfer rates. This study provides a reference method for preparing stable and highly active electrocatalysts for hydrogen evolution.

Keywords: metal-organic framework; transition-metal phosphides (TMPs); N/P co-doped carbon; electrocatalyst; hydrogen evolution reaction

Citation: Zhong, J.; Zhang, T.; Tian, J.; Gao, W.; Wang, Y. Nickel Foam-Supported FeP Encapsulated in N, P Co-Doped Carbon Matrix for Efficient Electrocatalytic Hydrogen Evolution. *Inorganics* **2024**, *12*, 291. https://doi.org/10.3390/inorganics12110291

Academic Editors: Torben R. Jensen and Roberto Nisticò

Received: 14 September 2024
Revised: 23 October 2024
Accepted: 4 November 2024
Published: 7 November 2024

1. Introduction

With the increasing severity of environmental pollution and energy shortages, the development of sustainable clean energy has gradually gained the attention of various countries [1–6]. Over the past few decades, numerous renewable green energy sources, such as biomass, solar, wind, geothermal, and hydrogen, have been explored. Among these renewable energy sources, hydrogen energy stands out due to its high energy density, and clean, pollution-free characteristics [7,8]. Noble metal-based materials are typically considered efficient catalysts for electrochemical water splitting to produce hydrogen. However, their limited abundance on earth and high costs restricts the commercial application of noble metal-based catalysts. Using abundant non-noble metals on earth as catalysts for water electrolysis to produce hydrogen has become a research hotspot [9]. Various catalysts have been developed to advance non-noble metal-based hydrogen evolution reaction (HER) catalysts, including transition metal carbides [10,11], transition metal sulfides [12,13], transition metal nitrides [14,15], transition metal oxides [16,17], and transition metal phosphides [18,19].

Among the various transition metal-based catalysts, transition metal phosphides have garnered significant attention for their excellent electrocatalytic activity and affordability. Among all the other transition metals, due to the low price and high Earth abundance of Fe, FeP has attracted much attention due to the electrocatalytic HER performance in consideration of other nature resources. FeP-based catalysts are reported to have high levels of activity in wide pH ranges in both acid and alkaline electrolytes. However, the HER activity of FeP is still inferior to that of Pt-based catalysts, so enhancing FeP-based electrocatalytic activity is a substantial challenge [20]. Previously, numerous researchers have investigated FeP-based electrocatalysts. Wei and his team synthesized spindle-shaped Co-doped FeP supported on three-dimensional reduced graphene oxide using a hydrothermal method [21]. The synthesized catalyst demonstrated a low overpotential of 110.8 mV at a current density of 10 mA cm^{-2} in an acidic environment. Similarly, Jiang's group constructed a hollow spherical heterostructure FeCo-P catalyst through a hydrothermal and phosphorization method [22]. The hollow spherical structure and the heterostructure between Co_2P and FeP endowed the catalyst with a low overpotential of only 131 mV at a current density of 10 mA cm^{-2} in 1.0 M KOH. Despite these advances, the catalytic activity and stability of FeP are significantly limited by the slow charge transfer rate, the low number of active sites, and its high sensitivity to oxygen during water electrolysis. Metal-organic frameworks (MOFs), consisting of metal ions or clusters linked to organic ligands, have gained significant attention for their large specific surface area and high porosity [23]. Utilizing MOFs as templates has proven to be an effective strategy for preparing porous carbon/transition metal phosphides, transition metal oxides, and transition metal sulfides, which exhibit excellent catalytic performance [24].

To improve the electrocatalytic activity and stability of FeP, it is crucial to design a structure that maximizes the exposure of active sites while preventing catalyst aggregation and surface oxidation. Thus, it is essential to develop an efficient structure that keeps FeP nanoparticles separated during the reaction. Recently, carbon materials have been widely used to encapsulate FeP nanoparticles [25]. Carbon prevents FeP nanoparticles from aggregating during phosphorization and protects FeP electrocatalysts from oxidation during the hydrogen evolution reaction (HER) [26]. However, the intrinsic activity of carbon materials requires enhancement. The low intrinsic activity of carbon diminishes the electrocatalytic performance of carbon-coated FeP nanoparticles. Recent studies have revealed that nitrogen (N) doping in carbon matrices can create additional active sites for transition metal phosphide-based catalysts during the HER, significantly enhancing their overall performance. Also, nickel foam exhibits excellent conductivity, facilitating efficient electron transport during electrochemical reactions, thus improving catalytic performance [27]. Furthermore, the robust structure of nickel foam ensures good catalyst stability. It also allows direct electrochemical performance testing without needing binders, significantly saving time. The interaction between the nickel foam substrate and the catalyst can generate synergistic effects, further enhancing overall catalytic efficiency [28].

Inspired by these developments, we synthesized an N, P-doped carbon-coated FeP catalyst supported on nickel foam (NF). This catalyst was prepared by immersing nickel foam in a Fe-MOF solution, enabling successful loading of the MOF onto the nickel foam, followed by phosphorization in a tube furnace under an Ar atmosphere. The resulting FeP@NPC/NF catalyst exhibited excellent HER catalytic activity and stability in alkaline and acid mediums.

2. Result and Discussion

To avoid interference from the nickel foam substrate and the X-ray diffractometer sample holder, X-ray diffraction spectra were collected for FeP@NPC-T (T = 400 °C, 450 °C, 500 °C) and Fe_2O_3@NC after removing the nickel foam substrate. The X-ray diffraction spectra of samples subjected to different heat treatment temperatures exhibit similar main features, as shown in Figure 1. The diffraction peaks of the FeP@NPC/NF-T samples match well with those of FeP (PDF#01-071-2262). The diffraction peaks observed at 30.9°, 32.8°, 34.5°, 35.5°, 37.2°, 46.3°, 46.9°, 48.3°, 50.4°, 56.1°, and 59.6° correspond to the (020), (011), (220), (120), (111), (121), (220), (211), (130), (221), and (002) crystal planes of FeP

(PDF#01-071-2262) [22], respectively. This indicates the successful synthesis of the FeP phase in the catalysts.

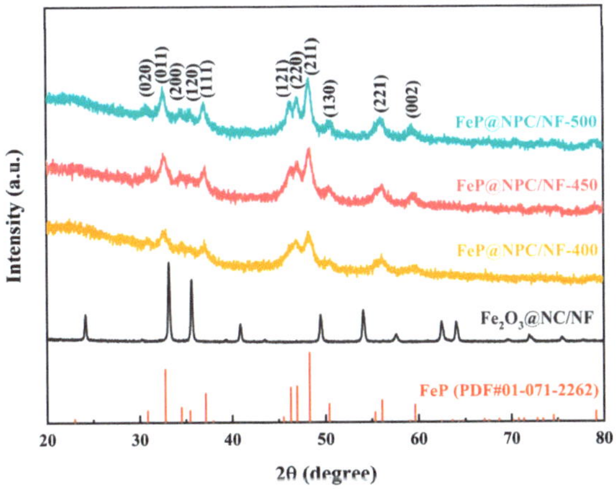

Figure 1. The XRD patterns of FeP@NPC/NF samples subjected to heat treatment at 400 °C, 450 °C, and 500 °C, along with the Fe_2O_3@NC/NF sample heat−treated at 450 °C.

The morphology and structure of FeP@NPC/NF-400, FeP@NPC/NF-450, and FeP@NPC/NF-500 were examined using scanning electron microscopy (SEM), as shown in Figure 2a–i. The nickel foam surface is porous and smooth, providing ample space for catalyst growth and gas release channels, which helps mitigate the decline in catalytic performance. After loading FeP@NPC nanoparticles onto the nickel foam, the surface becomes rougher. The size of the nanoparticles in the catalysts enhanced due to the increase in the annealing temperature from 400 °C to 500 °C. Figure 2g presents the EDS elemental mapping of the FeP@NPC/NF-450 sample, indicating the presence and uniform distribution of Ni, Fe, P, N, and C elements within the catalyst, further confirming the successful synthesis of the FeP@NPC/NF catalyst.

Transmission electron microscopy (TEM) characterization of FeP@NPC/NF-450 exhibits a large number of small nanoparticles encapsulated by the matrix, as shown in Figure 3. The formation of the matrix could be attributed to the heat treatment of 2-methylimidazole. The d spacing of the crystalline fringe is 0.241 nm, indicating the formation of (111) in the FeP phase [29]. The TEM results exhibit that the FeP nanoparticles are encapsulated in the N, P-doped carbon matrix.

To further investigate the chemical composition and electronic properties of the FeP@NPC/NF-450 catalyst, an X-ray photoelectron spectroscopy (XPS) analysis was performed. Figure 4a shows the XPS spectra of the FeP@NPC/NF-450 catalyst. The Ni 2p XPS spectrum of FeP@NPC/450/NF exhibits six characteristic peaks, as shown in Figure 4b. The two peaks at 853.33 eV and 870.65 eV, corresponding to Ni $2p_3/_2$ and Ni $2p_1/_2$, are attributed to the presence of metallic Ni^0, indicating that Ni primarily exists in its metallic form. The peaks at 856.81 eV and 874.01 eV are assigned to Ni-O, while the broad peaks at 861.93 eV and 880.32 eV correspond to satellite peaks [30,31]. The XPS spectrum of Fe 2p exhibits five characteristic peaks, as shown in Figure 4c. The two peaks at 706.27 eV and 719.07 eV, corresponding to Fe $2p_3/_2$ and Fe $2p_1/_2$, are attributed to the presence of Fe^{2+}, while the peaks at 709.65 eV and 723.51 eV are attributed to Fe^{3+}. The presence of both Fe^{2+} and Fe^{3+} is ascribed to the formation of Fe-O. The peak at 713.14 eV is associated with the formation of Fe-P bonds [32,33], consistent with the XRD results (Figure 4c). Figure 4d shows the high-resolution P 2p spectrum, which can be deconvoluted into three peaks: 129.85 eV (P $2p_{3/2}$) and 130.74 eV (P $2p_{3/2}$) corresponding to the formation of Fe-P bonds, and 134.50 eV attributed to P-O bonds formed by surface oxidation [34,35]. The

N 1s spectrum can be resolved into three primary peaks, which correspond to graphitic N at 401.87 eV, pyrrolic N at 400.34 eV, and pyridinic N at 399.36 eV, as previously reported [36,37]. The incorporation of nitrogen into the carbon framework serves to both improve the catalyst's conductivity and augment the quantity of active sites conducive to hydrogen generation [38]. The C 1s XPS spectrum (Figure 4f) shows a strong peak at 284.79 eV, corresponding to C=C bonds, along with peaks at 286.74 eV and 288.65 eV, representing C-N=O and O-N=O bonds, respectively, indicating successful nitrogen doping in the carbon matrix [39,40]. These analyses confirm the successful synthesis of the FeP@NPC/NF-450 catalyst.

Figure 2. SEM images of FeP@NPC/NF−400 (**a–c**), FeP@NPC/NF−450 (**d–f**), and FeP@NPC/NF−500 (**g–i**) at different magnifications; elemental mapping images of FeP@NPC/NF−450 (**j**).

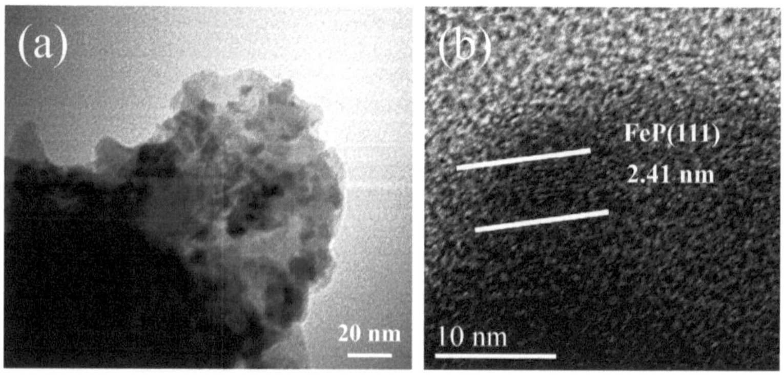

Figure 3. TEM images of (**a**) FeP@NPC/NF−450 and (**b**) high-resolution TEM of FeP@NPC/NF−450.

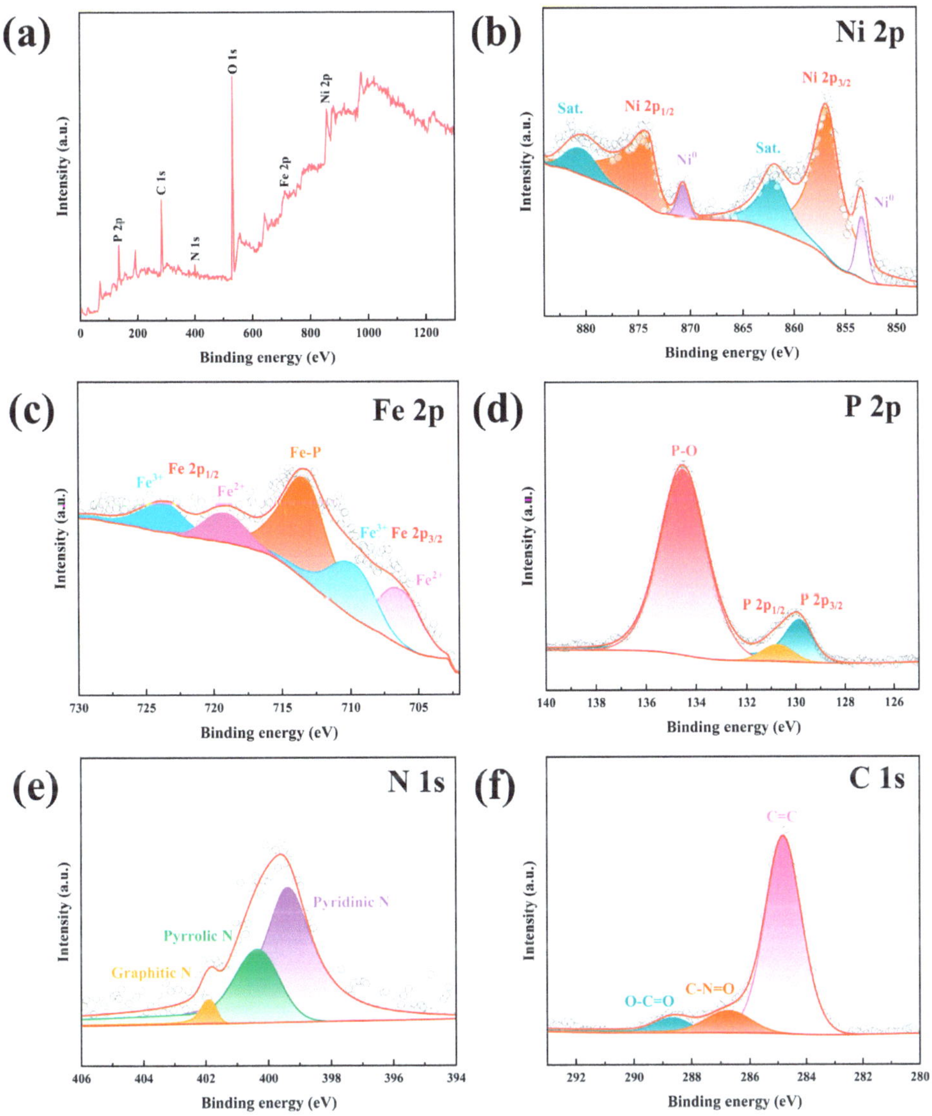

Figure 4. (**a**) FeP@NPC/NF−450 XPS survey spectrum; high-resolution XPS spectrum of FeP@NPC/NF−450: (**b**) Ni 2p, (**c**) Fe 2p, (**d**) P 2p, (**e**) N 1s, (**f**) C 1s.

The electrocatalytic performance of FeP@NPC/NF-450 during the hydrogen evolution reaction (HER) was tested in a 1.0 M KOH solution. The HER performance levels of Pt sheets, FeP@NPC/NF-400, FeP@NPC/NF-500, Fe_2O_3@NC/NF, and nickel foam were also analyzed for comparison. The linear sweep voltammetry (LSV) curves of the catalysts are shown in Figure 5a. The Pt sheet exhibited the lowest overpotential of 35.1 mV at a current density of 10 mA cm^{-2}, demonstrating the highest HER performance [41]. Among the FeP@NPC/NF catalysts synthesized at different phosphorization temperatures and Fe_2O_3@NC/NF and nickel foam, FeP@NPC/NF-450 showed the highest HER activity. At a current density of 10 mA cm^{-2}, the overpotential of FeP@NPC/NF-450 was only 106.1 mV, significantly lower than that of FeP@NPC/NF-400 (138.9 mV), FeP@NPC/NF-

500 (142.9 mV), Fe_2O_3@NC/NF (201.1 mV), and nickel foam (177.8 mV). FeP nanoparticles are protected by the N, P-doped carbon matrix from aggregation, the mass transfer in the big bulk particles will be hindered. The HER properties will decrease with the increase in the annealing temperature to 500 °C. However, when the annealing temperature is under 450 °C, the crystal structure of FeP is weak in the XRD results, indicating that some of the Fe-MOF precursor is transferring to the FeP phase. According to the LSV results, FeP is the active site of the catalyst; the weak crystal structure of FeP will decrease the number of active sites during the reaction, thus hindering the HER properties of the catalyst.

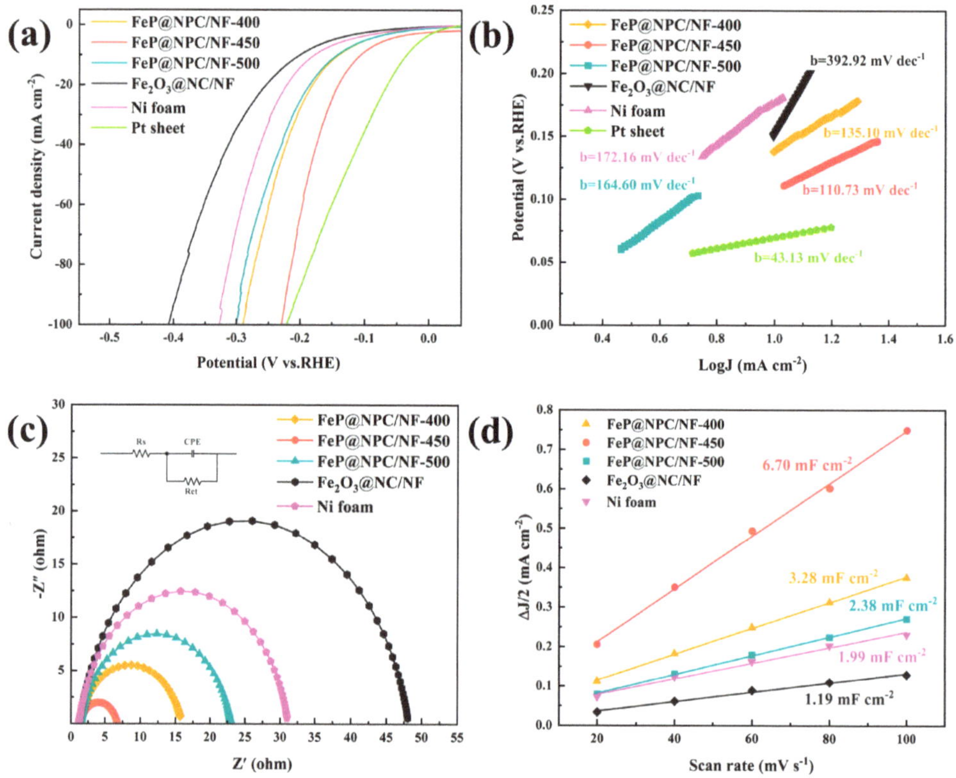

Figure 5. Electrochemical tests of catalysts in 1 M KOH: (**a**) LSV curve and (**b**) Tafel slope of FeP@NPC/NF−400, FeP@NPC/NF−450, FeP@NPC/NF−500, Fe_2O_3@NC/NF, Ni foam, and Pt sheet; (**c**) Nyquist plot and (**d**) capacitive current difference versus scan rate of FeP@NPC/NF−400, FeP@NPC/NF−450, FeP@NPC/NF−500, Fe_2O_3@NC/NF, and Ni foam.

To investigate the kinetics of the catalysts, their Tafel slopes were analyzed (Figure 5b). The Tafel slope of FeP@NPC/NF-450 was only 110.73 mV dec^{-1}, significantly lower than those of FeP@NPC/NF-400 (135.10 mV dec^{-1}), FeP@NPC/NF-500 (164.60 mV dec^{-1}), Fe_2O_3@NC/NF (392.92 mV dec^{-1}), and Ni foam (172.16 mV dec^{-1}). This indicated that FeP@NPC/NF-450 exhibits the highest efficiency for hydrogen production, so we conducted a further analysis of the catalysts' kinetics during the hydrogen evolution reaction (HER) process. Nyquist plots were obtained for FeP@NPC/NF-400, FeP@NPC/NF-450, FeP@NPC/NF-500, Fe_2O_3@NC/NF, and Ni foam, as depicted in Figure 5c. The efficiency of electron charge transfer within the catalysts is manifested by the diameter of the semicircles in the Nyquist plots, which is indicative of the charge transfer resistance (Rct) encountered during the reaction. The resistance at the interface between the electrolyte and the catalyst is termed as Rs [42]. Notably, FeP@NPC/NF-450 demonstrated the lowest internal resistance of 5.42 Ω cm^{-2}, signifying its superior electron transfer efficiency compared to the other catalysts. This enhanced performance can be ascribed to the N- and P-doped carbon matrix, along with the nickel foam substrate, which effectively mitigates the aggregation of FeP nanoparticles during the annealing process. As a result, these structural features facilitate accelerated charge transfer between the active sites and the electrolyte.

To delve deeper into the hydrogen evolution reaction (HER) performance of the catalysts during hydrogen production, the double-layer capacitance (C_{dl}) was determined by conducting cyclic voltammetry scans at various sweep rates, spanning from 20 to 100 mV s^{-1}. The C_{dl} values of FeP@NPC/NF-400, FeP@NPC/NF-450, FeP@NPC/NF-500, Fe_2O_3@NC/NF, and nickel foam was 3.28 mF cm^{-2}, 6.70 mF cm^{-2}, 2.38 mF cm^{-2}, 1.19 mF cm^{-2}, and 1.99 mF cm^{-2}, respectively, as shown in Figure 5d. FeP@NPC/NF-450 exhibited the largest double-layer capacitance, significantly higher than that of the other catalysts studied. This suggests that encapsulation in the N and P-doped carbon matrix prevents the aggregation of FeP nanoparticles and facilitates the redistribution of electrons among N, P, and C, promoting the exposure of more active sites.

Assessing the stability of FeP@NPC/NF-450 during the hydrogen production process is of utmost importance. The stability of FeP@NPC/NF-450 was assessed by 1000 cyclic voltammetry scans and chronoamperometry tests. As shown in Figure 6a, after long-term testing over 1000 cycles, there was no significant change in the polarization curve compared to the catalyst before cycling. As depicted in Figure 6b, the current density of FeP@NPC/NF-450 remained almost unchanged after 48 h at 106.1 mV in the chronoamperometry test. These results indicate that the catalyst's structure remains stable in the reaction and Fe atoms do not leach out from the materials. The stability of FeP@NPC/NF-450 can be attributed to the encapsulation by the N and P co-doped carbon, which prevents the oxidation and corrosion of the FeP nanoparticles in 1.0 M KOH electrolyte [25].

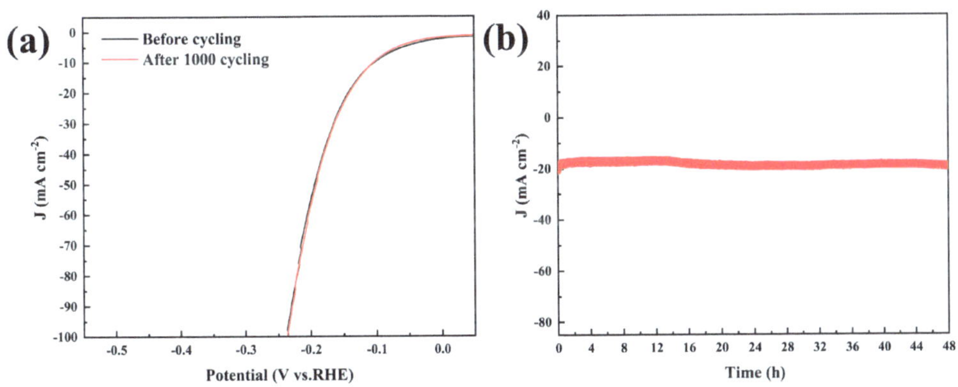

Figure 6. (**a**) Stability of FeP@NPC/NF−450 before and after 1000 cycles, and (**b**) chronoamperometric curve of FeP@NPC/NF−450 at 106.1 mV for 48 h in 1.0 M KOH.

Figure 7 showcases the hydrogen evolution reaction (HER) performance of the synthesized catalysts in an acidic electrolyte (0.5 M H_2SO_4). The linear sweep voltammetry (LSV) curves (Figure 7a) demonstrate that the HER catalytic activity of the FeP@NPC/NF-450 catalyst surpasses that of the other studied catalysts, second only to the Pt sheet. At a current density of 10 mA cm^{-2}, the overpotentials of the FeP@NPC/NF-400, FeP@NPC/NF-450, FeP@NPC/NF-500, Fe_2O_3@NC/NF, Ni foam, and Pt sheet catalysts were 86.8 mV, 68.3 mV, 135.9 mV, 229.5 mV, 134.6 mV, and 33.5 mV, respectively. Figure 7b shows the Tafel slopes of the prepared catalysts, with the Tafel slope of FeP@NPC/NF-450 being 57.90 mV dec^{-1}, significantly lower than those of FeP@NPC/NF-400 (72.81 mV dec^{-1}), Mo-FeP@NPC/NF-

500 (97.63 mV dec^{-1}), Fe_2O_3@NC/NF (210.42 mV dec^{-1}), and Ni foam (140.95 mV dec^{-1}), indicating its superior catalytic kinetics during the HER process.

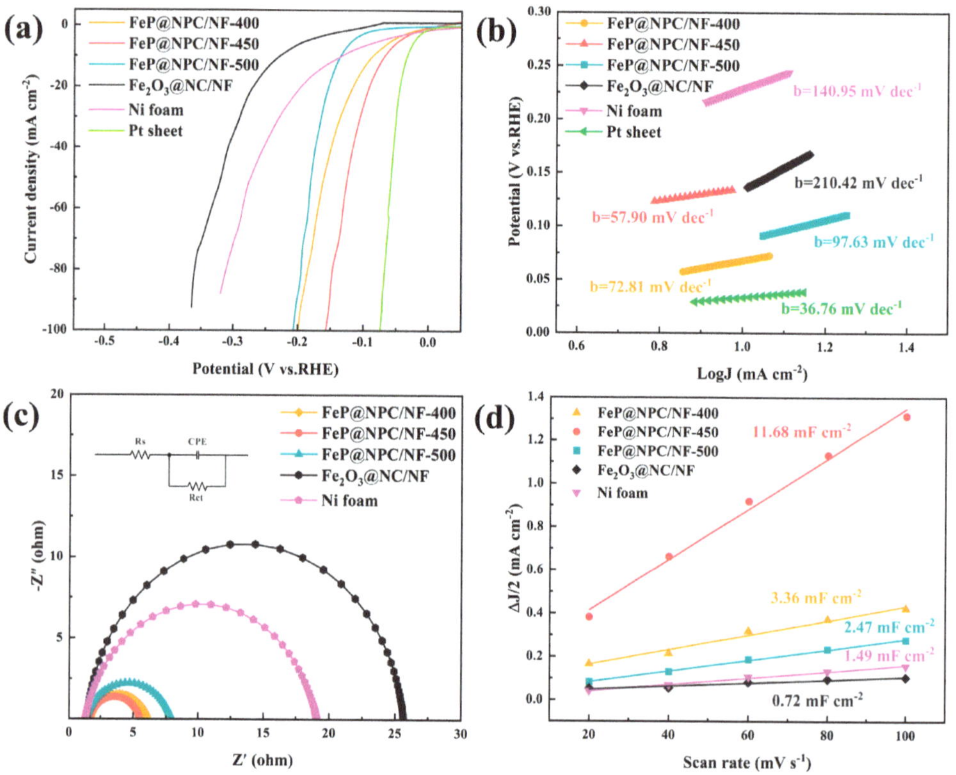

Figure 7. Electrochemical tests of catalysts in 0.5 M H_2SO_4: (**a**) LSV curves of FeP@NPC/NF-400, FeP@NPC/NF−450, FeP@NPC/NF−500, Fe_2O_3@NC/NF, nickel foam, and Pt sheet; (**b**) Tafel slopes; (**c**) Nyquist plots of FeP@NPC/NF−400, FeP@NPC/NF−450, FeP@NPC/NF−500, Fe_2O_3@NC/NF, and nickel foam; (**d**) capacitive current difference versus scan rate.

To evaluate the electron transfer dynamics of the synthesized catalysts, electrochemical impedance spectroscopy (EIS) measurements were performed. The charge transfer resistance (R_{ct}) of the FeP@NPC/NF-450 catalyst was 3.85 Ω cm^{-2}, lower than that of FeP@NPC/NF-400 (4.41 Ω cm^{-2}), FeP@NPC/NF-500 (6.63 Ω cm^{-2}), Fe_2O_3@NC/NF (24.24 Ω cm^{-2}), and Ni foam (17.78 Ω cm^{-2}), suggesting a faster electron transfer rate during hydrogen production. To further investigate the electrochemical double-layer capacitance (C_{dl}) of the catalysts, cyclic voltammetry (CV) curves were recorded at different scan rates. As shown in Figure 7d, the C_{dl} value of FeP@NPC/NF-450 was 11.68 mF cm^{-2}, significantly higher than those of FeP@NPC/NF-400 (3.36 mF cm^{-2}), FeP@NPC/NF-500 (2.47 mF cm^{-2}), Fe_2O_3@NC/NF (0.72 mF cm^{-2}), and Ni foam (1.49 mF cm^{-2}), indicating a larger electrochemically active surface area and more available active sites for hydrogen production.

The synthesized FeP@NPC/NF-450 catalyst exhibits excellent stability in 0.5 M H_2SO_4, as shown in Figure 8. After 1000 cycles, the LSV curve shows almost no change. Simultaneously, after 48 h of chronoamperometry testing, the change in current density is negligible, demonstrating its remarkable catalytic stability.

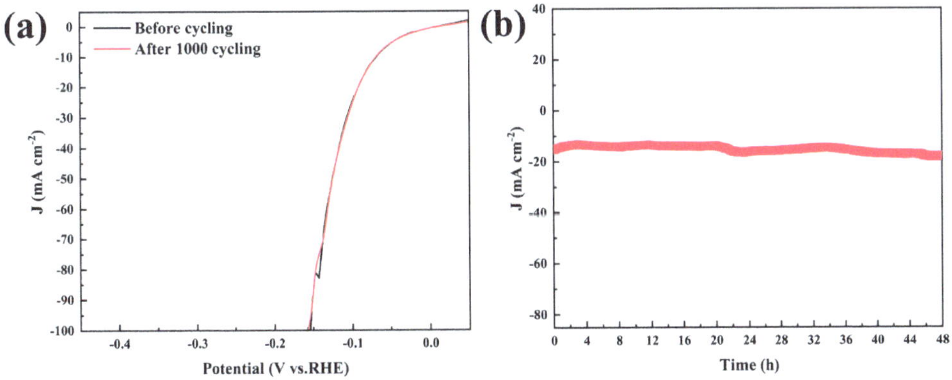

Figure 8. (**a**) Stability of FeP@NPC/NF−450 before and after 1000 cycles, and (**b**) chronoamperometric curve of FeP@NPC/NF−450 at 68.3 mV for 48 h in 0.5 M H_2SO_4.

3. Experimental Section

3.1. Chemicals

All chemicals were used as received without further purification. Hydrochloric acid was purchased from Shanghai Titan Technology Co., Ltd. Sodium hypophosphite (NaH_2PO_2) was obtained from Aladdin Biochemical Technology Co., Ltd. (Shanghai, China). 2-Methylimidazole ($C_4H_6N_2$) was purchased from Adamas-beta (Shanghai, China) Chemical Reagent Co., Ltd. Ethanol (C_2H_5OH) was acquired from Jiangsu Qiangsheng Functional Chemical Co., Ltd. (Changshu, China) Iron(III) chloride ($FeCl_3$) was purchased from Shanghai Runjie Chemical Reagent Co., Ltd. (Shanghai, China).

3.2. Pretreatment of Nickel Foam

First, the nickel foam was cut into small pieces of 40 mm × 25 mm. The cut nickel foam was immersed into 1 M HCl, deionized water, and anhydrous ethanol to ultrasonicate for 20 min. After the ultrasonication, a hair dryer was used to blow dry the pieces for spare.

3.3. Preparation of Fe-MOF Precursor on Nickel Foam

Firstly, 2 mmol of $FeCl_3$ was added to 50 mL of deionized water and stirred until completely dissolved. Then, 10 mmol of 2-methylimidazole was added, and the mixture was stirred for an additional 30 min to obtain the Fe-MOF precursor [18]. The pretreated nickel foam was then immersed in the Fe-MOF precursor solution and left to stand for 24 h. After 24 h, the nickel foam was removed and gently rinsed with deionized water and anhydrous ethanol to remove impurities and organic residues. Finally, the nickel foam was dried in an oven at 60 °C to obtain the Fe-MOF/NF precursor.

3.4. Preparation of FeP@NPC/NF

$FeCl_3$ was dissolved in the DI water to form uniform Fe^{2+} solution. 2-methylimidazole was dissolved in the $FeCl_3$ solution to prepared Fe-MOF precursor. Ni foam was soaked in the above-mentioned precursor to prepare Fe-MOF/NF precursor. Then, the prepared precursor was transferred to FeP@NPC/NF under phosphorization reaction.

The FeP@NPC/NF samples were prepared using a high-temperature phosphorization method under argon protection. The Fe-MOF/NF precursor and 0.4 g of sodium hypophosphite were placed separately at the downstream and upstream positions of a porcelain boat, respectively. The porcelain boat was then transferred to a tube furnace. Before heating, the tube furnace was evacuated and purged with argon gas three times. Under an argon atmosphere, the furnace was heated at 2 °C/min to 450 °C and maintained at this temperature for 2 h. The furnace was then cooled at a rate of 5 °C/min to 200 °C, followed by natural cooling to room temperature to obtain the phosphorized FeP@NPC/NF catalyst at 450 °C, designated as FeP@NPC/NF-450. For comparison, the phosphorization of Fe-MOF/NF was conducted at 400 °C and 500 °C under otherwise identical conditions, resulting in FeP@NPC/NF-400 and FeP@NPC/NF-500, respectively. Additionally, Fe-MOF/NF was thermally treated at 450 °C without adding a phosphorus source under the same conditions, yielding Fe_2O_3@NC/NF.

3.5. Preparation of FeP Powder Material Without Ni Foam

After Ni foam had been immersed for 24 h in the Fe-MOF precursor solution, powders were collected from the baker. The collected powder was washed and dried, then annealed under the same condition of FeP@NPC/NF, the prepared sample was labeled as FeP@NPC.

3.6. Material Characterization

The crystal structure of the prepared powder material (sample without Ni foam) was analyzed using powder X-ray diffraction (XRD, Rigaku SmartLab, Cambridge, MA, USA) with Cu $K_{\alpha 1}$ radiation (λ = 1.5406 Å) in the 2θ range of 20° to 80°. The microstructure was further investigated using scanning electron microscopy (FE-SEM, Hitachi SU8600, Tokyo, Japan). The surface chemical composition of the catalyst was determined by X-ray photoelectron spectroscopy (XPS, ThermoFisher Nexsa, Waltham, MA, USA).

3.7. Electrochemical Measurements

The HER performance was analyzed using a standard three-electrode system with a CHI 660E electrochemical workstation. The reference and counter electrodes were saturated calomel electrode (SCE) and a Pt sheet. The working area of the working electrode was 1 cm × 1 cm, and the HER was conducted in a 1 M KOH solution and 0.5 M H_2SO_4 solution. All electrode potentials (vs. SCE) were calibrated to a reversible hydrogen electrode (RHE) and IR-corrected (solution resistance), namely, $E_{(RHE)} = E_{(SCE)} + 0.244 + (0.0599 \times pH) - iR_s$ (at 25 °C) [43,44]. The linear sweep voltammetry (LSV) rate was 3 mV s^{-1}, and the electrochemical impedance spectroscopy (EIS) was analyzed at a voltage of 106.1 mV (1 M KOH) and 68.3 mV (0.5 M H_2SO_4) in the frequency range of 100 KHz to 10 MHz.

According to the formula given in [45],

$$ECSA = C_{dl}/C_s$$

The electrochemical active surface area (ECSA) can be analyzed through the double-layer capacitance (C_{dl}), and the C_{dl} of catalysts was evaluated based on the cyclic voltammogram (CV). The CV curves of the catalysts were tested at scan rates from 20 to 100 mV s^{-1}. The slope of the fitted linear regression function ($\Delta j = j_{anodic} - j_{cathodic}$) vs. scan rate) equals the value of C_{dl}.

4. Conclusions

In conclusion, we have devised a novel structure aimed at augmenting the activity and stability of FeP nanoparticles. A simple phosphidation method was employed to synthesize nitrogen and phosphorus co-doped carbon-coated FeP nanoparticles supported on nickel foam (FeP@NPC/NF). The N, P co-doped carbon serves to prevent the aggregation of FeP nanoparticles by encapsulating them, thereby increasing the number of active sites and shielding the FeP nanoparticles from corrosion and oxidation during the hydrogen production process. The multi-controllable network structure of the nickel foam offers a larger specific surface area during the hydrogen evolution reaction (HER) and facilitates gas release. The results demonstrate that FeP@NPC/NF-450 exhibits excellent HER activity in acidic and alkaline environments. In 1.0 M KOH and 0.5 M H_2SO_4, the hydrogen evolution overpotentials at a current density of 10 mA cm^{-2} were 68.3 mV and 106.1 mV, respectively, with corresponding Tafel slopes of 57.90 mV dec^{-1} and 110.73 mV dec^{-1}. Moreover, it showed long-term stability for 48 h in both acidic and alkaline media, surpassing the performance of recently reported FeP-based HER catalysts. This study provides a simple approach for improving the activity and stability of FeP-based electrocatalysts, presenting a promising application in green energy conversion, where highly active HER catalysts are essential.

Author Contributions: J.Z.: Investigation, Data curation, Writing—original draft. T.Z.: Methodology, Formal Analysis, Supervision. J.T.: Formal Analysis W.G.: Writing—review and editing. Y.W.: Supervision. All authors have read and agreed to the published version of the manuscript.

Funding: This research was funded by [Jiangsu Province Natural Science Foundation for Youths] grant number [No. BK20241021].

Data Availability Statement: Data are contained within the article.

Acknowledgments: The authors would like to acknowledge the assistance of the staff at the Jiangsu University of Science and Technology. Ting Zhang was supported by the "Jiangsu Province Excellent Postdoctoral Program".

Conflicts of Interest: The authors declare that they have no known competing financial interests or personal relationships that could have appeared to influence the work reported in this paper.

References

1. Xu, Y.; Wang, R.; Zheng, Y.; Zhang, L.; Jiao, T.; Peng, Q.; Liu, Z. Facile preparation of self-assembled Ni/Co phosphates composite spheres with highly efficient HER electrocatalytic performances. *Appl. Surf. Sci.* **2020**, *509*, 145383. [CrossRef]
2. Yi, Z.; Zeng, Y.; Wu, H.; Chen, X.; Fan, Y.; Yang, H.; Tang, Y.; Yi, Y.; Wang, J.; Wu, P. Synthesis, surface properties, crystal structure and dye-sensitized solar cell performance of TiO_2 nanotube arrays anodized under different parameters. *Results Phys.* **2019**, *15*, 102609. [CrossRef]
3. Zhang, T.; He, Z.; Yin, L.; Gao, W.; Wang, Y. CoNi alloy nanoparticles confined by N-doped carbon matrix with tailored d-Band center for electrocatalytic hydrogen evolution. *Fuel* **2024**, *365*, 131176. [CrossRef]
4. Gao, P.; Yue, C.; Zhang, J.; Bao, J.; Wang, H.; Chen, Q.; Jiang, Y.; Huang, S.; Hu, Z.; Zhang, J.J. Construction of unique NiCoP/FeNiCoP hollow heterostructured ellipsoids with modulated electronic structure for enhanced overall water splitting. *J. Colloid Interface Sci.* **2024**, *666*, 403–415. [CrossRef] [PubMed]
5. Vijayakumar, E.; Ramakrishnan, S.; Sathiskumar, C.; Yoo, D.J.; Balamurugan, J.; Noh, H.S.; Kwon, D.; Kim, Y.H.; Lee, H. MOF-derived CoP-nitrogen-doped carbon@ NiFeP nanoflakes as an efficient and durable electrocatalyst with multiple catalytically active sites for OER, HER, ORR and rechargeable zinc-air batteries. *Chem. Eng. J.* **2022**, *428*, 131115. [CrossRef]
6. Yuan, S.; Wu, Y.; Chen, W.; Yin, L.; Hao, J.; Wang, Y. Defects engineering of nickel tellurium electrocatalyst to boost urea oxidation reaction: Electrodeposition mechanism and performance. *J. Alloys Compd.* **2024**, *997*, 174944. [CrossRef]
7. Wu, H.; Chen, Z.; Wang, Y.; Cao, E.; Xiao, F.; Chen, S.; Du, S.; Wu, Y.; Ren, Z. Regulating the allocation of N and P in codoped graphene via supramolecular control to remarkably boost hydrogen evolution. *Energy Environ. Sci.* **2019**, *12*, 2697–2705. [CrossRef]
8. Zhang, T.; Zhong, J.; Gao, W.; Wang, Y. Interface Engineering Induced N, P-Doped Carbon-Shell-Encapsulated FeP/NiP_2/Ni_5P_4/NiP Nanoparticles for Highly Efficient Hydrogen Evolution Reaction. *Coatings* **2024**, *14*, 817. [CrossRef]
9. Zhang, T.; Yang, T.; Qu, G.; Huang, S.; Cao, P.; Gao, W. Phase control and stabilization of 1T-MoS_2 via black TiO_{2-x} nanotube arrays supporting for electrocatalytic hydrogen evolution. *J. Energy Chem.* **2022**, *68*, 71–77. [CrossRef]
10. Zhang, Q.; Jiang, Z.; Tackett, B.M.; Denny, S.R.; Tian, B.; Chen, X.; Wang, B.; Chen, J.G. Trends and descriptors of metal-modified transition metal carbides for hydrogen evolution in alkaline electrolyte. *ACS Catal.* **2019**, *9*, 2415–2422. [CrossRef]
11. Zhao, R.; Chen, Y.; Xiang, H.; Guan, Y.; Yang, C.; Zhang, Q.; Li, Y.; Cong, Y.; Li, X. Two-dimensional ordered double-transition metal carbides for the electrochemical nitrogen reduction reaction. *ACS Appl. Mater. Interfaces* **2023**, *15*, 6797–6806. [CrossRef] [PubMed]
12. Shen, W.; Cui, J.; Chen, C.; Zhang, L.; Sun, D. Metal-organic framework derived transition metal sulfides grown on carbon nanofibers as self-supported catalysts for hydrogen evolution reaction. *J. Colloid Interface Sci.* **2024**, *659*, 364–373. [CrossRef] [PubMed]
13. Kapuria, A.; Mondal, T.K.; Shaw, B.K.; Su, Y.-K.; Saha, S.K. Polysulfide functionalized reduced graphene oxide for electrocatalytic hydrogen evolution reaction and supercapacitor applications. *Int. J. Hydrogen Energy* **2023**, *48*, 17014–17025. [CrossRef]
14. Majhi, K.C.; Yadav, M. Palladium oxide decorated transition metal nitride as efficient electrocatalyst for hydrogen evolution reaction. *J. Alloys Compd.* **2021**, *855*, 157511. [CrossRef]
15. Gujral, H.S.; Singh, G.; Yang, J.H.; Sathish, C.; Yi, J.; Karakoti, A.; Fawaz, M.; Ramadass, K.; Ala'a, H.; Yu, X. Mesoporous titanium carbonitride derived from mesoporous C_3N_5 for highly efficient hydrogen evolution reaction. *Carbon* **2022**, *195*, 9–18. [CrossRef]
16. Li, Y.; Yu, Z.G.; Wang, L.; Weng, Y.; Tang, C.S.; Yin, X.; Han, K.; Wu, H.; Yu, X.; Wong, L.M. Electronic-reconstruction-enhanced hydrogen evolution catalysis in oxide polymorphs. *Nat. Commun.* **2019**, *10*, 3149. [CrossRef]
17. Hu, F.; Yu, D.; Ye, M.; Wang, H.; Hao, Y.; Wang, L.; Li, L.; Han, X.; Peng, S. Lattice-Matching formed mesoporous transition metal oxide heterostructures advance water splitting by active Fe–O–Cu bridges. *Adv. Energy Mater.* **2022**, *12*, 2200067. [CrossRef]

18. Zhang, T.; Yang, T.; Li, B.; Wei, S.; Gao, W. MOF-derived formation of ultrafine FeP nanoparticles confined by N/P Co-doped carbon as an efficient and stable electrocatalyst for hydrogen evolution reaction. *Appl. Surf. Sci.* **2022**, *597*, 153662. [CrossRef]
19. Zhou, Q.; Wang, D. 3D nanoporous NiCoP as a highly efficient electrocatalyst for the hydrogen evolution reaction in alkaline electrolyte. *New J. Chem.* **2022**, *46*, 7490–7496. [CrossRef]
20. Liu, H.; Shi, S.; Wang, Z.; Han, Y.; Huang, W. Recent Advances in Metal–Gas Batteries with Carbon-Based Nonprecious Metal Catalysts. *Small* **2022**, *18*, 2103747. [CrossRef]
21. Wei, S.; Xing, P.; Tang, Z.; Wang, Y.; Dai, L. Spindle-shaped cobalt-doped iron phosphide anchored on three-dimensional graphene electrocatalysis for hydrogen evolution reactions in both acidic and alkaline media. *J. Power Sources* **2023**, *555*, 232414. [CrossRef]
22. Jiang, H.; Zhao, Z.; Li, G.; Wang, M.; Chen, P.; Liu, X.; Tu, X.; Hu, Y.; Shen, Z.; Wu, Y. Hollow Spherical Heterostructured FeCo-P Catalysts Derived from MOF-74 for Efficient Overall Water Splitting. *Adv. Sci.* **2024**, *11*, 2306919. [CrossRef] [PubMed]
23. Cai, G.; Yan, P.; Zhang, L.; Zhou, H.-C.; Jiang, H.-L. Metal–organic framework-based hierarchically porous materials: Synthesis and applications. *Chem. Rev.* **2021**, *121*, 12278–12326. [CrossRef] [PubMed]
24. Rezaee, S.; Shahrokhian, S. Ruthenium/Ruthenium oxide hybrid nanoparticles anchored on hollow spherical Copper-Cobalt Nitride/Nitrogen doped carbon nanostructures to promote alkaline water splitting: Boosting catalytic performance via synergy between morphology engineering, electron transfer tuning and electronic behavior modulation. *J. Colloid Interface Sci.* **2022**, *626*, 1070–1084. [PubMed]
25. Zheng, Z.; Wu, H.-H.; Liu, H.; Zhang, Q.; He, X.; Yu, S.; Petrova, V.; Feng, J.; Kostecki, R.; Liu, P.; et al. Achieving fast and durable lithium storage through amorphous FeP nanoparticles encapsulated in ultrathin 3D P-doped porous carbon nanosheets. *ACS Nano* **2020**, *14*, 9545–9561. [CrossRef] [PubMed]
26. Zhang, T.; Zhong, J.; Ding, Y.; Gao, W.; Wang, Y. Mo-doped FeP nanoparticles encapsulated in N, P doped carbon matrix with optimized d-band center for highly efficient and stable hydrogen evolution reaction. *New J. Chem.* **2024**, *48*, 11302–11309. [CrossRef]
27. Sivanantham, A.; Ganesan, P.; Shanmugam, S. Hierarchical $NiCo_2S_4$ nanowire arrays supported on Ni foam: An efficient and durable bifunctional electrocatalyst for oxygen and hydrogen evolution reactions. *Adv. Funct. Mater.* **2016**, *26*, 4661–4672. [CrossRef]
28. Chen, H.; Qiao, S.; Yang, J.; Du, X. NiMo/$NiCo_2O_4$ as synergy catalyst supported on nickel foam for efficient overall water splitting. *Mol. Catal.* **2022**, *518*, 112086. [CrossRef]
29. Wang, Y.; Wang, X.; Zhang, L.; Zhang, Y.; Xu, Z.; Lu, L.; Huang, J.; Yin, L.; Zhu, W.; Zhuang, Z. Insights into the effect of precursors on the FeP-catalyzed hydrogen evolution reaction. *Inorg. Chem.* **2022**, *61*, 2954–2961. [CrossRef]
30. Zhang, H.; Wu, Y.; Wang, X.; Li, C.; Xiao, Z.; Liu, Y.; Deng, Y.; Li, Z.; Wang, L. The construction of defect-rich CoP@CoP@$(Co/Ni)_2P$ triple-shell hollow nanospheres with boosted electrocatalytic hydrogen evolution performances over a wide pH range. *Chem. Eng. J.* **2023**, *463*, 142448. [CrossRef]
31. Qian, M.; Cui, S.; Jiang, D.; Zhang, L.; Du, P. Highly efficient and stable water-oxidation electrocatalysis with a very low overpotential using FeNiP substitutional-solid-solution nanoplate arrays. *Adv. Mater.* **2017**, *29*, 1704075. [CrossRef] [PubMed]
32. Ma, R.; Cui, X.; Xu, X.; Wang, Y.; Xiang, G.; Gao, L.; Lin, Z.; Yang, Y. Collaborative integration of ultrafine Fe_2P nanocrystals into Fe, N, P-codoped carbon nanoshells for highly-efficient oxygen reduction. *Nano Energy* **2023**, *108*, 108179. [CrossRef]
33. Xu, L.-H.; Che, P.-C.; Zhang, X.-J.; Cosnier, S.; Shan, D. FeP nanoparticles highly dispersed on N, P-doped petaloid carbon nanosheet: Interface engineering and boosted intrinsic ORR activity. *Appl. Surf. Sci.* **2023**, *620*, 156770. [CrossRef]
34. Zhai, W.; Chen, Y.; Liu, Y.; Ma, Y.; Vijayakumar, P.; Qin, Y.; Qu, Y.; Dai, Z. Covalently bonded Ni sites in black phosphorene with electron redistribution for efficient metal-lightweighted water electrolysis. *Nanomicro Lett.* **2024**, *16*, 115. [CrossRef]
35. Yang, X.; Wang, F.; Jing, Z.; Chen, M.; Wang, B.; Wang, L.; Qu, G.; Kong, Y.; Xu, L. A General "In Situ Etch-Adsorption-Phosphatization" Strategy for the Fabrication of Metal Phosphides/Hollow Carbon Composite for High Performance Liquid/Flexible Zn–Air Batteries. *Small* **2023**, *19*, 2301985. [CrossRef]
36. Xing, G.; Zhang, G.; Wang, B.; Tong, M.; Tian, C.; Wang, L.; Fu, H. Strengthening oxygen reduction activity based on the cooperation of pyridinic-N and graphitic-N for atomically dispersed Fe sites. *J. Mater. Chem. A* **2023**, *11*, 9493–9503. [CrossRef]
37. Shin, C.-H.; Ted, H.Y.; Lee, H.-Y.; Lee, B.-J.; Kwon, S.; Goddard, W.A., III; Yu, J.-S. Ru-loaded pyrrolic-N-doped extensively graphitized porous carbon for high performance electrochemical hydrogen evolution. *Appl. Catal. B* **2023**, *334*, 122829. [CrossRef]
38. Zhou, W.; Zhou, J.; Zhou, Y.; Lu, J.; Zhou, K.; Yang, L.; Tang, Z.; Li, L.; Chen, S. N-doped carbon-wrapped cobalt nanoparticles on N-doped graphene nanosheets for high-efficiency hydrogen production. *Chem. Mater.* **2015**, *27*, 2026–2032. [CrossRef]
39. Xie, J.; Fu, W.; Zhang, X.; Song, G.; Wu, H.; Zhou, M. Cobalt species immobilized on nitrogen-doped ordered mesoporous carbon for highly efficient catalytic ozonation of atrazine. *Sep. Purif.* **2024**, *335*, 126077. [CrossRef]
40. Lv, Q.; Si, W.; He, J.; Sun, L.; Zhang, C.; Wang, N.; Yang, Z.; Li, X.; Wang, X.; Deng, W.; et al. Selectively nitrogen-doped carbon materials as superior metal-free catalysts for oxygen reduction. *Nat. Commun.* **2018**, *9*, 3376. [CrossRef]
41. Chang, Y.-H.; Lin, C.-T.; Chen, T.-Y.; Hsu, C.-L.; Lee, Y.-H.; Zhang, W.; Wei, K.-H.; Li, L.-J. Highly efficient electrocatalytic hydrogen production by MoS_x grown on graphene-protected 3D Ni foams. *Adv. Mater.* **2013**, *25*, 756–760. [CrossRef] [PubMed]
42. Papiya, F.; Pattanayak, P.; Kumar, P.; Kumar, V.; Kundu, P.P. Development of highly efficient bimetallic nanocomposite cathode catalyst, composed of Ni: Co supported sulfonated polyaniline for application in microbial fuel cells. *Electrochim. Acta* **2018**, *282*, 931–945. [CrossRef]

43. Ye, J.; Zang, Y.; Wang, Q.; Zhang, Y.; Sun, D.; Zhang, L.; Wang, G.; Zheng, X.; Zhu, J. Nitrogen doped FeS_2 nanoparticles for efficient and stable hydrogen evolution reaction. *J. Energy Chem.* **2021**, *56*, 283–289. [CrossRef]
44. Wu, Y.; Sun, Z.; Wang, Y.; Yin, L.; He, Z.; Zhang, Z.; Hayat, M.D.; Zang, Q.; Lian, J. Cyclic voltammetric deposition of binder-free Ni-Se film on Ni foams as efficient bifunctional electrocatalyst for boosting overall urea-water electrolysis. *J. Alloys Compd.* **2023**, *937*, 168460. [CrossRef]
45. Cheng, C.; Ao, W.; Ren, H.; Fan, Z.; Xu, T.; Dai, L.; Yin, P. Amorphous versus crystalline CoS_x anchored on CNTs as heterostructured electrocatalysts toward hydrogen evolution reaction. *Sci. China Mater.* **2023**, *66*, 1383–1388. [CrossRef]

inorganics

MDPI

Article

Hierarchically Porous Titanosilicate Hollow Spheres Containing TS-1 Zeolite Precursors for Oxidative Desulfurization

Yao Wang, Hongda Yu, Huan Wang and Tiehong Chen *

Smart Sensing Interdisciplinary Science Center, Institute of New Catalytic Materials Science, School of Materials Science and Engineering, Nankai University & Cangzhou Bohai New Area Green Chemical Institute, Nankai University, Tianjin 300350, China
* Correspondence: chenth@nankai.edu.cn

Abstract: The environmental and health impacts of sulfur compounds in fuel oil have prompted considerable research interest in oxidative desulfurization (ODS) technology. Tetrahedrally coordinated titanium has been demonstrated to exhibit excellent activity in the context of oxidative desulfurization processes. However, further improving the catalytic property of the tetrahedrally coordinated titanium remains a challenging endeavor. In the context of ODS processes conducted at near room temperatures, the improvement of conversion remains a subject of considerable challenge. In this study, hierarchically porous titanosilicate hollow spheres were synthesized by using TS-1 zeolite precursors as Ti and Si sources to obtain the catalyst with only tetrahedrally coordinated titanium. The synthesized materials were characterized through transmission electron microscopy (TEM), Fourier-transform infrared spectroscopy (FT-IR), ultraviolet–visible diffuse reflectance spectroscopy (UV-Vis), and nitrogen adsorption analysis. These techniques confirmed the formation of hollow spherical hierarchically porous structures with Ti species uniformly incorporated in tetrahedral coordination and the presence of five-member rings of TS-1 zeolite. As a result, the hierarchically porous titanosilicate hollow spheres demonstrated excellent catalytic performance in ODS, achieving complete dibenzothiophene (DBT) removal within 15 min and a high turnover frequency (TOF) of up to 123 h^{-1} at 30 °C.

Keywords: hierarchically porous; titanosilicate; hollow spheres; oxidative desulfurization

Academic Editors: Torben R. Jensen and Roberto Nisticò

Received: 19 December 2024
Revised: 21 January 2025
Accepted: 22 January 2025
Published: 25 January 2025

Citation: Wang, Y.; Yu, H.; Wang, H.; Chen, T. Hierarchically Porous Titanosilicate Hollow Spheres Containing TS-1 Zeolite Precursors for Oxidative Desulfurization. *Inorganics* **2025**, *13*, 37. https://doi.org/10.3390/inorganics13020037

1. Introduction

Fuel oil is widely utilized as energy source; however, its high quantity of sulfur remains a problem that needs to be solved [1]. During the combustion of gasoline and diesel, sulfur compounds such as dibenzothiophene (DBT), 4,6-dimethyldibenzothiophene (4,6-DMDBT), and 4-methylbenzothiophene (4-MDBT) produce sulfur oxides, which are harmful to human health and contribute to environmental pollution [2]. In response, stricter sulfur content regulations have been introduced worldwide, attracting research interest for development of effective petroleum desulfurization technologies [3–6]. Despite the recent advancements in the field, the efficient removal of sulfur compounds from fuel oils using minute quantities of catalysts at ambient temperatures is still rare.

Hydrodesulfurization (HDS) is the most prevalent industrial technology utilized for sulfur removal due to its high efficiency [7]. However, producing ultra-low sulfur fuels with HDS requires high temperatures, high pressures, and significant hydrogen consumption, which increases operational costs and complicates process control [8]. To overcome these limitations, alternative non-hydrodesulfurization methods, such as biological desulfurization [9], adsorption desulfurization [10,11], extraction desulfurization [12], and oxidative desulfurization (ODS) [13] have been explored. Among them, ODS eliminates the need for hydrogen, enhancing operational safety and reducing costs. The two types of ODS are non-catalytic and catalytic; while non-catalytic ODS is compatible with oils that contain high levels of sulfur, it necessitates elevated reaction temperatures and extended reaction times [14]. In contrast, catalytic ODS has emerged as particularly promising, as it efficiently removes sulfur compounds under mild conditions, including ambient pressure and reaction temperatures below 100 °C [15]. More importantly, aromatic sulfur containing organic compounds (ASOCs) including DBT, 4,6-DMDBT, and 4-MDBT are not easily eliminated through the HDS process because of steric hindrance; in contrast, ASOCs can be efficiently removed through the ODS process under mild conditions [16].

The ODS process means that sulfur compounds are catalytically oxidized into more polar sulfone derivatives, which can then be separated from hydrocarbons through adsorption or extraction [17]. The most common oxidants are hydrogen peroxide (H_2O_2) and tert-butyl hydroperoxide (TBHP). H_2O_2 is often favored for its environmental friendliness and cost-effectiveness; however, the reaction occurs in a biphasic water/oil system. In this setup, sulfur compounds remain in the oil phase, while the catalyst and H_2O_2 are in the aqueous phase, causing mass transfer limitations that hinder the oxidation rate [18]. In comparison, TBHP, being oil-soluble, enables the reaction to proceed in a single-phase system, eliminating interfacial mass transfer limitations and enhancing reaction performance [19–21].

Various catalysts have shown significant potential in oxidative desulfurization (ODS), including transition metal oxides [22,23], TS-1 zeolite [24,25], and mesoporous silica [26,27]. Careo et al. [23] investigated ODS reactions in synthetic diesel using vanadium oxide catalysts supported on titanium, analyzing the impact of operating conditions on sulfur removal efficiency. Similarly, Du et al. [24] synthesized hierarchical porous TS-1 (HTS) using a soft-template method. The hierarchical pore structure of HTS enabled superior dibenzothiophene (DBT) removal compared to conventional TS-1, which relies solely on microporous structures and demonstrated lower efficiency under identical conditions. Furthermore, the use of mesoporous materials has been shown to enhance ODS performance due to their large pore channels, which improve the accessibility of sulfur compounds. For instance, Cho et al. [28] synthesized Ti-SBA-15 through a grafting method, achieving good catalytic activity [29,30]. In summary, some ODS catalysts have been reported; however, their activities remain unsatisfactory and introduce single tetrahedrally coordinated Ti to the catalysts, which is still a challenge.

Previous studies [31,32] have demonstrated that aging a solution containing appropriate amounts of silicon sources and template agents, such as TPAOH or TEAOH, at controlled temperatures produces a transparent solution rich in zeolite primary and secondary structural units, along with nanoclusters formed through their aggregation. Adding Ti species to the above precursors makes it more likely that material precursors containing tetrahedrally coordinated Ti will be obtained. In addition, due to their substantial surface area and comparatively short channels, which are conducive to molecular diffusion, hierarchically porous nanoparticles are frequently used as catalytic support and a variety of templates have been shown to facilitate the synthesis of hierarchically porous structures such as PEO-*b*-PMMA and CTAB [33], P123 and SDS [34], and CTAB-PAA [26,35].

Herein, hierarchically porous titanosilicate hollow spheres (Ti-HHS) containing TS-1 zeolite precursors were synthesized using a polyelectrolyte–surfactant mesoporous complex template as a structure-directing agent. Ti-HHS exhibited hollow spherical mesoporous structures with Ti species uniformly incorporated in tetrahedral coordination and the presence of five-member rings of TS-1 zeolite, and gave rise to excellent catalytic performance in ODS, achieving complete dibenzothiophene (DBT) removal with a high turnover frequency (TOF) of up to 123 h^{-1} at 30 °C.

2. Results and Discussion

2.1. Characterization of Ti-HHS

The synthesis of the Ti-HHS mechanism is outlined in Figure 1. The small-angle XRD patterns of the calcined Ti-HHS-1, Ti-HHS-2, and Ti-HHS-3 samples are presented in Figure 2a. These patterns display a single broad diffraction peak around 2.0°, consistent with the worm-like pore structures observed in the TEM images. Figure 2b shows the wide-angle XRD patterns of the same samples. A broad diffraction peak centered at 23° is observed, corresponding to the characteristic peak of amorphous silica. No diffraction peaks associated with zeolite crystal structures or titanium species are detected.

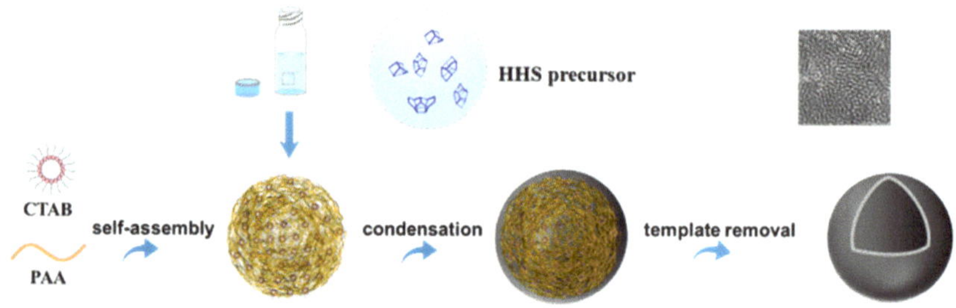

Figure 1. Schematic representation of the synthesis mechanism of hierarchically porous titanosilica hollow spheres.

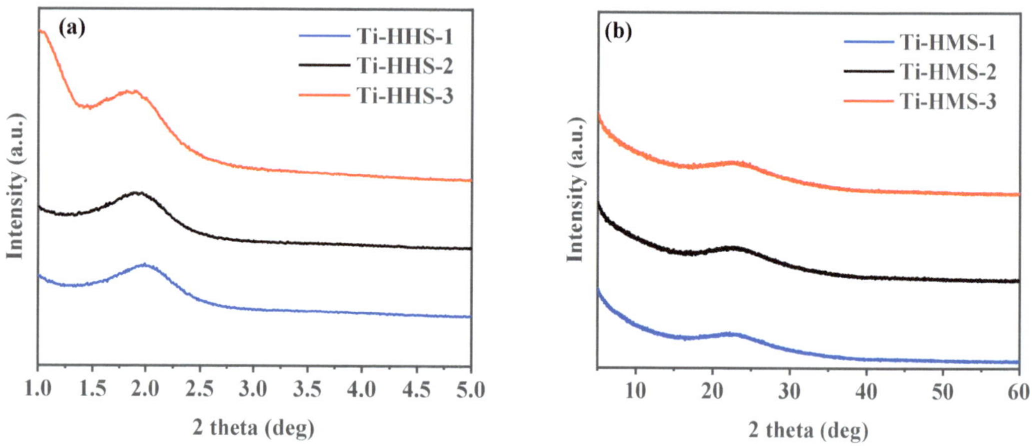

Figure 2. Small-angle XRD patterns (**a**) and wide-angle XRD patterns (**b**) of hierarchically porous titanosilicate hollow spheres with various Ti contents: Ti-HHS-1, Ti-HHS-2, and Ti-HHS-3.

The SEM and TEM images of Ti-HHS-3 are shown in Figures 3 and 4, respectively. The SEM images (Figure 3) reveal that Ti-HHS-3 has a uniform spherical morphology with particle sizes ranging around several hundreds of nanometers and exhibits good particle dispersion. Figure 3b shows a rough particle surface with visible macropores. The TEM images in Figure 4a confirm that the hollow spherical shell structure is consistent across the sample, supporting the observations from the SEM images. Figure 4b indicates that the shell thickness ranges from 20 to 40 nanometers. Figure 4d, an enlarged view of

Figure 4c, highlights the uniformly distributed wormhole structure within the particles, further emphasizing the hierarchical porous characteristics of the sample [36,37].

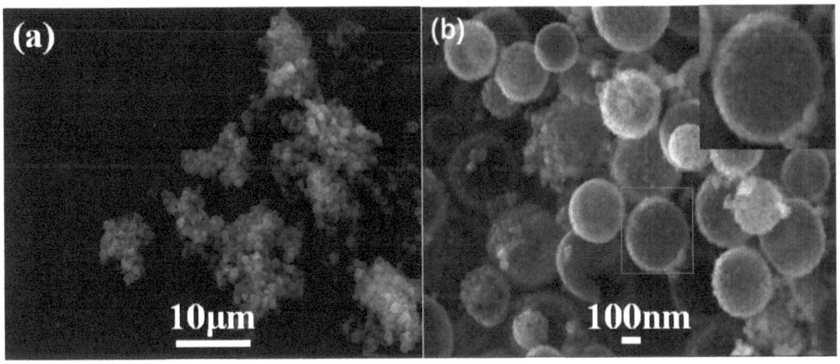

Figure 3. SEM images of Ti-HHS-3. (**a**) ×2000, and (**b**) ×50,000, the red square indicates the original location of the enlarged image, which is situated in the upper right corner.

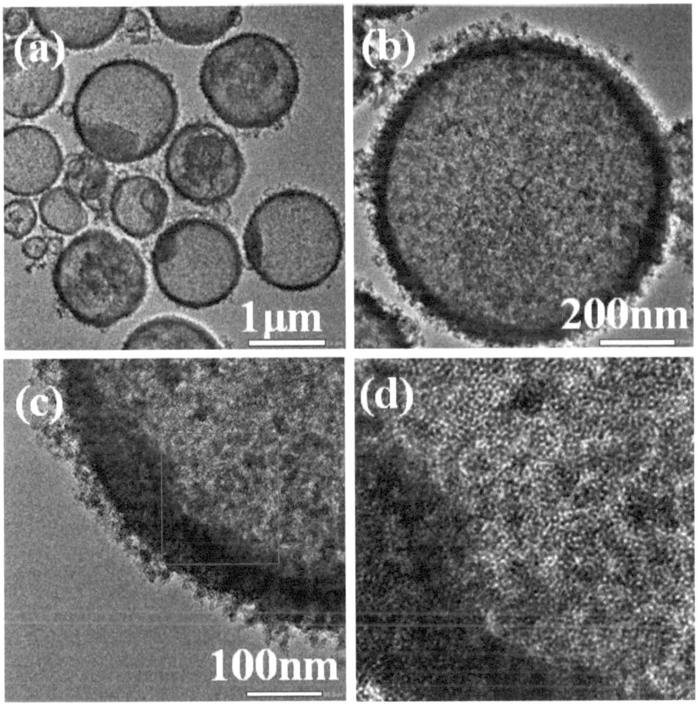

Figure 4. TEM images of Ti-HHS-3. (**a**) 1 μm, (**b**) 200 nm, (**c**) 100 nm, and (**d**) enlarged view of the red square in figure (**c**).

The EDS mapping for Ti-HHS-3, presented in Figure 5, characterizes the Ti distribution within the sample. The mapping confirms a uniform distribution of Ti throughout the hollow sphere structure, with no evidence of local aggregation or clustering. These results validate the successful incorporation of Ti into all three samples. The titanium content in the sample was determined by ICP-AES; the Ti content of Ti-HHS-1, Ti-HHS-2, and Ti-HHS-3 are 0.55 wt%,1.46 wt%, and 2.60 wt%, respectively.

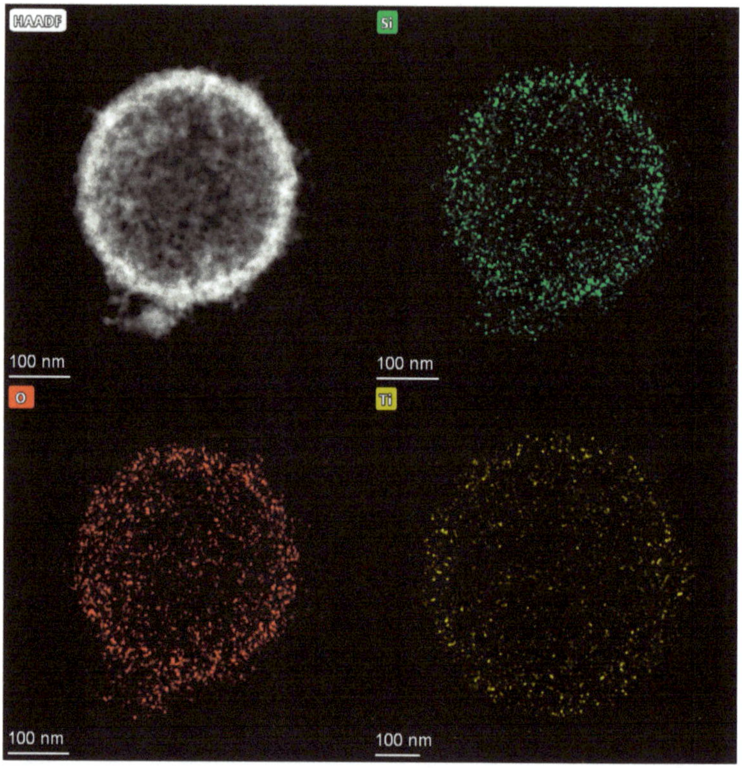

Figure 5. EDS–mapping images of Ti-HHS-3.

Figure 6a,b present the nitrogen adsorption–desorption isotherms and the corresponding pore size distribution curves for the hierarchically porous titanosilicate hollow sphere samples. All three samples exhibit typical Type IV isotherms with an H3-type hysteresis loop, characterized by two adsorption steps in the relative pressure (P/P_0) range of 0.25–0.35 and 0.85–0.95 in Figure 6a. The first step corresponds to a peak at approximately 3 nm in Figure 6b, indicating the presence of wormhole-like mesopores within the samples [36,37]. The second step corresponds to the larger pores within the shell template by phase-separated PAA [35]. Furthermore, the nearly overlapping adsorption–desorption curves suggest highly accessible mesopores, consistent with the thin-shell structure of the hollow spheres. According to Table 1, the three samples show relatively large specific surface areas and pore volumes. While the surface area and pore volume of Ti-HHS-3 are lower than Ti-HHS-1 and Ti-HHS-2, which probably owes to different additions of Ti species [38].

Table 1. The detailed structural parameters of the samples.

Samples	Mesopore Size (nm)	S_{BET} (m^2/g)	V_{MIC} (cm^3/g)	V_{MES} (cm^3/g)
Ti-HHS-1	2.8	710	0.005	1.022
Ti-HHS-2	2.8	647	0.006	0.915
Ti-HHS-3	2.8	600	0.009	0.941

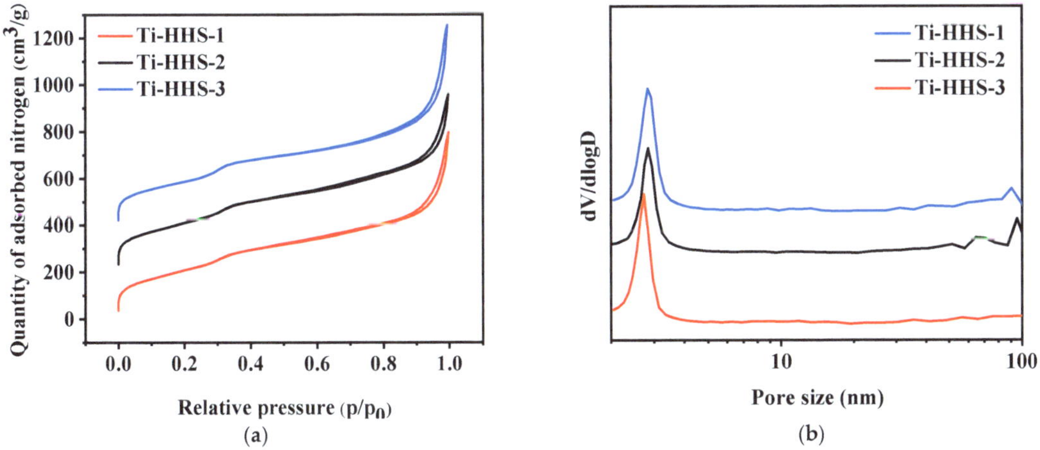

Figure 6. Nitrogen adsorption–desorption isotherms (**a**) and pore size distribution curves (**b**) for Ti-HHS-1, Ti-HHS-2, and Ti-HHS-3.

As shown in the FT-IR spectra in Figure 7, all three samples exhibit characteristic absorption peaks at 810 cm^{-1} and 1053 cm^{-1}. These absorption peaks correspond to the symmetric and asymmetric Si–O–Si stretching vibrations, respectively, which are typical features of amorphous silica in infrared spectra [39]. Additionally, a minor absorption peak appears at 560 cm^{-1} in the Ti-HHS-1, Ti-HHS-2, and Ti-HHS-3 samples. This peak is associated with the stretching vibrations of double five-membered rings in MFI structures [40], suggesting the presence of primary and secondary zeolite structural units and indicating short-range order within the samples. Each sample also shows a distinct absorption peak at 960 cm^{-1}, attributed to Si–O–Ti vibrations [39,41], confirming the incorporation of Ti into the structure. Furthermore, the intensity of the 960 cm^{-1} peak decreases with lower Ti content, particularly in the Ti-HHS-1 sample, where the peak at 960 cm^{-1} is less prominent.

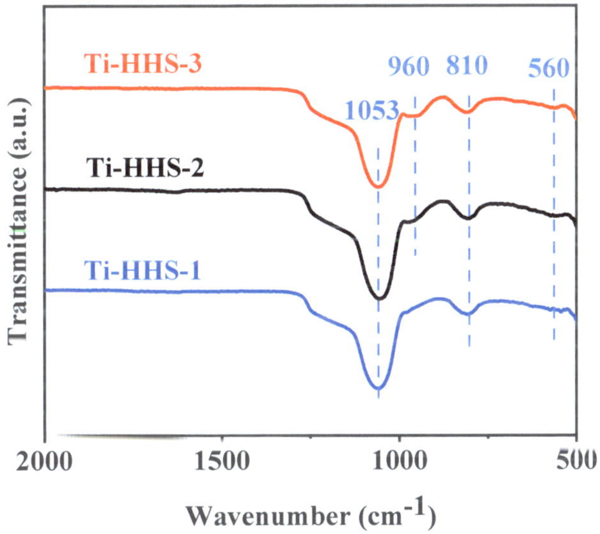

Figure 7. FT-IR spectra for the Ti-HHS-1, Ti-HHS-2, and Ti-HHS-3.

UV–visible absorption spectroscopy was employed to investigate the coordination state of Ti species in the hierarchically porous titanosilicate hollow spheres. As shown in Figure 8, all three samples exhibit a narrow, strong absorption peak at 215 nm, with no additional peaks observed at longer wavelengths. The absorption peak near 220 nm is characteristic of the tetrahedral coordination of Ti species within the framework [42]. These results confirm that the doped Ti exists exclusively in tetrahedral coordination within the framework, with no evidence of non-framework octahedral Ti species or separate crystalline TiO_2 phases [43,44].

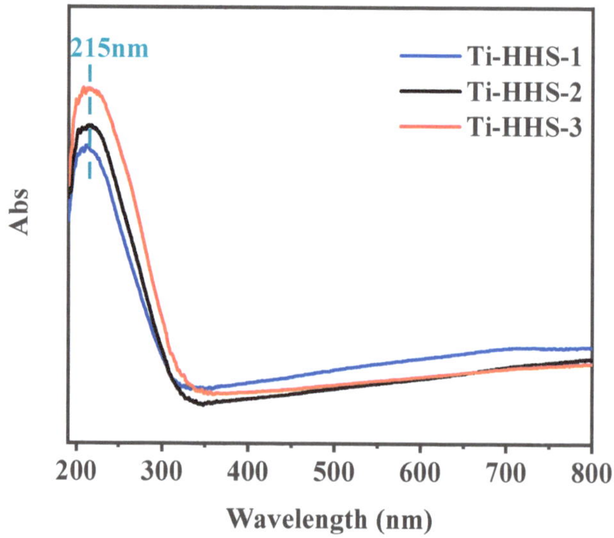

Figure 8. UV-vis spectra for the Ti-HHS-1, Ti-HHS-2, and Ti-HHS-3.

2.2. *ODS Performance of Ti-HHS*

In the ODS reaction at 30 °C, both the catalyst and oxidant are crucial for effective desulfurization. As shown in Figure 9a, the sulfur compound conversion rate remains low in the absence of an oxidant, and no significant conversion occurs without a catalyst. However, the simultaneous introduction of the Ti-HHS-3 catalyst and the oxidant (TBHP) results in rapid DBT conversion, reaching 100% within 15 min. Previous studies [32,45] have demonstrated that tetrahedral Ti sites exhibit higher oxidation activity compared to Ti species in amorphous silica. Structural characterization of Ti-HHS-3 confirms the presence of Ti species primarily in tetrahedral coordination, which significantly enhances its activity. As illustrated in Figure 9b, the Ti-HHS-3 catalyst, synthesized from TS-1 precursor, exhibits substantially higher ODS activity than TiO_2/HHS prepared via direct the impregnation of TiO_2 with an equivalent Si/Ti molar ratio. The catalytic ODS process is shown in Figure 10, and, as reported in the literature, sulfoxides exhibited a high polarity and tended to be absorbed onto the silanol groups in silica via hydrogen bonding, achieving a one-pot oxidation–adsorption process with the hierarchically porous titanosilicate [26].

The ODS activity of catalysts with varying Ti content (Ti-HHS-1, Ti-HHS-2, and Ti-HHS-3) was evaluated under identical conditions. As shown in Figure 11, Ti-HHS-3 exhibits the highest catalytic performance, achieving 100% DBT conversion within 15 min. In contrast, Ti-HHS-1 and Ti-HHS-2 achieve only 71% and 94% conversion after 30 min, respectively. These results highlight the critical role of Ti content in enhancing catalytic activity. At the 2 min mark, Ti-HHS-3 achieves a TOF of up to 123 h^{-1}, which could be due to the tetrahedrally coordinated Ti of the TS-1 precursors in Ti-HHS-3. As illustrated in

Table 2, an analysis of the extant literature reveals a range of catalysts with different levels of performance in the oxidation of DBT. It is evident that Ti-HHS-3 exhibits higher activity than the majority of reported catalysts, especially the TOF, although it should be noted that Ti-HHS-3 operates at near room temperature.

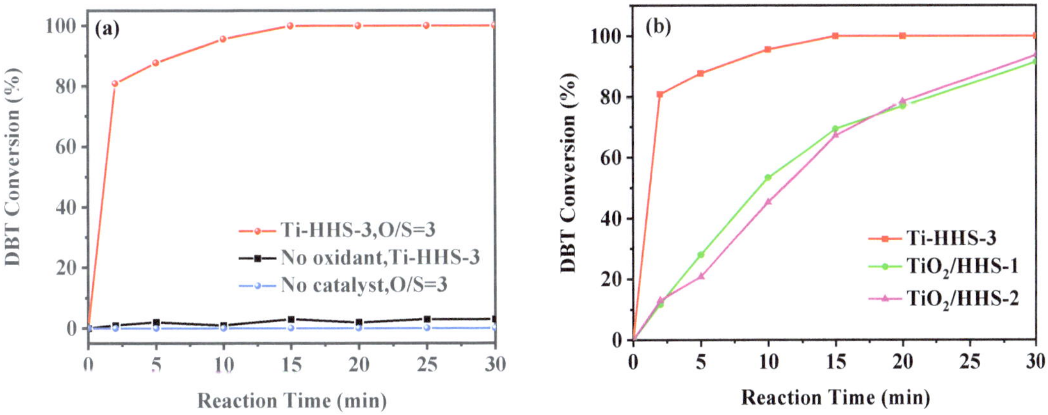

Figure 9. (**a**) DBT ODS activity of the titanosilicate hollow spheres under various conditions. (**b**) Comparison with impregnated catalysts Ti/HHS-1 and Ti/HHS-2 (reaction conditions: 20 g of 500 ppm model gasoline, 20 mg catalyst, 30 °C, O/S = 3).

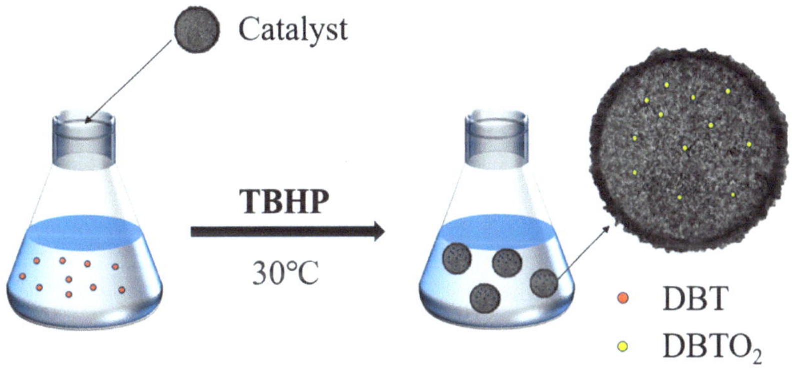

Figure 10. The flow chart of the ODS process.

Table 2. Comparison of catalytic ODS performance of DBT with different catalysts.

Catalysts	Ti Content (wt%)	S Content (ppm)	Oxidant	Temperature (°C)	TOF (h^{-1})	Ref.
Ti-HHS-3	2.6	500	TBHP	30	123	This work
Ti-B-M-DA	5.5	1000	TBHP	60	58.8	[46]
Ti-SBA-2	5.5	500	TBHP	40	18.8	[47]
Meso-TS-1	0.98	174	H_2O_2	60	3.7	[48]
NSTS-10	2.7	500	TBHP	25	14.9	[26]
Ti-HMS	0.88	584	H_2O_2	70	1.8	[49]
Ti-MCM-41S	1.26	1740	TBHP	80	18.2	[19]

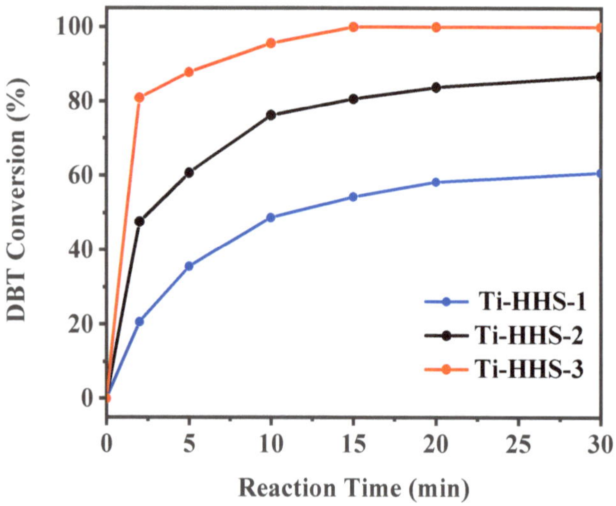

Figure 11. DBT ODS activity of Titanosilicate hierarchically porous hollow spheres with varying Ti content (reaction conditions: 20 g of 500 ppm model gasoline, 20 mg).

As shown in Figure 12, with an O/S ratio of 1, the DBT conversion rate reaches 95.51% within 30 min, but complete conversion is not achieved. A closer examination of the conversion rate over time reveals a plateau after 15 min, indicating that this ratio is insufficient for full DBT conversion. When the O/S ratio is increased to 3, DBT conversion reaches 100% within 15 min. With further increases in the oxidant ratio to 6 and 9, the conversion of DBT did not change too much. These findings suggest that an optimal O/S ratio exists for efficient DBT conversion while minimizing oxidant consumption. Accordingly, an O/S ratio of 3 is identified as optimal, and all subsequent investigations in this study are based on this ratio. It was reported that octane could not be oxidized by TBHP below 80 °C [50]; thus, in our experiments, the oxidation of model oil by TBHP could be neglected.

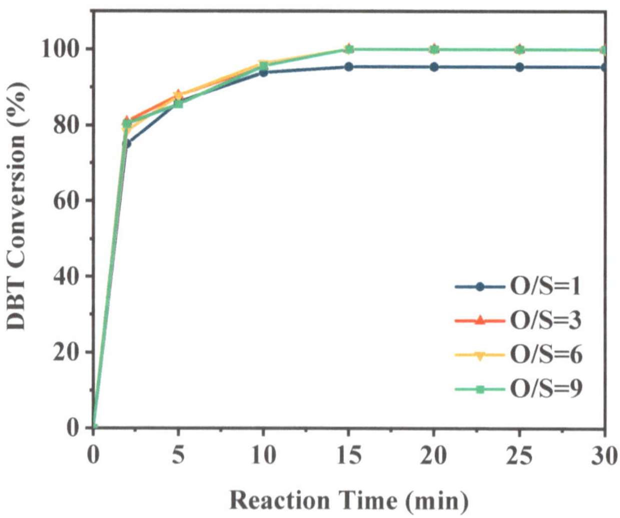

Figure 12. Effect of O/S molar ratio on ODS reaction (reaction conditions: 20 g of 500 ppm model gasoline, 20 mg Ti-HHS-3 catalyst, 30 °C).

In industrial applications, the concentration of sulfur compounds in fuel oil can vary significantly. To more accurately simulate the sulfur concentrations found in actual fuel oils, Ti-HHS-3 was selected as the catalyst to conduct the ODS reaction on model gasolines with different sulfur contents. The model gasolines were prepared with sulfur concentrations of 500 ppm, 1000 ppm, and 2000 ppm, as shown in Figure 13. As the DBT concentration increased, the conversion rate of the sulfur compounds significantly decreased over the same period. Notably, when the DBT concentration reached 2000 ppm, complete conversion could not be achieved within 30 min. This may be attributed to the relatively low Ti doping in Ti-HHS-3, resulting in insufficient catalytic active sites to meet the demands of the reaction at high concentrations, thereby causing a significant decline in catalytic activity.

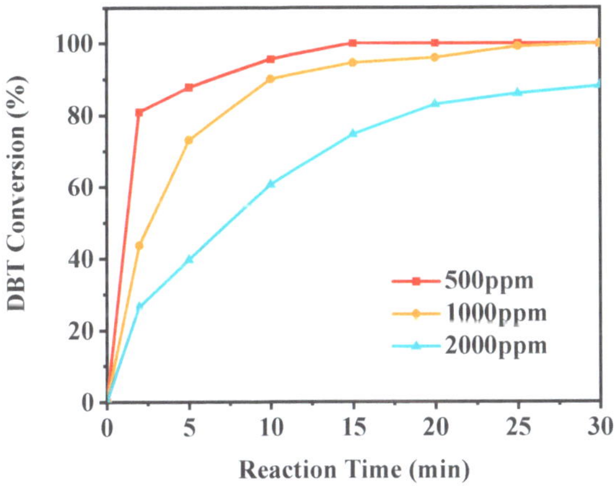

Figure 13. Conversion of different DBT concentrations over Ti-HHS-3 (reaction conditions: 20 g of model gasoline, 20 mg catalyst, 30 °C, O/S = 3).

The reaction temperature is one of the critical parameters influencing catalyst activity. As shown in Figure 14, complete DBT conversion requires 15 min at 30 °C. When the reaction temperature is increased to 50 °C, the time for complete DBT conversion decreases to 10 min. Further increasing the temperature to 80 °C reduces the conversion time to just 2 min. These results demonstrate that the ODS reaction catalyzed by Ti-HHS-3 is highly temperature-dependent, with reaction rates significantly increasing at higher temperatures. As the reaction temperature increases, the conversion efficiency of sulfur compounds improves, highlighting the potential for temperature optimization in practical applications. Higher temperatures likely enhance the reactivity of the catalyst's active sites, reducing the activation energy and thereby accelerating the reaction.

Fuel oils commonly used in daily life include petrol and diesel. Compared to gasoline, diesel contains longer alkane chains and boils in the range of 180–350 °C. To model diesel fuel, this study used long-chain alkanes, specifically tetradecane. The types of sulfur compounds can vary significantly. To further investigate the ODS performance of Ti HHS-3, other sulfur compounds, specifically 4-MDBT and 4,6-DMDBT, were selected to evaluate their ODS performance with Ti-HHS-3 as the catalyst in model diesel. The experimental results (Figure 15) demonstrate that Ti-HHS-3 exhibits excellent catalytic activity in both short-chain alkanes (n-octane) and long-chain alkanes (n-tetradecane), with no significant difference in catalytic performance between the two alkanes. However, the presence of methyl groups in the molecular structure of 4-MDBT and 4,6-DMDBT introduces steric hindrance, limiting interaction between the reactive Ti sites on the catalyst and the sulfur compounds [51,52]. Consequently, the conversion rates for 4-MDBT and

4,6-DMDBT are lower than that of DBT at the 2 min, 5 min, and 10 min mark, highlighting the impact of molecular structure on ODS efficiency and providing valuable insights for the desulfurization of diesel oils.

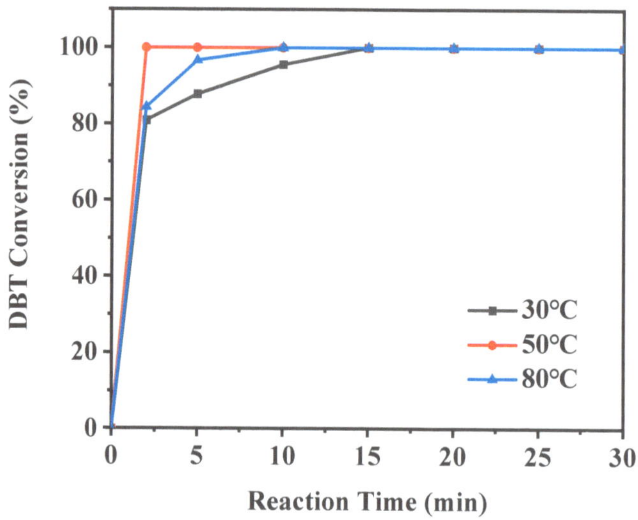

Figure 14. Effect of reaction temperature on DBT conversion (Reaction conditions: 20 g of 500ppm model gasoline, 20 mg catalyst, O/S = 3).

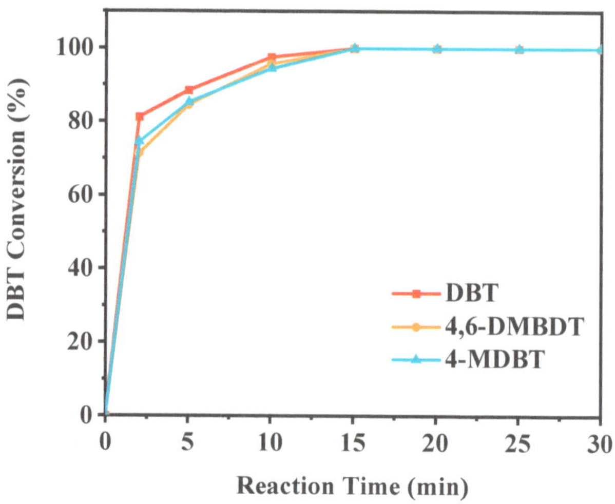

Figure 15. Conversion of different sulfur compounds in model diesel (reaction conditions: 20 g of 500 ppm model oil, 20 mg catalyst, O/S = 3).

In addition to the factors mentioned above, the effect of stirring speed on the ODS reaction should also be examined (as shown in Figure 16). For consistency, the stirring speed was maintained at 300 r/min in other experiments. The results show that the catalytic activity of the ODS reaction remained largely unaffected by changes in stirring speed. Specifically, whether the stirring speed is set at 300 r/min or adjusted to 100 r/min or 500 r/min, the ODS reaction proceeds similarly. This indicates that, under the experimental conditions used, stirring speed only slightly impacts on the catalytic activity of the ODS reaction.

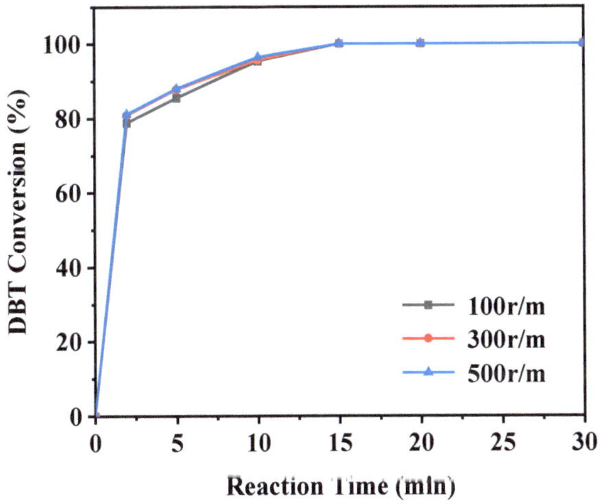

Figure 16. Effect of stirring speed on DBT conversion (reaction conditions: 20 g of 500 ppm model gasoline, 20 mg catalyst, O/S = 3).

In previous research [46], the ODS reaction typically involves the decomposition of the oxidant on the catalyst surface, producing active radicals. Electron paramagnetic resonance (EPR) characterization is conducted to confirm the generation of active radicals within the Ti-HHS-3 catalytic system. By using 5,5-dimethyl-1-pyrroline-N-oxide (DMPO) as a radical trap at room temperature, the addition of the oxidant TBHP resulted in a typical radical/DMPO adduct signal (Figure 17). Based on the literature [53], the average spacing of the four stable peaks with a height ratio of 1:2:2:1 is 14.9 Gs, which indicates hydroxyl radicals were produced in the catalytic system. Hydroxyl radicals, as highly reactive species, can effectively oxidize DBT, thus facilitating desulfurization. The EPR spectrum confirms hydroxyl radical production during Ti-HHS-3 catalyzes DBT oxidation, supporting a radical-based oxidation mechanism for the ODS reaction.

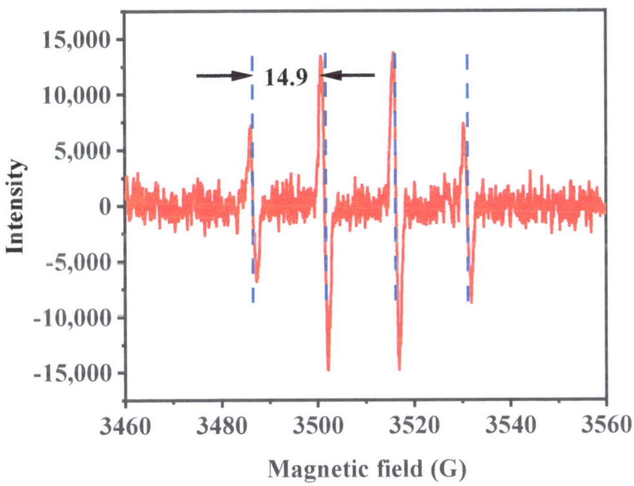

Figure 17. EPR spectrum of Ti-HHS-3. (The red lines indicate the EPR signals and the blue lines demonstrate the spacing of the signal peaks.).

3. Materials and Methods

3.1. Materials

Cetyltrimethylammonium bromide (CTAB, 99%), tetraethyl orthosilicate (TEOS, ≥99%), tetrabutyl titanate (TEOT, 99%), ammonium hydroxide (25–28 wt%), dibenzothiophene (DBT, ≥99 wt%), 4-methyl dibenzothiophene (4-MDBT, ≥99 wt%), 4,6-dimethyl dibenzothiophene (4,6-DMDBT, ≥99 wt%), n-octane (≥99 wt%), and n-tetradecane (≥99 wt%) were purchased from Aladdin Biochemical Technology (Shanghai, China). Anhydrous ethanol (≥99 wt%) was obtained from Tianjin Bohua Chemical Reagent (Tianjin, China). Tert-butyl hydroperoxide (TBHP, 5.0–6.0 mol/L in decane) was acquired from Sigma-Aldrich (Shanghai, China). Tetrapropylammonium hydroxide (TPAOH, 25 wt%) was purchased from Meryer Chemical Technology (Shanghai, China), and polyacrylic acid (PAA, 25 wt%, MW = 240,000) was obtained from Thermo Fisher Scientific (Shanghai, China).

3.2. Synthesis of Ti-HHS

To prepare the HHS precursor solution, mix 6 mL of TPAOH solution with 12 mL of deionized water and stir for 15 min until homogeneous. Then, add 5.6 mL of TEOS and a specified amount of TEOT (0.075 mL, 0.15 mL, and 0.3 mL) to achieve Si/Ti molar ratios of 116, 58, and 29, respectively. Continue stirring for 2 h to allow the hydrolysis of Si and Ti sources, resulting in a clear solution. Place the solution in a sealed container and let it stand at 45 °C for 3 days; the resulting clear solution is the HHS precursor solution. Here, the transparent solution containing zeolite structural units was prepared as the precursor for hierarchically porous titanosilicate hollow spheres (HHS).

Dissolve 0.56 g of CTAB in 20 mL of deionized water, stirring at 60 °C until CTAB is fully dissolved. After the solution becomes clear, bring the solution to room temperature, then add 3.0 g of PAA, and stir for 20 min until the solution is clear again. While stirring, add 2.0 g of ammonium hydroxide, which will immediately turn the solution milky white; continue stirring for another 20 min. Then, add 8.8 mL of precursor solution with varying Si/Ti molar ratios and stir at room temperature for 2 h. Transfer the emulsion to a 50 mL autoclave and let it stand at 80 °C for 2 days. Collect the product by centrifugation, wash it twice with deionized water and anhydrous ethanol, and dry it at 60 °C for 12 h. Calcine at 550 °C for 6 h to remove the template. Samples with varying Ti contents, obtained by adding 0.075 mL, 0.15 mL, and 0.3 mL of the Ti source, are designated as Ti-HHS-1, Ti-HHS-2, and Ti-HHS-3, respectively. When 0 mL of TEOT is added and the other synthesis steps are as above, a pure silicon hollow (HHS) sphere sample is obtained. A surfactant-polymer composite was used as the structure-directing agent, which became colloidal particles via self-assembly between anionic PAA and cationic CTAB. Under alkaline conditions, a series of Ti-HHS catalysts with varying titanium contents (Ti-HHS-1, Ti-HHS-2, and Ti-HHS-3) were successfully synthesized. The HHS precursor, with its larger volume and greater polymerization capability compared to monomeric or oligomeric silicates, is unable to penetrate the interior of the PAA-CTAB template. Consequently, co-assembly of the silicon source with the template and phase separation of PAA primarily occur at the surface of the PAA-CTAB composite, resulting in the formation of a thin shell. During calcination, the template is removed, leaving behind a hollow spherical structure.

For preparation of TiO_2-supported samples, a certain amount of tetrabutyl titanate was dissolved in 1 mL of deionized water, 0.1 g of pure silica hollow spheres was added to the above solution after continuous stirring for 2 h at room temperature, and then it was put into a vacuum oven and dried at 40 °C for 10 h. The solid samples were taken and ground and then roasted at 400 °C for 2 h in air, which yielded TiO_2/HHS-1 and TiO_2/HHS-2, with Si/Ti ratios of 58 and 29, respectively.

3.3. Characterization

X-ray diffraction (XRD) patterns were recorded on a Rigaku Smart Lab 3 kW instrument (Tokyo, Japan) using Cu Kα radiation (40 kV, 40 mA). Wide-angle diffraction patterns were measured in the 2θ range of 5–50° with a scan speed of 8° per minute, while small-angle diffraction patterns were measured in the 2θ range of 1–5° with a scan speed of 0.1° per minute. Scanning electron microscopy (SEM) images were taken using a JSM-7800F (JEOL, Tokyo, Japan). The SEM instrument operates under vacuum conditions, and samples were gold-sputtered prior to testing. Transmission electron microscopy (TEM) images were obtained with a JEM-2800 (JEOL, Japan). Nitrogen adsorption–desorption isotherms were obtained using a JW-TB220A analyzer from JWGB Sci & Tech Co., Ltd. (Beijing, China). Samples were pretreated at 350 °C under nitrogen flow for 4 h prior to measurement. The specific surface area was calculated using the Brunauer–Emmett–Teller (BET) model (JWGB, Beijing, China); the microporous pore volume (V_{mic}) and the mesoporous pore volume (V_{mes}) were obtained by DFT method. Infrared spectra were collected on a Bruker VECTOR 22 spectrometer (Billerica, MA, USA), with samples prepared by pressing into KBr pellets; the position and shape of absorption peaks were analyzed to characterize whether Si-O-Ti structures were present in the silicate materials. The diffuse reflectance UV–visible (UV-vis) spectra of the dehydrated samples relative to $BaSO_4$ were measured using a Shimadzu (UV-2450) spectrophotometer (Tokyo, Japan) over a wavelength range of 500–2000 nm at room temperature. Electron paramagnetic resonance (EPR) spectra were recorded at room temperature using a Bruker EMX-plus X-band spectrometer (Billerica, MA, USA). Ti content was determined by inductively coupled plasma emission spectrometry (ICP) on a Thermo Scientific iCAP 7400 (Waltham, MA, USA) atomic emission spectrometer.

3.4. Catalytic Tests

Dissolve 0.1 g DBT, 4,6-DMDBT, or 4-MDBT in 199.9 g of n-octane or n-tetradecane, stirring for 10 min, then seal and store as 500 ppm model gasoline. Place 20 g of model gasoline in a 50 mL Erlenmeyer flask and transfer it to a 30 °C water bath, stirring until the temperature stabilizes. Add 20 mg of the calcined hierarchically porous titanosilica hollow spheres as the catalyst and stir for another 10 min. Subsequently, add a specific amount of tert-butyl hydroperoxide (TBHP, 5.5 M in decane) solution to the system according to the predetermined oxygen-to-sulfur ratio. Start timing as soon as the oxidant is added. The stirring speed is maintained as 300 r/min. Samples are collected at various reaction times, and the catalyst is removed by filtering through a syringe with a membrane filter before analyzing reactant concentrations using a gas chromatograph (GC-7800, Grockway, Tengzhou, China). The signal detection during the analysis is performed using a flame ionization detector (FID). All catalysts were dehydrated at 150 °C for 2 h prior to use. The major ODS reaction conditions are listed in Table 3.

Table 3. The conditions for ODS reactions.

ODS Reaction Runs	Raw Material: Catalyst Ratio	Raw Material: Oxidant Ratio	O/S Ratio	Temperature (°C)	S Content (ppm)	Stirring Speed (r/min)
1	1000	1363	3	30	500	300
2	1000	4089	1	30	500	300
3	1000	454	9	30	500	300

3.5. Analysis

The GC-7800 equipped with a flame ionization detector (FID) was used to analyze the DBT, 4,6-DMDBT, and 4-MDBT concentration in the samples after the catalytic oxidation reaction, using high purity nitrogen as the carrier gas at a flow rate of 1 mL/min.

The conversion of DBT, 4,6-DMDBT, and 4-MDBT were calculated to evaluate the performance of the catalyst Ti-HHS. The reaction equation for oxidative desulfurization is based on the following equation:

$$\eta = [(C_0 - C)/C_0] \times 100\% \quad (1)$$

where η is the DBT, 4,6-DMDBT, and 4-MDBT conversion rate, and C_0 and C correspond to the initial and final concentration of DBT in model gasoline, respectively (Units: ppm).

The oxidant/sulfur molar ratio (O/S) of the reaction was evaluated using the following equation:

$$\alpha = n_o/n_s \quad (2)$$

where α is the O/S, and n_o and n_s denote the molar mass of oxidant added to the ODS and the molar mass of sulfur compound in the model gasoline, respectively.

4. Conclusions

In summary, a zeolite TS-1 precursor was successfully used as the source of Ti and Si, with a PAA-CTAB composite serving as the template, to synthesize hierarchically porous hollow spheres, Ti-HHS-x, under mildly alkaline conditions. The Ti-HHS-3 sample, with the highest Ti content, exhibited remarkable catalytic activity in the ODS reaction at room temperature. Using only 20 mg of catalyst, a complete conversion of DBT in 20 g of 500 ppm model gasoline was achieved at 30 °C with a TOF as high as 123 h^{-1}. This outstanding catalytic performance can be attributed to the tetrahedral coordination Ti of TS-1 sub-units and the hierarchically porous hollow structure of the material itself.

Author Contributions: Experiments, Y.W. and H.Y.; data analysis, Y.W., H.Y., and H.W.; writing—review and editing, funding acquisition and supervision, T.C. All authors have read and agreed to the published version of the manuscript.

Funding: This research received no external funding.

Data Availability Statement: The original contributions presented in this study are included in the article. Further inquiries can be directed to the corresponding author.

Conflicts of Interest: The authors declare no conflicts of interest.

References

1. Haruna, A.; Merican Aljunid Merican, Z.; Gani Musa, S.; Abubakar, S. Sulfur Removal Technologies from Fuel Oil for Safe and Sustainable Environment. *Fuel* **2022**, *329*, 125370. [CrossRef]
2. Si, Y.; Jiang, F.; Qiang, L.; Teng, X.; Gong, C.; Tang, Q. A Visible-Light-Responsive Molecularly Imprinted Polyurethane for Specific Detection of Dibenzothiophene in Gasoline. *Anal. Methods* **2022**, *14*, 1254–1260. [CrossRef] [PubMed]
3. Song, C.; Ma, X. New Design Approaches to Ultra-Clean Diesel Fuels by Deep Desulfurization and Deep Dearomatization. *Appl. Catal. B Environ.* **2003**, *41*, 207–238. [CrossRef]
4. Gao, S.; Yu, G.; Abro, R.; Abdeltawab, A.A.; Al-Deyab, S.S.; Chen, X. Desulfurization of Fuel Oils: Mutual Solubility of Ionic Liquids and Fuel Oil. *Fuel* **2016**, *173*, 164–171. [CrossRef]
5. Jiang, Z.; Lü, H.; Zhang, Y.; Li, C. Oxidative Desulfurization of Fuel Oils. *Chin. J. Catal.* **2011**, *32*, 707–715. [CrossRef]
6. Khalid, H.; Umar, A.; Saeed, M.H.; Nazir, M.S.; Akhtar, T.; Ikhlaq, A.; Ali, Z.; Hassan, S.U. Advances in Fuel Oil Desulfurization: A Comprehensive Review of Polyoxometalate Catalysts. *J. Ind. Eng. Chem.* **2024**, *141*, 32–45. [CrossRef]
7. Pham, D.D.; Nguyen, T.M.; Ho, T.H.; Le, Q.V.; Nguyen, D.L.T. Advancing Hydrodesulfurization in Heavy Oil: Recent Developments, Challenges, and Future Prospects. *Fuel* **2024**, *372*, 132082. [CrossRef]

8. Javadli, R.; de Klerk, A. Desulfurization of Heavy Oil. *Appl. Petrochem. Res.* **2012**, *1*, 3–19. [CrossRef]
9. Kabir, S.F.; Zheng, R.; Delgado, A.G.; Fini, E.H. Use of Microbially Desulfurized Rubber to Produce Sustainable Rubberized Bitumen. *Resour. Conserv. Recycl.* **2021**, *164*, 105144. [CrossRef]
10. Liu, Y.; Wang, H.; Zhao, J.; Liu, Y.; Liu, C. Ultra-Deep Desulfurization by Reactive Adsorption Desulfurization on Copper-Based Catalysts. *J. Energy Chem.* **2019**, *29*, 8–16. [CrossRef]
11. Mahmood, Q.A.; Humadi, J.I.; Algawi, R.J.; Nawaf, A.T.; Ahmed, I.A. Adsorption Desulfurization of Simulated Diesel Fuel Using Graphene Oxide. *Chem. Chem. Technol.* **2024**, *18*, 436–441. [CrossRef]
12. Kianpour, E.; Azizian, S.; Yarie, M.; Zolfigol, M.A.; Bayat, M. A Task-Specific Phosphonium Ionic Liquid as an Efficient Extractant for Green Desulfurization of Liquid Fuel: An Experimental and Computational Study. *Chem. Eng. J.* **2016**, *295*, 500–508. [CrossRef]
13. Wang, P.; Jiang, L.; Zou, X.; Tan, H.; Zhang, P.; Li, J.; Liu, B.; Zhu, G. Confining Polyoxometalate Clusters into Porous Aromatic Framework Materials for Catalytic Desulfurization of Dibenzothiophene. *ACS Appl. Mater. Interfaces* **2020**, *12*, 25910–25919. [CrossRef] [PubMed]
14. Pyshyev, S.; Korchak, B.; Miroshnichenko, D.; Vytrykush, N. Influence of Water on Noncatalytic Oxidative Desulfurization of High-Sulfur Straight-Run Oil Fractions. *ACS Omega* **2022**, *7*, 26495–26503. [CrossRef]
15. Sahraei, S. Assessment of Reaction Parameters in the Oxidative Desulfurization Reaction. *Energy Fuels* **2023**, *37*, 15373–15393. [CrossRef]
16. Ma, C.; Dai, B.; Liu, P.; Zhou, N.; Shi, A.; Ban, L.; Chen, H. Deep Oxidative Desulfurization of Model Fuel Using Ozone Generated by Dielectric Barrier Discharge Plasma Combined with Ionic Liquid Extraction. *J. Ind. Eng. Chem.* **2014**, *20*, 2769–2774. [CrossRef]
17. Mirshafiee, F.; Movahedirad, S.; Sobati, M.A.; Alaee, R.; Zarei, S.; Sargazi, H. Current Status and Future Prospects of Oxidative Desulfurization of Naphtha: A Review. *Process Saf. Environ. Prot.* **2023**, *170*, 54–75. [CrossRef]
18. García-Gutiérrez, J.L.; Laredo, G.C.; García-Gutiérrez, P.; Jiménez-Cruz, F. Oxidative Desulfurization of Diesel Using Promising Heterogeneous Tungsten Catalysts and Hydrogen Peroxide. *Fuel* **2014**, *138*, 118–125. [CrossRef]
19. Chica, A., Corma, A., Dómine, M.E. Catalytic Oxidative Desulfurization (Ods) of Diesel Fuel on a Continuous Fixed-Bed Reactor. *J. Catal.* **2006**, *242*, 299–308. [CrossRef]
20. Tang, Q.; Lin, S.; Cheng, Y.; Liu, S.; Xiong, J.-R. Ultrasound-Assisted Oxidative Desulfurization of Bunker-C Oil Using Tert-Butyl Hydroperoxide. *Ultrason. Sonochem.* **2013**, *20*, 1168–1175. [CrossRef]
21. Ding, Y.; Wang, J.; Liao, M.; Li, J.; Zhang, L.; Guo, J.; Wu, H. Deep Oxidative Desulfurization of Dibenzothiophene by Novel Pom-Based Il Immobilized on Well-Ordered Kit-6. *Chem. Eng. J.* **2021**, *418*, 129470. [CrossRef]
22. Bakar, W.A.W.A.; Ali, R.; Kadir, A.A.A.; Mokhtar, W.N.A.W. Effect of Transition Metal Oxides Catalysts on Oxidative Desulfurization of Model Diesel. *Fuel Process. Technol.* **2012**, *101*, 78–84. [CrossRef]
23. Caero, L.C.; Hernández, E.; Pedraza, F.; Murrieta, F. Oxidative Desulfurization of Synthetic Diesel Using Supported Catalysts: Part I. Study of the Operation Conditions with a Vanadium Oxide Based Catalyst. *Catal. Today* **2005**, *107–108*, 564–569. [CrossRef]
24. Du, Q.; Guo, Y.; Wu, P.; Liu, H.; Chen, Y. Facile Synthesis of Hierarchical Ts-1 Zeolite without Using Mesopore Templates and Its Application in Deep Oxidative Desulfurization. *Microporous Mesoporous Mater.* **2019**, *275*, 61–68. [CrossRef]
25. Lv, G.; Deng, S.; Zhai, Y.; Zhu, Y.; Li, H.; Wang, F.; Zhang, X. P123 Lamellar Micelle-Assisted Construction of Hierarchical Ts-1 Stacked Nanoplates with Constrained Mesopores for Enhanced Oxidative Desulfurization. *Appl. Catal. A Gen.* **2018**, *567*, 28–35. [CrossRef]
26. Wang, H.; Shi, C.; Chen, S.; Chen, R.; Sun, P.; Chen, T. Hierarchically Mesoporous Titanosilicate Single-Crystalline Nanospheres for Room Temperature Oxidative–Adsorptive Desulfurization. *ACS Appl. Nano Mater.* **2019**, *2*, 6602–6610. [CrossRef]
27. Wang, J.; Wu, W.; Ye, H.; Zhao, Y.; Wang, W.-H.; Bao, M. Moo3 Subnanoclusters on Ultrasmall Mesoporous Silica Nanoparticles: An Efficient Catalyst for Oxidative Desulfurization. *RSC Adv.* **2017**, *7*, 44827–44833. [CrossRef]
28. Cho, K.-S.; Lee, Y.-K. Effects of Nitrogen Compounds, Aromatics, and Aprotic Solvents on the Oxidative Desulfurization (Ods) of Light Cycle Oil over Ti-Sba-15 Catalyst. *Appl. Catal. B Environ.* **2014**, *147*, 35–42. [CrossRef]
29. Hao, L.; Sun, L.; Su, T.; Hao, D.; Liao, W.; Deng, C.; Ren, W.; Zhang, Y.; Lü, H. Polyoxometalate-Based Ionic Liquid Catalyst with Unprecedented Activity and Selectivity for Oxidative Desulfurization of Diesel in [Omim]Bf4. *Chem. Eng. J.* **2019**, *358*, 419–426. [CrossRef]
30. Hori, H.; Ogi, K.; Fujita, Y.; Yasuda, Y.; Nagashima, E.; Matsuki, Y.; Nomiya, K. Oxidative Removal of Dibenzothiophene and Related Sulfur Compounds from Fuel Oils under Pressurized Oxygen at Room Temperature with Hydrogen Peroxide and a Phosphorus-Free Catalyst: Sodium Decatungstate. *Fuel Process. Technol.* **2018**, *179*, 175–183. [CrossRef]
31. Ramaswamy, V.; Shah, P.; Lazar, K.; Ramaswamy, A.V. Synthesis, Characterization and Catalytic Activity of Sn-Sba-15 Mesoporous Molecular Sieves. *Catal. Surv. Asia* **2008**, *12*, 283–309. [CrossRef]
32. Han, Y.; Xiao, F.-S.; Wu, S.; Sun, Y.; Meng, X.; Li, D.; Lin, S.; Deng, F.; Ai, X. A Novel Method for Incorporation of Heteroatoms into the Framework of Ordered Mesoporous Silica Materials Synthesized in Strong Acidic Media. *J. Phys. Chem. B* **2001**, *105*, 7963–7966. [CrossRef]

33. Wei, J.; Yue, Q.; Sun, Z.; Deng, Y.; Zhao, D. Synthesis of Dual-Mesoporous Silica Using Non-Ionic Diblock Copolymer and Cationic Surfactant as Co-Templates. *Angew. Chem. Int. Ed.* **2012**, *51*, 6149–6153. [CrossRef] [PubMed]
34. Chen, D.; Li, Z.; Yu, C.; Shi, Y.; Zhang, Z.; Tu, B.; Zhao, D. Nonionic Block Copolymer and Anionic Mixed Surfactants Directed Synthesis of Highly Ordered Mesoporous Silica with Bicontinuous Cubic Structure. *Chem. Mater.* **2005**, *17*, 3228–3234. [CrossRef]
35. Wang, J.-G.; Zhou, H.-J.; Sun, P.-C.; Ding, D.-T.; Chen, T.-H. Hollow Carved Single-Crystal Mesoporous Silica Templated by Mesomorphous Polyelectrolyte−Surfactant Complexes. *Chem. Mater.* **2010**, *22*, 3829–3831. [CrossRef]
36. Gopalakrishnan, A.; Raju, T.D.; Badhulika, S. Green Synthesis of Nitrogen, Sulfur-Co-Doped Worm-Like Hierarchical Porous Carbon Derived from Ginger for Outstanding Supercapacitor Performance. *Carbon* **2020**, *168*, 209–219. [CrossRef]
37. Li, Y.; Wu, Z.; Liu, Y.; Zhang, K.; Luo, S.; Li, W.; Liu, S. Worm-Like Ordered Mesoporous Carbon from Liquefied Wood: Morphological Manipulation by Varying Hydrothermal Temperature. *Aggregate* **2024**, *5*, e570. [CrossRef]
38. Macina, D.; Opioła, A.; Rutkowska, M.; Basąg, S.; Piwowarska, Z.; Michalik, M.; Chmielarz, L. Mesoporous Silica Materials Modified with Aggregated Transition Metal Species (Cr, Fe and Cr-Fe) in the Role of Catalysts for Selective Catalytic Oxidation of Ammonia to Dinitrogen. *Mater. Chem. Phys.* **2017**, *187*, 60–71. [CrossRef]
39. Ren, J.; Li, Z.; Liu, S.; Xing, Y.; Xie, K. Silica–Titania Mixed Oxides: Si–O–Ti Connectivity, Coordination of Titanium, and Surface Acidic Properties. *Catal. Lett.* **2008**, *124*, 185–194. [CrossRef]
40. Liu, J.; Wang, J.; Zhang, Y.; Zheng, W.; Yao, Y.; Liu, Q.; Zhang, X.; Yang, Y.; Wang, X. Improved C–H Activation in Propane Dehydrogenation Using Zeolite-Stabilized Co–O Moieties. *ACS Catal.* **2023**, *13*, 14737–14745. [CrossRef]
41. Liang, Z.; Yang, Y.; Zhang, Y.; Li, S.; Zhang, W.; Zhang, L.; Chan, S.H. Synergistic Photocatalysis of Mesoporous Confinement Effect and Si-O-Ti Interface for Organic Pollutants Degradation. *Surf. Interfaces* **2024**, *51*, 104715. [CrossRef]
42. Bhaumik, A.; Tatsumi, T. Organically Modified Titanium-Rich Ti-Mcm-41, Efficient Catalysts for Epoxidation Reactions. *J. Catal.* **2000**, *189*, 31–39. [CrossRef]
43. Capel-Sanchez, M.C.; Campos-Martin, J.M.; Fierro, J.L.G. Removal of Refractory Organosulfur Compounds Via Oxidation with Hydrogen Peroxide on Amorphous Ti/SiO_2 Catalysts. *Energy Environ. Sci.* **2010**, *3*, 328–333. [CrossRef]
44. de la Peña O'Shea, V.A.; Capel-Sanchez, M.; Blanco-Brieva, G.; Campos-Martin, J.M.; Fierro, J.L.G. The Usefulness of Time-Dependent Density Functional Theory to Describe the Electronic Spectra of Ti-Containing Catalysts. *Angew. Chem.* **2003**, *115*, 6031–6034. [CrossRef]
45. Su, J.; Xiong, G.; Zhou, J.; Liu, W.; Zhou, D.; Wang, G.; Wang, X.; Guo, H. Amorphous Ti Species in Titanium Silicalite-1: Structural Features, Chemical Properties, and Inactivation with Sulfosalt. *J. Catal.* **2012**, *288*, 1–7. [CrossRef]
46. Leng, K.; Li, X.; Ye, G.; Du, Y.; Sun, Y.; Xu, W. Ti-Containing Hierarchical Beta with Highly Active Sites for Deep Desulfurization of Fuels under Mild Conditions. *Catal. Sci. Technol.* **2016**, *6*, 7615–7622. [CrossRef]
47. Shi, C.; Wang, W.; Liu, N.; Xu, X.; Wang, D.; Zhang, M.; Sun, P.; Chen, T. Low Temperature Oxidative Desulfurization with Hierarchically Mesoporous Titaniumsilicate Ti-Sba-2 Single Crystals. *Chem. Commun.* **2015**, *51*, 11500–11503. [CrossRef]
48. Fang, Y.; Hu, H. Mesoporous Ts-1: Nanocasting Synthesis with Cmk-3 as Template and Its Performance in Catalytic Oxidation of Aromatic Thiophene. *Catal. Commun.* **2007**, *8*, 817–820. [CrossRef]
49. Hulea, V.; Fajula, F.; Bousquet, J. Mild Oxidation with H_2O_2 over Ti-Containing Molecular Sieves—A Very Efficient Method for Removing Aromatic Sulfur Compounds from Fuels. *J. Catal.* **2001**, *198*, 179–186. [CrossRef]
50. Soobramoney, L.; Bala, M.D.; Friedrich, H.B. Coordination Chemistry of Co Complexes Containing Tridentate Sns Ligands and Their Application as Catalysts for the Oxidation of N-Octane. *Dalton Trans.* **2014**, *43*, 15968–15978. [CrossRef]
51. Stanger, K.J.; Angelici, R.J. Silica-Catalyzed Tert-Butyl Hydroperoxide Oxidation of Dibenzothiophene and Its 4,6-Dimethyl Derivative: A Route to Low-Sulfur Petroleum Feedstocks. *Energy Fuels* **2006**, *20*, 1757–1760. [CrossRef]
52. Serrano, D.P.; Sanz, R.; Pizarro, P.; Moreno, I.; Medina, S. Hierarchical Ts-1 Zeolite as an Efficient Catalyst for Oxidative Desulphurization of Hydrocarbon Fractions. *Appl. Catal. B Environ.* **2014**, *146*, 35–42. [CrossRef]
53. Teng, Z.; Yang, H.; Zhang, Q.; Cai, W.; Lu, Y.-R.; Kato, K.; Zhang, Z.; Ding, J.; Sun, H.; Liu, S.; et al. Atomically Dispersed Low-Valent Au Boosts Photocatalytic Hydroxyl Radical Production. *Nat. Chem.* **2024**, *16*, 1250–1260. [CrossRef] [PubMed]

inorganics

MDPI

Article

Studies of Various Batch Adsorption Parameters for the Removal of Trypan Blue Using Ni-Zn-Bi-Layered Triple Hydroxide and Their Isotherm, Kinetics, and Removal Mechanism

Ganesan Sriram [1,*], Nimisha Baby [1], Karmegam Dhanabalan [1], Muthuraj Arunpandian [1], Karuppaiah Selvakumar [2], Thangarasu Sadhasivam [1] and Tae Hwan Oh [1,*]

1 School of Chemical Engineering, Yeungnam University, Gyeongsan 38541, Republic of Korea
2 Department of Physiology, Saveetha Dental College and Hospitals, Saveetha Institute of Medical and Technical Science (SIMATS), Saveetha University, Chennai 600077, India
* Correspondence: sriramyu@yu.ac.kr (G.S.); taehwanoh@ynu.ac.kr (T.H.O.)

Abstract: The present study addressed the removal of Trypan blue (TB) from water using a novel Ni-Zn-Bi-layered triple hydroxide (NZB LTH or NZB) synthesized through the co-precipitation technique. The physiochemical properties of NZB were analyzed before and after TB adsorption using XRD, BET, FESEM, FTIR-ATR, Raman, and XPS. Studies on adsorption indicate that 80 mg of NZB has a maximum TB removal effectiveness of around 96.7% at natural pH (~4.5–5.0). This study found that NZB has a maximum adsorption capacity (q_{max}) of 5.3 mg·g^{-1} at dye concentrations ranging from 5 to 30 mg·L^{-1}. When combined with various anionic dye mixtures, NZB's selectivity studies showed that it is highly selective for the removal of TB and is also effective at removing cationic dyes. When compared to Na_2SO_4 and NaCl salts, NZB had a lower dye removal percentage for TB removal in the presence of Na_2SO_3. In an adsorption process, the interaction between the TB and NZB in an aqueous solution is caused by hydrogen bonding and electrostatic interactions, which are investigated in the adsorption mechanism. In comparison with ethanol and methanol, the recyclability investigation of NZB revealed the notable removal of TB using 0.1 M NaOH for the desorption. Therefore, the present investigation suggests that NZB is an appropriate adsorbent for the removal of TB from an aqueous solution.

Keywords: synthesis; co-precipitation; dye removal; anionic dye; LTH; adsorption capacity; selectivity; salt-effect; reusability

Citation: Sriram, G.; Baby, N.; Dhanabalan, K.; Arunpandian, M.; Selvakumar, K.; Sadhasivam, T.; Oh, T.H. Studies of Various Batch Adsorption Parameters for the Removal of Trypan Blue Using Ni-Zn-Bi-Layered Triple Hydroxide and Their Isotherm, Kinetics, and Removal Mechanism. *Inorganics* **2024**, *12*, 296. https://doi.org/10.3390/inorganics12110296

Academic Editor: Eleonora Aneggi

Received: 14 October 2024
Revised: 9 November 2024
Accepted: 15 November 2024
Published: 19 November 2024

1. Introduction

Dye contamination of water necessitates treatment by removing the dyes with diverse adsorbents or methods before they discharge into local drainage systems or natural water bodies [1–3]. Various industries, such as textiles, cosmetics, ink, plastics, paper, food, pharmaceuticals, pulp, leather, printing, paints, and rubber, use dyes because they are easily soluble in water and give concentrated color to the water [4,5]. Trypan blue (TB) is a water-soluble azo dye that is used for biological staining, vitreoretinal surgery, and measuring cell viability, and it is also used in the food coloring dyeing process [6–8]. However, TB is toxic to cells for a short period of exposure time and carcinogenic to humans and animals, if exhaled or ingested, it causes eye agitation, skin irritation, and digestive problems. Excessive levels of TB may harm aquatic life. Due to the toxic effects of TB, the European Chemicals Agency has determined that TB is carcinogenic to human beings and requested that its usage in testing facilities be restricted [9]. Discharging untreated water polluted with this dye may have detrimental effects on natural water resources, disrupt the aquatic lifecycle, and pose substantial risks to human health, including toxicological and carcinogenic effects. To prevent contamination of the environment, the TB discharges from various sectors should be removed from the water. This concern of dye contaminations in water has led to the development of many water treatments, such as advanced oxidation processes, ion exchange, deep eutectic solvent, coagulation-flocculation, adsorption, electrocoagulation, wet-oxidation, sedimentation, arc discharge, membrane bioreactors, phytoremediation, reverse osmosis, sequencing batch reactors, membrane filtration, etc. [8,10–17]. Among these, adsorption is considered to be one of the easiest, low-cost, most rapid, and effective approaches to removing dyes from water using adsorbents [18–20].

The process of adsorption is capable of efficiently removing many kinds of pollutants [21–25]. In addition, the application of this technique in the removal of dyes from water-based solutions may be seen visibly in an instant. Adsorbents with a substantial surface area, extensive porous architectures, and abundant surface hydroxyl groups have the potential to expeditiously remove dyes. Consequently, an extensive study has been conducted to develop adsorbents for the goal of removing harmful dyes and other pollutants. These adsorbents include porous carbon derived from biomass, activated carbon from bamboo fibers, carbon xerogel, silica xerogel, agricultural wastes, hydrogels, metal-organic frameworks (MOFs), biochar, calcium silicate, aerogel, zeolite, polymers, natural adsorbents, layered double hydroxides (LDHs), montmorillonite, layered triple hydroxides (LTHs), red mud waste, nanoparticles, and various other adsorbents [12,18,26–41]. Among these adsorbents, LTHs possess a high positive charge, nanosheet configuration, high porosity, abundant surface functional groups, and a moderate surface area. Additionally, they are synthesized using low-cost chemicals and inexpensive methods, making them a promising and cost-effective option for effectively removing anionic-based dyes. The advantages of LTHs means they can be used to effectively treat water contaminated with anionic-based dyes.

Recently, nanomaterials have been explored for environmental applications because of their distinctive features, primarily their high surface-to-volume ratio. Accordingly, because of their layered structure and three cationic constituents, LTHs are one of the nanomaterials that are becoming more popular among researchers working with water research [42–44]. Conversely, LDHs, which are a mixture of two cationic precursors, would be less efficient than LTHs due to the possibility of reduced surface area and low functional groups [45–48]. As a result, three cationic structures with LTHs could be able to remove dye more effectively than LDHs. Due to their unique characteristics, the applications for LTHs include water purification, supercapacitors, batteries, electrolysis, photocatalysts, electrocatalysts, and water splitting [49–61]. The LTH structure has notable qualities, such as a large number of active sites and high interlayer space, as well as a remarkable ability for efficient ion exchange. Thus, water research communities are focusing on these LTHs to use them as active adsorbents for water purification.

It has been reported in several studies that LTHs are used for the removal of dyes. Accordingly, Kazemi et al. synthesized a MgZnAl LTH using the hydrothermal method to remove acid yellow 76 (AY76) at pH 4 [38]. Because of the electrostatic interaction and hydrogen bonding between the LTHs and AY76, the maximum adsorption capacity (q_{max}) of the LTHs was found to be 277.8 $mg \cdot g^{-1}$, and they had a high reusability of 90.3% after four cycles. Recently, Abdel-Hady et al. produced ZnMgAl LTHs via co-precipitation to remove crystal violet (CV) at a pH of 9 [62]. However, the q_{max} of the prepared LTHs adsorbent was only 64.8 mg g^{-1}, which was a result of its ineffectiveness at removing the positively charged CV dye. In contrast, after five cycles, the reusability of ZnMgAl LTHs was determined to be moderate at 64.0%. However, the potential mechanisms that facilitated adsorption between ZnMgAl LTHs and CV exhibited substantial efficacy, such as mesoporous filling, surface diffusion, hydrogen bonding, electrostatic interaction, and π–π interaction. Using the co-precipitation method, Bahadi et al. synthesized a polyethyleneimine-functionalized graphene oxide composite with MgFeAl LTHs (PEI@GO/MgFeAl) to remove acid red 1 (AR1) dye at pH 6 [43]. AR1 adsorbed on PEI@GO/MgFeAl exhibited a q_{max} of 225.7 $mg \cdot g^{-1}$, and an effective reusability of 80.7% was observed after five cycles. The adsorption of AR1 was achieved by the mechanism of electrostatic interactions, pore-filling, hydrogen bonding, and π–π interactions. Similarly, co-precipitation was employed to synthesize MgNiAl-CO_3, which was then calcined at 500 °C to yield MgNiAl-C to remove methyl orange (MO) at pH 8 [63]. Because of the calcination effect, MgNiAl-C (375.4 $mg \cdot g^{-1}$) had a higher q_{max} than MgNiAl-CO_3 (118.5 $mg \cdot g^{-1}$) for MO removal. Similarly, co-precipitation was used to produce NiMgAl LTHs with variable percentages (20% and 30%) of $Ni(NO_3)_2 \cdot 6H_2O$, which were then calcined at 400 °C to obtain NiMgAl-layered triple oxides (LTOs) to remove MO at pH 9 [64]. This work demonstrated increased $Ni(NO_3)_2 \cdot 6H_2O$ loading into MgAl, as well as improved adsorption performance following calcination. After calcination, 30%NiMgAl LTO (322.5 $mg \cdot g^{-1}$) had a higher q_{max} compared to 20%NiMgAl LTO (270.2 $mg \cdot g^{-1}$), 20%NiMgAl LTHs (250.0 $mg \cdot g^{-1}$), and 30%NiMgAl LTHs (232.5 $mg \cdot g^{-1}$). In contrast to the 30%NiMgAl LTHs, which only removed 58% of the MO after three cycles, the 30%NiMgAl LTO was able to remove 72% of the MO. On the other hand, samples of LTHs still showed a good q_{max}, and they offer the following advantages over LTO: a low-temperature process, no calcination or annealing is necessary, a time-saving procedure, high hydroxyl groups on their surface that are crucial for the removal of dyes through hydrogen bonding and electrostatic interaction, mild-temperature drying following synthesis does not impact structure or morphology, the ability to synthesize LTHs using a single method, and a high crystalline structure. For this reason, the use of LTH-based adsorbents for dye removal via the adsorption process has the potential to achieve a high adsorption capacity for dye removal. According to the above studies, LTHs are often synthesized via co-precipitation, which is a low-cost approach, and this type of method is advantageous for the adsorption process since it produces inexpensive and efficient sorbents. The pH of the solution is crucial for the adsorption process and should not harm the environment. Using a pH of the dye solution between 4 and 8 for the adsorption process would be environmentally friendly. However, some studies performed at pH 9 could be harmful. Further, it is necessary to test the reusability of the adsorbent in the adsorption process, which helps to reduce the material waste or cost of the process. Therefore, it is necessary to desorb the dye molecules from the adsorbent surface by either chemical or thermal methods. It would be preferable if the adsorbent could be reused for at least five cycles, ensuring a high removal efficiency of over 90%. Additionally, the q_{max} is dependent on the volume and dosage of the adsorbents, as well as on the concentration of the dye molecules. For the higher q_{max}, experiments could use a high volume of dye solution, a low adsorbent dosage, and a higher dye concentration.

This is the first time NZB has been synthesized using co-precipitation to remove TB from an aqueous solution. Several characterization approaches were employed to compare before and after dye adsorption. To investigate the dye removal effectiveness and potential mechanisms of the synthesized adsorbent, a systematic batch adsorption procedure was carried out, accounting for the effects of pH, dosage, dye concentrations, time, isotherms, and kinetic studies. Additionally, the study involves the process of selectively removing TB by combining various dye mixtures, the salt effect, and exploring the potential for reusing the materials. The potential of NZB as an effective adsorbent in the treatment of aqueous solutions contaminated with both anionic and cationic dyes is demonstrated by this study. Figure 1 presents a diagram of the synthesis of NZB and its ensuing removal of TB.

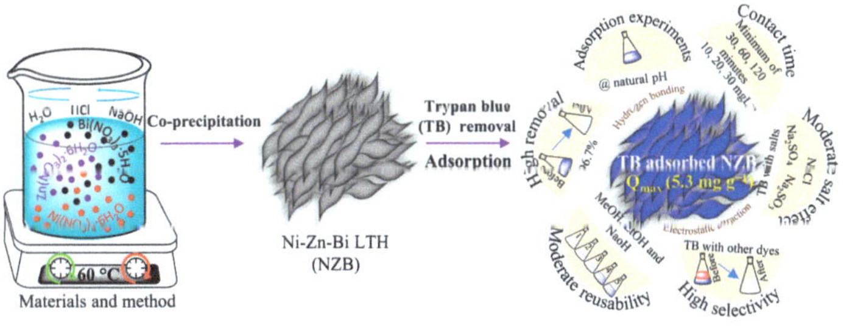

Figure 1. Schematic diagram of the synthesis of NZB and its adsorption properties regarding TB.

2. Results and Discussion

The adsorbent's surface area could potentially assist in achieving a faster and higher level of dye removal. Thus, an analysis was conducted on the surface area of the NZB using the N_2 adsorption–desorption technique and the Barrett–Joyner–Halenda (BJH) method to assess the porosity of the adsorbent before TB adsorption. As shown in Figure 2a, according to the IUPAC classification, the NZB shows a Type IV isotherm curve with a characteristic H3 hysteresis loop opening at 0.45 (P/P_0) and closing at 1.0 (P/P_0), demonstrating the presence of a mesoporous structure on the surfaces. This loop represents the capillary condensation properties occurring in the mesopores. The NZB's surface area was determined to be low, at 17.08 $m^2 \cdot g^{-1}$, whereas the pore volume and pore diameter were found to be 0.0436 $cm^3 \cdot g^{-1}$ and 5.9 nm, respectively. The lower surface area might be owing to the presence of moisture, which can block the pores and reduce surface area. However, the layered triple hydroxide (LTH) type of materials may have high positive charges on their external surfaces, and this could play a crucial role in the removal of a large number of anionic dye molecules rather than their surface area. A similar result was obtained in the report of the N_2 adsorption–desorption isotherm study on the HKUST-1 and HHK-10 before and after water vapor adsorption. The surface area of these MOF materials was 1510 and 1284 $m^2 \cdot g^{-1}$, respectively. After water vapor adsorption, the surface area decreased to 607 and 687 $m^2 \cdot g^{-1}$. As a result, the presence of moisture could reduce the material's surface area [65]. In light of this, NZB has the potential to function as an effective adsorbent in the process of water treatment for the removal of TB.

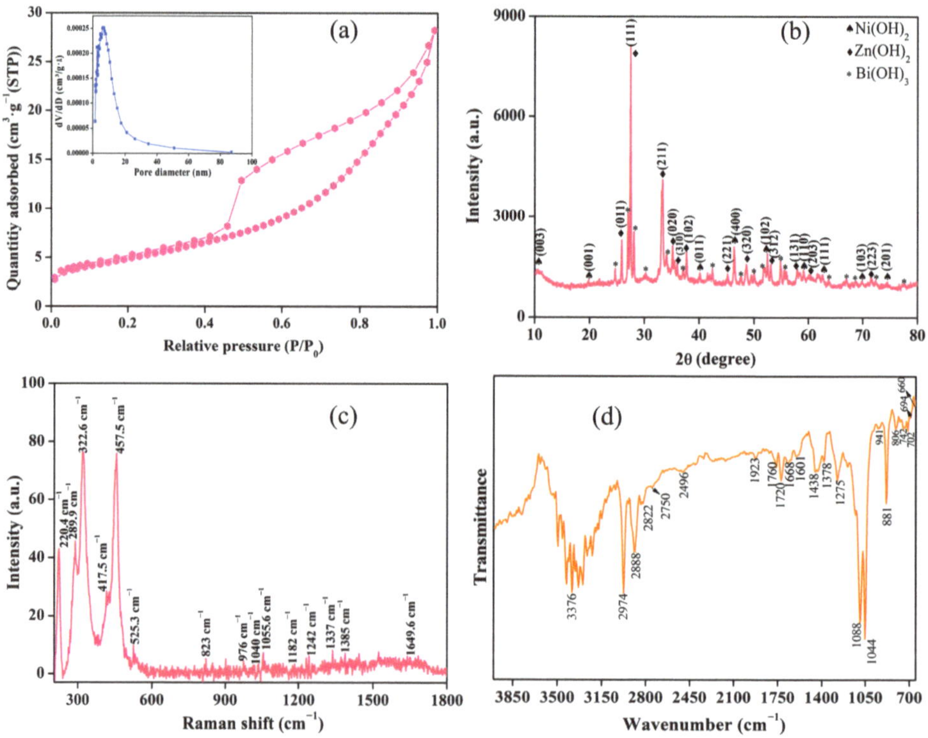

Figure 2. Characterization of NZB LTH before dye adsorption: (**a**) the nitrogen (N_2) adsorption–desorption isotherm curve and porosity measurement by Barrett–Joyner–Halenda (BJH) plot (inset); (**b**) XRD pattern; (**c**) Raman spectrum; and (**d**) FTIR-ATR spectrum.

X-ray diffraction (XRD) is a non-destructive method for material structural investigation. Figure 2b presents the XRD analysis of the structural matter of NZB before TB adsorption. The XRD pattern of NZB revealed a crystalline structure. Accordingly, $Ni(OH)_2$ was confirmed by the presence of characteristic peaks at 2θ, including 11.7°, 19.2°, and 38.2° and 46.4°, 52.4°, 59.2°, 62.5°, 69.8°, and 74.3°, denoted as (003), (001), and (011) and (400), (102), (110), (111), (103), and (201), respectively [3,66–68]. The appearance of peaks like (011), (111), (211), (020), (310), (102), (221), (320), (312), (131), (203), and (223) confirms $Zn(OH)_2$, which corresponds to 25.8°, 27.3°, 33.3°, 35.2°, 36.0°, 37.6°, 45.1°, 48.7°, 53.0°, 57.9°, 60.3°, and 71.4°, respectively [69–71]. Furthermore, the earlier reports confirmed that all remaining characteristic peaks were indeed $Bi(OH)_3$ [72–74]. However, some noise in the diffraction pattern can be seen due to the presence of moisture on the NZB. The presence of characteristic peaks of respective hydroxides such as $Ni(OH)_2$, $Zn(OH)_2$, and $Bi(OH)_3$ in the diffraction pattern confirms the LTH structure in NZB. Furthermore, Raman spectroscopy was utilized to examine and govern the NZB's functional composition. Figure 2c shows the Raman spectrum of the adsorbent, NZB, which covers from 200 to 1800 cm^{-1}. The Raman shift at 220.4 cm^{-1} for Bi-OH, 289.9 and 322.6 cm^{-1} for Zn-OH, and 417.5, 457.7, and 525.3 cm^{-1} for Ni-OH showed the existence of M-OH (Metal-hydroxide) stretching vibration in the adsorbent [70,75,76]. The peaks at 976, 823, 1040, 1337–1385, and 1649 cm^{-1} are ascribed to NO_3^- in-plane bending, symmetric stretching, antisymmetric stretching, and out-of-plane bending vibration, respectively [77,78]. Additional peaks at 1055.6 cm^{-1} and 1182 cm^{-1} to 1242 cm^{-1} were due to the symmetric and antisymmetric vibration of carbonate ions (CO_3^{2-}) [78]. The observation of these peaks verifies the LTHs' composition of $Ni(OH)_2$, $Zi(OH)_2$, and $Bi(OH)_3$ in their structure.

The functional groups in NZB were found using FTIR-ATR. Figure 2d thus reveals the IR peaks of NZB in the wavenumber range of 650 to 4000 cm^{-1}. The wider peak at 3200–3500 cm^{-1} was produced by stretching vibrations of interlayer hydroxyl (-OH) groups from the NZB. The peaks at 2974, 2750, 2496, 1923, 1088, and 1438 cm^{-1} indicate strong and mild C-H stretching vibrations, while the peak from 2822 to 2882 cm^{-1} represents CH_2 stretching [3,79]. The characteristic peaks at 1378 and 1275 cm^{-1} correspond to the symmetric stretching vibrations of NO_3^- ion and 1044 cm^{-1} to the symmetric stretching vibration of CO_3^{2-} ion, respectively [3,79–81]. The symmetric stretching vibration of the carboxylate molecule results in an IR peak regarding 1720–1767 cm^{-1} [82]. The peak between 1601 and 1668 cm^{-1} is caused by the bending vibration of water molecules [83]. The peaks at 941 and 881 cm^{-1} correspond to the metal-oxygen and metal-hydroxyl groups in LTHs [84]. The absorption peaks in the 806–660 cm^{-1} range may be related to the vibrational modes of O-M-O and M-O-M, where M denotes Ni, Zn, and Bi [80]. The presence of the peaks mentioned above served as evidence that several functional groups are present in NZB LTHs.

Furthermore, this study analyzed the NZB's morphology and elemental composition using FESEM and EDX. Accordingly, Figure 3a–c shows the morphology of the NZB at various scale ranges, such as 5 μm, 1 μm, and 500 nm. The observed layered structure and predicted flat sheet-type morphologies were noted. The rough surface was produced and the stacked sheet layer of NZB was identified at 500 nm scale. However, particles or grains were seen on the NZB's surface. This could be because the particles were crushed into a fine powder after the material was made. The sizes of these particles or grains could be between 100 and 500 nm. These sheet structures could have more active sites and surface areas, which might be beneficial for removing a considerable amount of dye molecules from water. The elemental mapping result shows that the NZB's distribution of Ni, Zn, Bi, and O is uniform (Figure 3d–g). These mapping pictures show that NZB has a higher concentration of Ni than Zn, Bi, and O elements, which is confirmed by the quantity of each chemical used in the preparation. Figure 3h depicts the NZB's elemental composition. The EDX spectrum revealed that the weight percentage of Ni K (53.78%) was larger than Zn K (29.46%), Bi M (6.35%), and O K (10.4%), and no further elements were identified, suggesting that NZB is produced only from the nickel-zinc-bismuth composition. Similarly, the proportion of NZB's atomic weight % was identified as Ni K > Zn K > O K > Bi M.

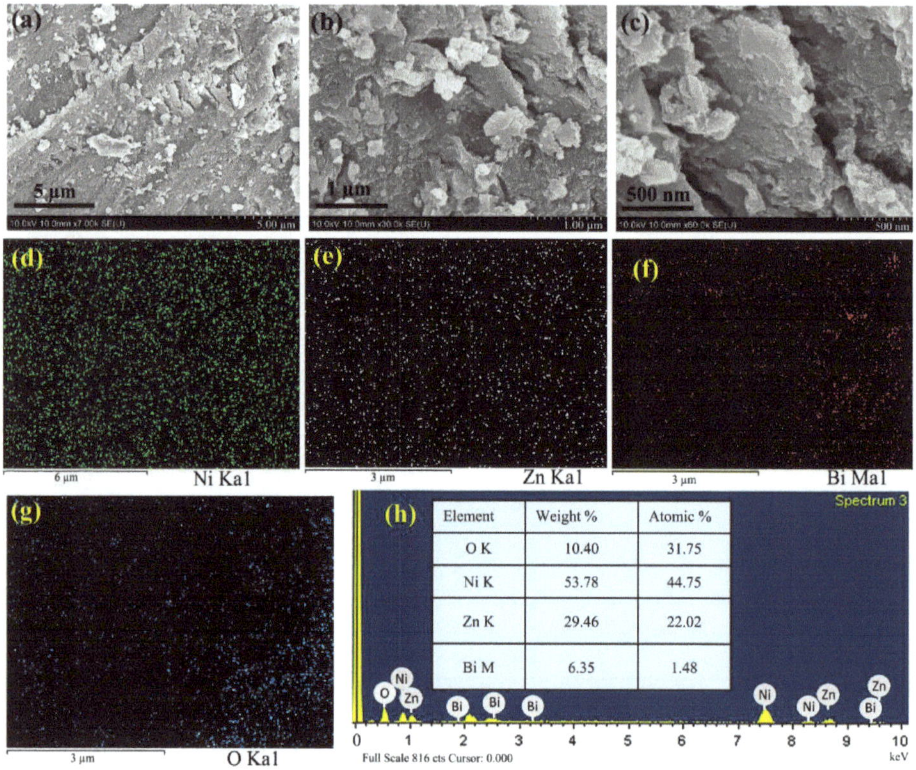

Figure 3. Characterization of NZB before dye adsorption: (**a**–**c**) FESEM images at different scales of 5 µm, 1 µm, and 500 nm; (**d**–**g**) elemental mapping; and (**h**) EDX spectrum with the table of element weights and atomic percentages.

XPS was used to determine the existence of the elemental composition on the produced adsorbent. The Ni-Zn-Bi LTH (NZB) of the full XPS survey spectrum is depicted in Figure 4a. The spectrum was analyzed within the binding energy range of 0 to 1200 eV, and the presence of NZB elements, including Bi 4f, C 1s, O 1s, Ni 2p, and Zn 2p, was confirmed at specific binding energies. Moreover, Figure 4b–f demonstrated the high resolution of each NZB component present at corresponding binding energies; the discussion of these peaks follows. The high-resolution spectrum of Bi 4f confirms two distinctive peaks at 157.7 eV and 163.1 eV, which are caused by Bi $4f_{7/2}$ and Bi $4f_{5/2}$, respectively. In the C1s peaks of the high-resolution spectrum, two peaks were observed at 284.2 eV and 288.3 eV. This was a result of the carbon bonding of the HCO_3^- and CO_3^{2-} within the interlayer of NZB LTHs [85]. In contrast, O 1s exhibits a single characteristic peak at 530.5 eV, which is attributed to the M-OH (M=Ni-Zn-Bi) bonding. Consequently, the presence of this peak demonstrates that the hydroxyl group of water is bonded to the surface of the NZB LTHs. The Ni 2p spectrum exhibits the two peaks of Ni $2p_{3/2}$ and Ni $2p_{1/2}$ at binding energies of 855.0 eV and 872.7 eV, respectively. Furthermore, the Ni 2p is accompanied by the formation of satellite peaks at 860.6 eV and 878.9 eV. Lastly, the binding energies of Zn $2p_{3/2}$ and Zn $2p_{1/2}$, respectively, assigned two peaks to the Zn 2p spectrum at 1020.9 eV and 1044.0 eV, indicating the presence of Zn^{2+} in NZB LTHs. Therefore, the presence of the aforementioned elements confirmed the synthesized Ni-Zn-Bi-layered triple hydroxide.

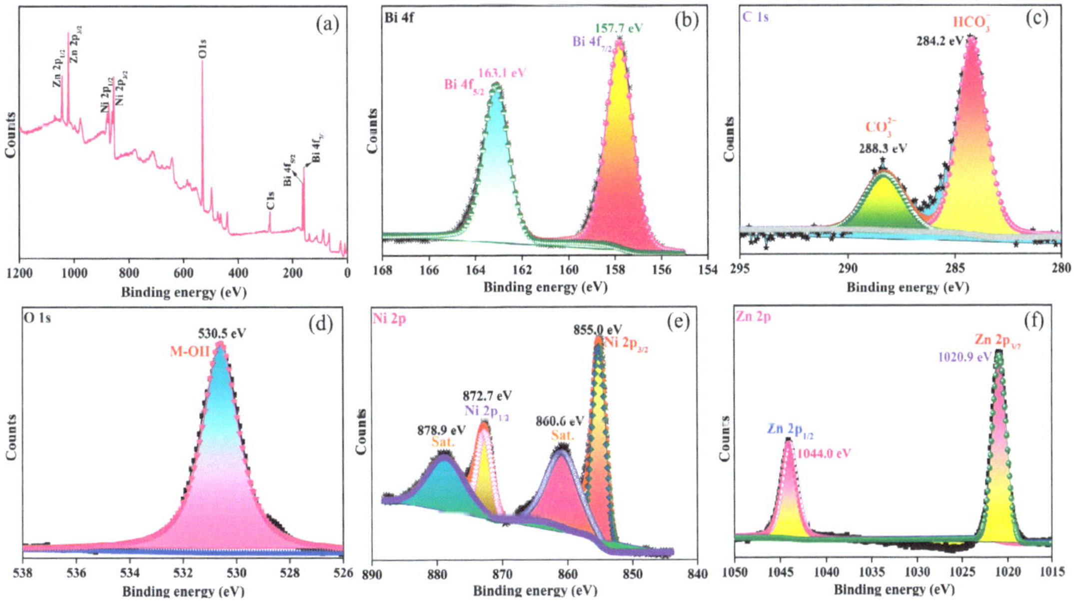

Figure 4. NZB LTH XPS spectra before dye adsorption: (**a**) Full survey spectrum and high-resolution spectrum of (**b**) Bi 4f, (**c**) C ls, (**d**) O 1s, (**e**) Ni 2p, and (**f**) Zn 2p.

3. Batch Adsorption

3.1. pH

The adsorption process for the removal of dye from water may be influenced by the pH of the solution. Thus, it is crucial to investigate TB adsorption at various pH levels. Figure 5a shows the percentage of dye removed at various pH levels using 80 mg of NZB and 15 mg·L^{-1} of TB concentration. The dye removal percentage dropped steadily from 97.6% to 12.3% as the pH increased from 2 to 10. The higher removal (97.6%) at pH 2 was due to the higher protonation (-OH^{2+}) of NZB surfaces that may be highly attracted with TB molecules through hydrogen bonding with chemical structures such as -OH, -NH_2, and -N=N- and also capable of high interaction with negatively charged sulfonate (-SO_3^-) groups (Figure S1). The decrease in removal efficiency from 96.2% to 12.3% from pH 4 to 10 may be attributed to the increasing deprotonation (-OH^{2-}) on NZB surfaces, which are unable to attract negatively charged groups in dye structures. Figure S2 represents the UV-Vis absorbance spectra of TB at various pHs, as well as spectra before and after TB adsorption by NZB, showing that the absorbance peak was substantially reduced in acidic conditions (2, 4, and natural pH (~4.5–5.0)), indicating strong adsorption. In contrast, the spectra obtained in the base media exhibit increased absorption peaks, indicating restricted adsorption. NZB surfaces may contain vast positive charges (-OH^+) in acidic environments, electrostatically attracting or hydrogen bonding with the TB dye. However, when the pH increases (natural (~4.5–5.0) and 6–10), they acquire negative charges (-OH^-), repelling the TB. Furthermore, at natural pH, the dye removal percentage was found to be 87.9%, which is significant. As a result, accomplishing an adsorption procedure in either an acidic or basic medium may affect the environment; hence, subsequent adsorption experiments were undertaken using natural pH (~4.5–5.0).

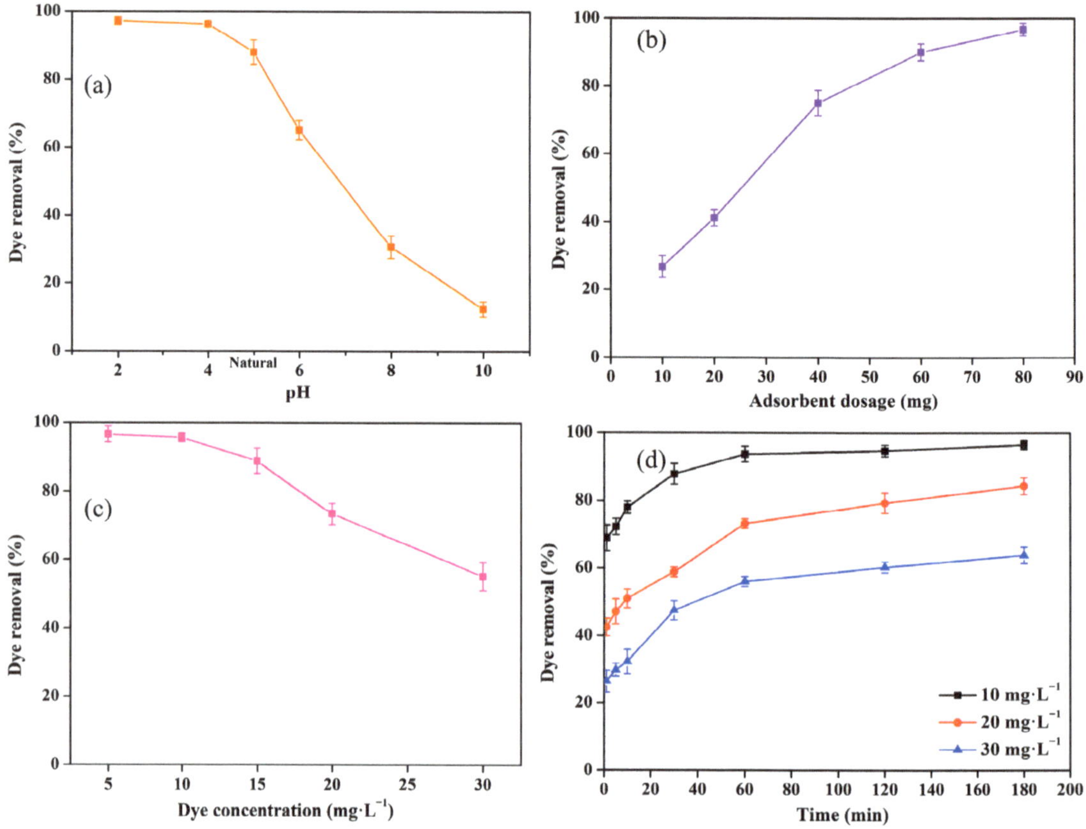

Figure 5. Batch adsorption parameters for dye (TB) removal using NZB LTH at room temperature: (**a**) effect of pH (Dosage: 80 mg, dye concentration: 15 mg·L^{-1}, volume: 25 mL, time: 1 h); (**b**) effect of adsorbent dosage (Dye concentration: 10 mg·L^{-1}, pH: natural (~4.5–5.0), volume: 25 mL, time: 1 h); (**c**) effect of dye concentration (Dosage: 80 mg, pH: natural (~4.5–5.0), volume: 25 mL, time: 1 h); and (**d**) effect of contact time (Dosage: 80 mg, pH: natural (~4.5–5.0), volume: 25 mL, dye concentration: 10, 20 and 30 mg·L^{-1}) at different dye concentrations.

3.2. Adsorbent Dosage

The adsorbent dosage may be important in the adsorption process for dye removal. Therefore, an experiment was performed to test the influence of NZB dosage on dye removal. The dosage ranged from 10 to 80 mg, with a dye concentration of 10 mg·L^{-1} at natural solution pH (~4.5–5.0). Figure 5b shows the dye removal percentage plotted against the adsorbent dosage. As the NZB's dosage increased from 10 to 80 mg, the dye removal percentage increased steadily from 26.6% to 96.7%. The results show that increasing the adsorbent dosage resulted in a considerable increase in adsorbent sites, which was attributed to a higher dye removal percentage. Figure S3 depicts absorbance spectra both before and after TB adsorption by NZB at various dosages; the UV-Vis absorbance peak significantly decreased when the NZB dosage increased from 10 to 80 mg. The ideal dosage of 80 mg was determined via these studies and utilized in subsequent investigations.

3.3. Dye Concentration

The increasing concentration of dye molecules may affect the adsorbent-based dye removal process. Thus, it is vital to examine the highly concentrated dye molecule in the adsorption process for dye removal effectiveness by the NZB; determining the q_{max} of the generated adsorbents is also critical. A batch experiment was carried out to assess the efficiency of an optimum dosage of NZB at various TB concentrations to compute the maximal dye removal from each concentration. Figure 5c illustrates the percentage of dye molecules removed at concentrations ranging from 5 to 30 mg·L^{-1}. As the TB concentration increased from 5 to 30 mg·L^{-1}, the dye removal percentage of NZB gradually decreased from 96.5% to 55.1%. The decrease in dye removal percentage was attributed to the NZB surface not having enough active sites or an inadequate adsorbent to adsorb large numbers of TB molecules at increasing concentrations. However, according to Figure 5c, a substantial removal percentage was observed owing to positive charges on the NZB surface caused by intensified driving forces and the diffusion rate of the TB dyes. Figure S4 shows absorbance spectra for various before and after dye concentrations, revealing that increasing the dye concentration increases the absorbance peak after dye adsorption by NZB, which suggests restricting the adsorption at higher concentrations.

3.4. Contact Time

To measure the adsorption rate at various dye concentrations, contact times extending from 0 to 180 min were utilized, since dye molecules may adsorb quickly or slowly on adsorbent surfaces. The optimal dosage of 80 mg of NZB was treated with 25 mL of TB concentrations ranging from 10 to 30 mg·L^{-1} to determine the dye removal percentage based on contact time along with natural pH (~4.5–5.0). Figure 5d depicts dye removal percentages vs. contact time at various dye concentrations. Accordingly, NZB removes 10 mg·L^{-1} of dye, with 68.8% and 72.0% efficiency during 1 and 30 min. After steadily increasing the dye removal percentage, it reached 93.7% after 60 min of contact, almost reaching saturation. A fast removal of a low-concentration dye from an aqueous solution may be ascribed to the presence of more active sites that may quickly adsorb dye molecules. For 10 mg·L^{-1}, the minimal contact time is 30 min. At a concentration of 20 mg·L^{-1}, the influence of time on the adsorbent was evaluated, with contact times up to 180 min. In this batch of contact time, the dye removal percentage for the NZB was 42.4% and 47.0% for 1 and 30 min, respectively. After 60 min, it increased to 73.0% and attained equilibrium. At a TB concentration of 30 mg·L^{-1}, NZB effectively removed 26.3%, 47.3%, and 55.9% of TB after 1, 30, and 60 min, respectively. After 120 min, it reached 60.2%, and no more improvements were seen. As a result, the removal of TB at low, moderate, and high concentrations takes at least 30, 60, and 120 min, respectively. The optimal amount of adsorbent may absorb dye molecules with a lower concentration than those with higher concentrations; raising the dye concentration to be absorbed by NZB necessitates a longer time need. It should be noted that NZB was able to remove TB efficiently, even at higher concentrations; however, it takes a longer time, as seen with the naked eye through UV-Vis absorbance spectra and supernatant TB solutions after adsorption at different contact times for various dye concentrations. Accordingly, Figure 6 depicts UV-Vis absorbance spectra for the influence of contact time, as well as related photographs of before and after dye adsorption for various dye concentrations. With an increase in the dye concentration, there is an increase in the contact time of the absorbance spectra, indicating that dye adsorption is fast at low concentrations but slow at higher concentrations. The adsorption process for dye concentrations of 10, 20, and 30 mg·L^{-1} needs a minimum contact time of 30, 60, and 120 min, as seen in the supernatant solution image.

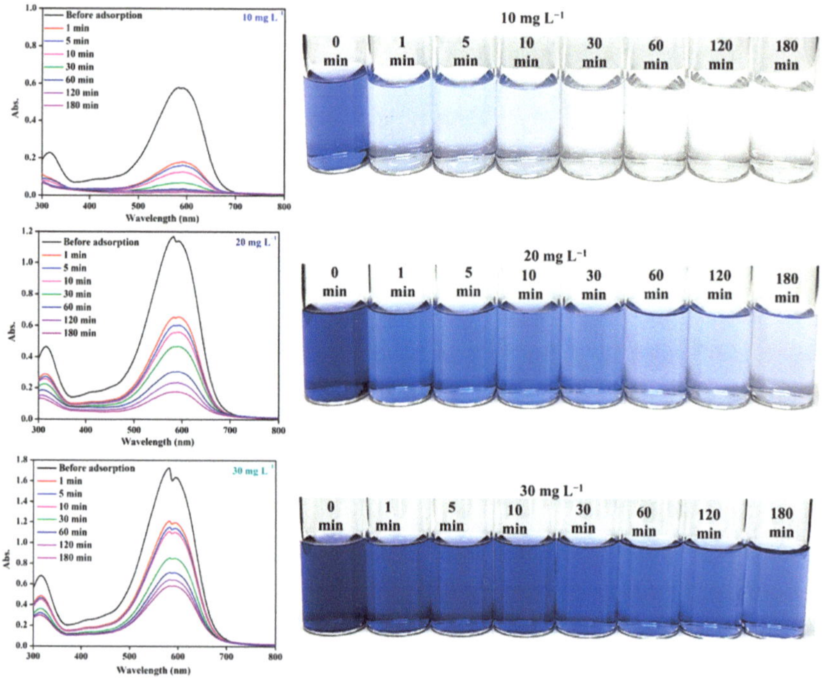

Figure 6. UV-Vis absorption spectra of the contact time before and after TB for dye concentrations of 10, 20, and 30 mg·L^{-1} adsorption by NZB LTH with their corresponding photograph of dye supernatants (Dye: TB, dosage: 80 mg, dye concentration: 10–30 mg·L^{-1}, pH: Natural (~4.5–5.0), volume: 25 mL, temperature: RT).

4. Characterization of the Adsorbent After TB Adsorption

To further understand TB adsorption on NZB surfaces, the adsorbent must be analyzed following dye adsorption by utilizing different characterization approaches. Figure S5a illustrates the N_2 adsorption–desorption BET Type IV isotherm curve, with an inset showing the BJH pore size distribution curve of NZB for porosity measurements (refer to the Supplementary Materials). The typical H3 loop Type IV isotherm curve remained unchanged after dye adsorption; however, the surface area of the NZB was reduced to 9.8 m^2·g^{-1}. The reduced surface area could be caused by the adsorbed dye particles blocking the pores on the adsorbents' surface. Following dye adsorption, pore volumes decreased while pore diameters increased by 0.0243 cm^3·g^{-1} and 6.56 nm, respectively. The decreasing pore volume and increasing pore diameter are due to the dye particles taking up a high place in the pore's depth and the chances of increasing the pore edges through the pore breathing effect [86]. As seen by the diffraction pattern in Figure S5b, dye adsorption weakened and decreased all of the NZB's crystalline peaks, indicating a high likelihood of TB adsorption on the NZB's surface. However, the diffraction pattern showed no other structural changes, suggesting that the adsorbent remained structurally strong after dye adsorption. Moreover, the Raman spectra reveal that the dye adsorption boosted the intensity of the peaks, caused shifts, led to the disappearance of most peaks, and resulted in the formation of new peaks, indicating the potential for dye adsorption on NZB surfaces (Figure S5c). Importantly, the distinctive peaks range from 812 to 836 cm^{-1} and 990 cm^{-1}, respectively, because of the benzene ring vibration and the bending vibration of the aromatic C=C structure in the dye. Furthermore, the peak at 1222 cm^{-1} was related to the stretching vibration of C-N in the dye structure. The peaks at 1346 and 1424 cm^{-1} were caused by the stretching vibration of S=O from sulfonic acid in the dye and the bending vibration of the -OH group from the TB dye, respectively. The peak ranging from 1519 to 1577 cm^{-1} was because of the stretching vibration of -N=N- in the dye structure [87]. The signal at 1691 cm^{-1} was found to be the bending vibration of the aromatic C-H from the dye structure.

Figure S5d shows the NZB of the IR spectrum following dye absorption, revealing the shifting of peaks and the appearance of new ones. As a result, 3422 and 3256 cm^{-1} appeared due to the dye molecules' -OH and N-H stretching vibrations, respectively. The stretching vibration of the azo group (-N=N-) was responsible for the peaks at 1642 and 1550 cm^{-1}. Additionally, the aromatic stretching vibration of C=C and the stretching vibration of SO_3^- in the dye structure caused characteristic peaks, such as at 1527 and 1349 cm^{-1}. Furthermore, 1326 and 1188 cm^{-1} were claimed to be the stretching vibration of the dye molecule's C-N group. Figure S5e shows FESEM images of NZB following dye treatment at various scales. It was found that the NZB's morphology exhibited irregularities and its surface had become uneven, resulting in the dye molecules settling over the layered structure. Moreover, there were no noticeable changes in the structure or form of the adsorbent surface, suggesting that it remained unchanged, even after the dye was absorbed. Then, XPS was utilized for confirming dye adsorption on NZB surfaces by detecting an extra N 1s peak of dye molecule binding energies at 399.1 and 401.5 eV, which were attributable to -N= and $-NH_2$, respectively (Figure S6). Also, the XPS peaks of Bi 4f shifted from 157.7 and 163.1 eV to 158.3 and 163.6 eV, Ni 2p shifted from 855.0 and 872.7 eV to 855.5 and 873.2 eV, and Zn 2p shifted from 1020.9 and 1044.0 eV to 1021.3 and 1044.5 eV, respectively. This is because dye molecules adsorbed on metals in the LTHs made the binding energies higher.

4.1. Adsorption Isotherm and Kinetics

It is important to determine the q_{max} of an adsorbent by using the Langmuir and Freundlich adsorption isotherms. Additionally, it is vital to determine the kind of adsorption process to understand the dye molecules that are adsorbed on the NZB surfaces. The Langmuir and Freundlich isotherms are linearly fitted in Figure 7a,b, while Table 1 contains the values of the isotherms. The values of the Langmuir and Freundlich isotherms were tabulated using Equations (S1)–(S3) (refer to the Supplementary Materials). Consequently, the q_{max} of NZB is 5.3 $mg \cdot g^{-1}$ for the dye concentrations used in this study, ranging from 5 to 30 $mg \cdot L^{-1}$. The linear fitting ($R^2 > 0.998$) of the Langmuir isotherm calculated R^2 value was greater than that of the Freundlich isotherm, signifying that the adsorption process could be conducted on a monolayer. The NZB surface exhibited a higher adsorption energy (K_L) for dye molecules, which was determined to be 2.06 $L \cdot mg^{-1}$. Equation S2 was employed to calculate the separation factor (R_L), and the value was found to be between 0 and 1, demonstrating that the removal of TB was quite promising for the adsorbent. The adsorbent's $1/n$ value (0.256) was lower than 1, demonstrating that the NZB's surface may exhibit heterogeneity and so can adsorb a substantial quantity of TB dye. Additionally, the Freundlich constant of adsorption capacity (K_F) value was used to determine the quantity of adsorbate adsorbed on adsorbent surfaces, which was 3.02 $mg \cdot g^{-1}$. Additionally, the experimental results were validated by the adsorption isotherm models (Figure 7c). The experimental (q_e) consequences for NZB are consistent with the Langmuir model when contrasted with the Freundlich model. This finding suggests that the mechanism of TB adsorption may be a monolayer.

Table 1. Adsorption isotherm findings of NZB LTH after TB adsorption.

Adsorbent	Langmuir				Freundlich		
	q_{max} ($mg \cdot g^{-1}$)	K_L ($L \cdot mg^{-1}$)	R_L	R^2	1/n	K_F ($mg \cdot g^{-1}$)	R^2
NZB LTH	5.3	2.06	0.088 to 0.015	0.998	0.256	3.022	0.792

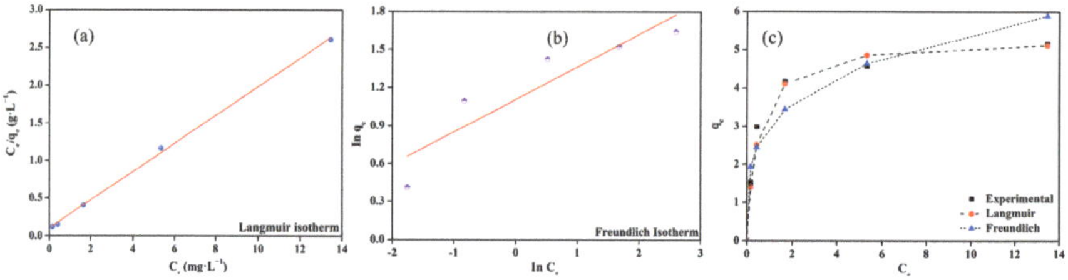

Figure 7. Adsorption isotherm fitting of (**a**) Langmuir, (**b**) Freundlich, and (**c**) Fitting of the experimental q_e data with calculated Langmuir and Freundlich q_e values of the NZB LTH for the TB removal.

Kinetic studies are essential for the identification of adsorption models and the comprehension of the adsorption process's rate. This study used Equations (S4)–(S7) to analyze the pseudo-first-order and pseudo-second-order models, and intraparticle diffusion (refer to the Supplementary Materials). Adsorption kinetics linear plots for TB concentrations of 10, 20, and 30 mg L^{-1} are depicted in Figure 8. Table 2 indicates that the rate of adsorption studies for all dye concentrations follows pseudo-second-order. This is because the experimental q_e value of all dye concentrations was matched with the calculated q_e value using pseudo-second-order when compared to pseudo-first-order. Moreover, the R^2 value of the correlation coefficient for the pseudo-second-order was greater than that for the pseudo-first-order. Consequently, these findings suggest that the adsorption of TB from water onto NZB is followed by a pseudo-second-order kinetic model [88]. Table 3 depicts the results of Weber's intraparticle diffusion, which was employed to ascertain the dye adsorption rate. The diffusion process is divided into two stages, as illustrated in Figure 8. The initial stage is responsible for the movement of dye molecules from the solution to the adsorbent surface, while the second stage is responsible for the diffusion of particles into the surface pores of the NZB. The intra-particle diffusion rate constants K_{int1} and K_{int2} are high and low for the first and second stages, respectively, for all dye concentrations. This is a result of the high driving force of dye molecules in water at the beginning of the process, which decreases to a low level at the end of the process on the pores of the NZB surfaces. Nevertheless, the K_{int1} and K_{int2} values are elevated as a result of the increased propelling forces of dye molecules in the water toward the pores of the adsorbent, which is a result of the concentration increase from 10 to 30 mg L^{-1}. Additionally, the gradual increase in the thickness of the dye particle's boundary layers on the adsorbent pores from the first stage to the second stage is indicative of the diffusion of dye molecules at low and high concentrations, respectively, as the values for all dye concentrations increase from C_1 to C_2. The concentration of dye molecules was increased from 10 to 30 mg L^{-1}, which increased the C_1 and C_2 values. This increase in the number of dye molecules contributed to the thickening of the boundary layers on the pores following pore filling. However, the C1 value decreased for 30 mg L^{-1}, which may be attributed to the low driving force toward the adsorbent's pores at the primary step, as a result of the possibility of agglomerated dye particles at a higher concentration. Consequently, a higher amount of dye particles (30 mg·L^{-1}) resulted in a higher thickness of the boundary layer on the adsorbent's surface in comparison to 10 and 20 mg L^{-1}.

Table 2. Adsorption kinetic models with the correlation coefficients of TB adsorption on NZB LTH.

Adsorbent		**Pseudo-First-Order**				**Pseudo-Second-Order**		
	C_0 (mg·L^{-1})	$q_{e, exp}$ (mg·g^{-1})	K_1 (min^{-1})	$q_{e, cal}$ (mg·g^{-1})	R^2	K_2 (g·mg^{-1}·min^{-1})	$q_{e, cal}$ (mg·g^{-1})	R^2
NZB LTH	10	2.99	0.022	0.70	0.945	0.165	3.03	0.999
	20	4.58	0.026	3.54	0.920	0.026	5.36	0.994
	30	5.16	0.031	4.34	0.912	0.020	6.13	0.996

Table 3. Intraparticle diffusion constants of TB adsorption on NZB LTH.

		1st Stage	2nd Stage	1st Stage	2nd Stage	1st Stage	2nd Stage
NZB LTH	10 mg·L^{-1}	K_{int1}	K_{int2}	C_1	C_2	$R^2{}_1$	$R^2{}_2$
		0.137	0.015	1.989	2.808	0.980	0.905
	20 mg·L^{-1}	K_{int1}	K_{int2}	C_1	C_2	$R^2{}_1$	$R^2{}_2$
		0.228	0.124	2.431	3.595	0.997	0.999
	30 mg·L^{-1}	K_{int1}	K_{int2}	C_1	C_2	$R^2{}_1$	$R^2{}_2$
		0.449	0.129	1.841	4.235	0.921	0.999

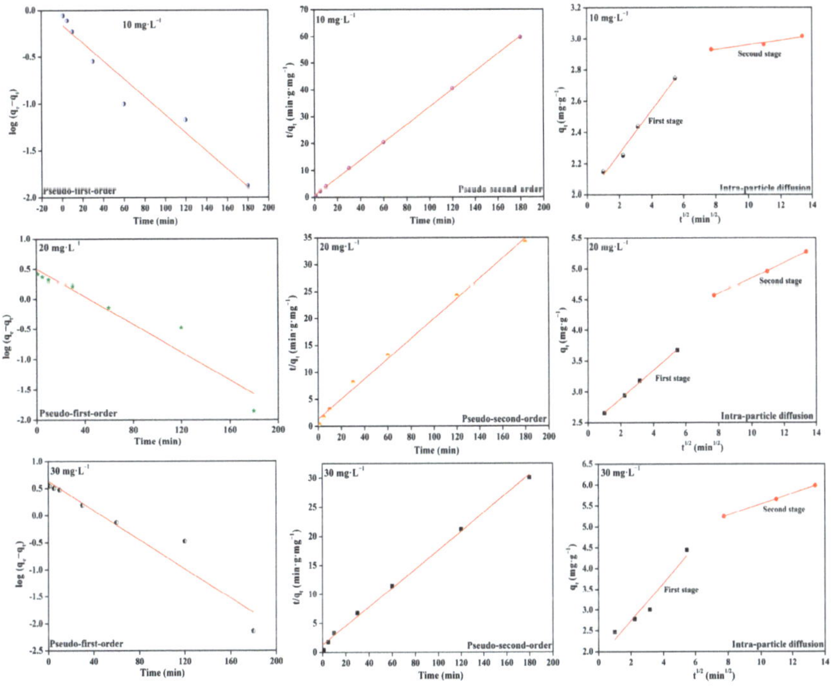

Figure 8. Adsorption kinetics at TB concentrations of 10, 20, and 30 mg·L^{-1} using pseudo-first-order, pseudo-second-order, and intra-particle diffusion models.

4.2. Mechanism

The adsorption mechanism might be used to understand the adsorbent's capacity to adsorb the dye in water. The decrease in surface area of NZB from 17.08 to 9.8 m^2·g^{-1} following dye adsorption occurred because the dye molecules blocked the pores of NZB by either electrostatic or hydrogen bonding. The pore volume (0.0436 to 0.0243 cm^3·g^{-1}) dropped while the pore diameters (5.9 to 6.56 nm) rose as a result of the pore-filling process during adsorption. Additionally, the pore diameters of the adsorbent increased owing to the pore breathing effect. The adsorption of dye molecules onto the surfaces of NZB may result in a reduction in peak intensity seen in the XRD pattern. The existence of a substantial number of positive charges on the surface of NZB may account for the electrostatic interaction with the negatively charged TB dye, which likely leads to the physisorption phenomenon. Moreover, the adsorption of dye molecules induces changes in the Raman spectra throughout the 600 to 1800 cm^{-1} region. The observed IR peaks at 3422 and 3256 cm^{-1} following adsorption may be attributed to the presence of -OH and N-H groups in the dye molecule. Furthermore, the presence of -N=N-, C=C, -SO_3^-, and C-N in the dye molecules resulted in the appearance of 1642, 1527, 1349, and 1326 cm^{-1} vibrations. The formation of these IR peaks may be attributed to the electrostatic or hydrogen bonding contact between the positively charged surface of NZB surfaces, which consist of metal nodes and hydroxyl groups. Following the adsorption of dye, the high-resolution spectrum of the NZB LTH reveals the presence of the N 1s peak, which is attributed to the existence of nitrogen in the azo and amino groups within the structure of TB. The binding energies of Bi 4f, Ni 2p, and Zn 2p elements were enhanced upon adsorption, as a result of dye molecules being absorbed onto NZB surfaces. This enhancement could be due to electrostatic interactions between the dye's -SO_3^- groups and the NZB's metal nodes. As seen in Figure 9, the primary mechanism of dye adsorption by the NZB involves a hydrogen bonding contact between the -OH^+ groups of the NZB and the sulfonate groups (-SO_3^-), amino groups (-NH_2), hydroxyl groups (-OH), and azo groups (-N=N-) of TB. On the other hand, the metal ions present in the surface LTHs, such as Ni^{2+}, Zn^{2+}, and Bi^{3+}, may be electrostatically attracted to the negatively charged -SO_3^- group in the dye structure. As a result, the adsorption process implies that NZB LTH and Trypan blue dye may bind together, enhancing their effectiveness in water remediation via hydrogen bonding and electrostatic interaction.

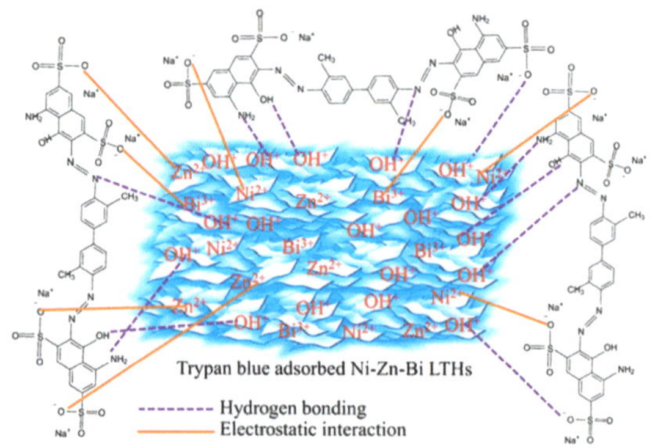

Figure 9. The possible mechanism of Trypan blue (TB) adsorption on the surfaces of NZB LTH.

4.3. Selectivity

It is critical to investigate dye removal from different anionic and cationic dye mixtures using an adsorbent to determine whether or not it is capable of removing additional dyes or simply the specified dye for industrial applications. To this extent, 100 mg of NZB LTH was employed to remove anionic dyes, including Eriochrome black T (EBT), Rose Bengal (RB), Methyl orange (MO), Congo red (CR), and Amido black 10B (AB), as well as cationic dyes, including Safranin O (SO), Crystal violet (CV), Rhodamine 6B (Rh6B), Methylene blue (MB), Brilliant green (BG), and Rhodamine 6G (Rh6G), from a mixture of 18 $mg \cdot L^{-1}$ concentration of Trypan blue (TB). The chemical structures of these dyes are illustrated in Figure S7 (refer to the Supplementary Materials). Because the absorbance value of 1.058 determined the concentration of TB, each cationic and anionic dye was adjusted to match the absorbance of TB for selectivity experiments. The UV-Vis absorbance spectra before and after adsorption of a variety of anionic (EBT + TB, MO + TB, RB + TB, CR + TB, and AB + TB) and cationic (SO + TB, CV + TB, Rh6B + TB, MB + TB, BG + TB, and Rh6G + TB) dyes with a mixture of TB are illustrated in Figure 10. The absorbance spectra of the dye mixture after adsorption indicate that NZB LTH is capable of selectively removing TB from a mixture of anionic dyes, such as EBT, MO, RB, and AB. Accordingly, EBT, which only has one negatively charged -SO_3^- group, may not be able to attract NZB as quickly as TB, which has four -SO_3^- and two primary amines (-NH_2). As a result of the competition between EBT and TB in the aqueous solution, NZB may have high selectivity for TB. In contrast to the four -SO_3^- groups and two -NH_2 groups in TB, the adsorption of one positively tertiary amine (-$N(CH_3)_2$) and negatively charged -SO_3^- group in MO dye to the positively charged NZB may be a weak contact via hydrogen bonding and electrostatic interaction. Therefore, in the presence of MO and TB, negatively charged TB attracts NZB more easily than MO. The chemical structure of RB has quite a few negative charges, such as one COO^- and one O^- group. As a consequence of the competition between RB and TB for NZB, TB has a higher adsorption because of its more negatively charged dye. Therefore, when combined with RB, NZB exhibits significant TB selectivity, whereas AB only contains two -SO_3^- and one -NH_2 group. Thus, when AB competes with TB for adsorption on NZB, the interaction via electrostatic and hydrogen bonding may be slower than TB, which has a faster adsorption owing to four negative charges and two amine groups. NZB hence exhibits notable selectivity for TB when combined with AB. The structure of CR includes two -SO_3^- and two -NH_2 groups. Despite having four negative charges and two positive charges, NZB has a significantly higher removed CR than TB. CR may be easier to remove than TB because it has a low molecular weight, stronger driving forces in water, and/or more hydrogen bonds and electrostatic interactions. In instances of cationic dyes containing TB, NZB was able to remove TB while also removing almost all cationic dyes except CV. The structures of cationic dyes like SO, Rh6B, MB, BG, and Rh6G mostly contain positive charge groups like -NH_2, -N-, and -$N(CH_3)_2$, and aromatic rings, which form dipole–dipole hydrogen bonds and Yoshida hydrogen bonds with the highly positively charged hydroxyl (-OH^+) groups of NZB. Despite the presence of tertiary amine in CV's structure, it is possible that the combination of CV and TB's larger molecular size inhibits its rapid adsorption on the surface of NZB through hydrogen bonds, resulting in the observation of a higher absorption peak. As a result, anionic TB has no competition with other cationic dyes for adsorption on positively charged NZB. Because of the high reduction in absorbance peaks of cationic dyes with TB following adsorption by NZB, the removal efficiency of the SO, CV, Rh6B, MB, BG, and Rh6G was determined to be 74.0%, 69.7%, 82.1%, 86.2%, 82.8%, and 77.0%, respectively. Figure 11 shows photographs of different anionic and cation dye mixtures with TB before and after adsorption. The absorption spectra indicate that NZB LTH effectively selectively removes TB when mixed with anionic dyes such as EBT, MO, RB, and AB, in contrast to CR. In the combination of TB with cationic dyes, NZB LTH completely removed TB and was also capable of removing almost all cationic dyes. Therefore, the present study has shown that NZB LTH is a viable material for the highly selective removal of TB when combined with an anionic dye mixture. Additionally, it is capable of removing cationic dyes.

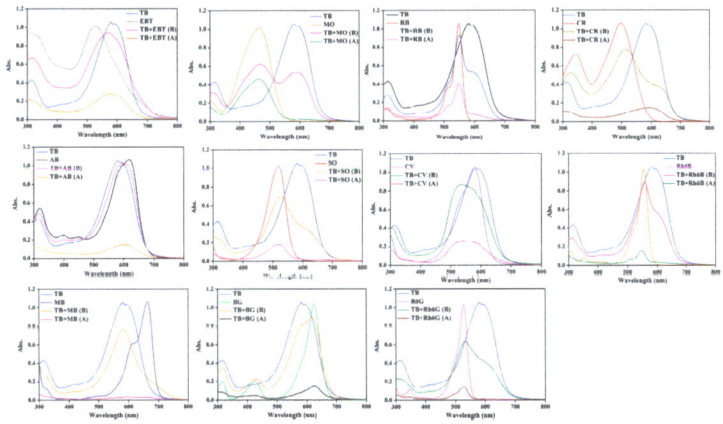

Figure 10. UV-Vis absorbance spectra of TB selectivity using NZB LTH (Dye: TB with other anionic and cationic; Dosage: 100 mg; TB concentration: 18 mg·L^{-1}; pH: Natural (~4.5–5.0); time: 1 h; volume: 12.5 mL (TB) + 12.5 mL (other dye) and temperature: Room temperature).

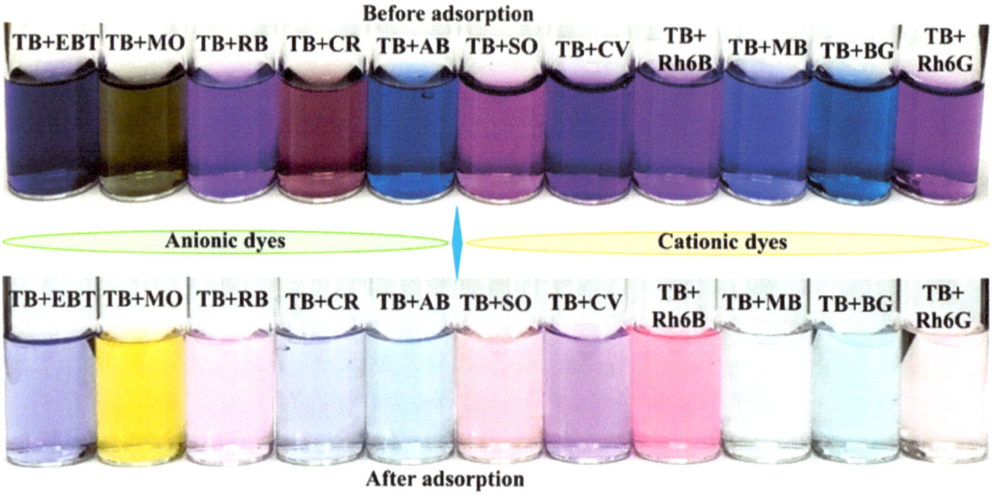

Figure 11. Trypan blue (TB) selectivity with anionic and cationic dyes using NZB LTH.

4.4. Salt Effect

Textiles and other industries that use coloring for products that produce wastewater may contain various types of salt that may affect the efficacy of adsorbents for dye removal using adsorption. As a result, it is critical to determine how the inorganic salts and their higher concentrations affect NZB LTH's dye removal percentage. In this work, sodium sulfate (Na_2SO_4), sodium sulfite (Na_2SO_3), and sodium chloride (NaCl) are the common salts used for TB removal. Figure 12 shows that the experiment without salt had a higher dye removal percentage (73.5%). But when Na_2SO_4 salt was added, the percentage of dye removal dropped to 35.5% at 0.01 M. This was because the negative charges of $SO_4{}^{2-}$ and $SO_3{}^{2-}$ in the water caused competition, and the positively charged adsorbent was better at attracting $SO_4{}^{2-}$ than the $SO_3{}^{2-}$ group of TB. Furthermore, an increase in salt concentrations from 0.03 M to 0.09 M led to a decrease in the dye removal percentage from 38.4% to 35.1%, primarily due to increased competition from $SO_4{}^{2-}$. The addition of Na_2SO_3 to the dye solution resulted in a drastic decrease in the dye removal percentage, reaching 18.0% at a concentration of 0.01 M. This was caused by the salt's high adsorption of the $SO_3{}^{2-}$ group onto the NZB surfaces, which acted as a shield for TB's $SO_3{}^{2-}$ group. When compared to Na_2SO_4, Na_2SO_3 may have a severe effect on the NZB's dye removal percentage. Increasing the salt content from 0.03 M to 0.05 M reduced the dye removal percentage from 17.4% to 10.3%, perhaps because of the NZB's strong attraction to the $SO_3{}^{2-}$ of salt. Subsequently, the removal percentage increased a little from 11.1% to 11.6% as the concentration was increased from 0.07 M to 0.09 M, mostly due to a higher salt effect. Furthermore, when the concentration of NaCl increased from 0.01 M to 0.09 M, the dye removal percentage increased from 39.7% to 43.1%, which might be attributed to the salt effect, in which the solubility of dye molecules rises with the increasing ion concentration of NaCl in an aqueous solution. Consequently, the percentage of dye removal rose as salt concentration increased. NaCl in the dye solution might therefore have a small effect in comparison to Na_2SO_4 and Na_2SO_3 salts. The UV-Vis absorbance spectra of the different salts' effects on TB adsorption by NZB LTH are shown in Figure 12. These findings demonstrate that as the concentration of salt in the dye solution rises, the high absorbance value could be seen in the following order: NaCl $<$ Na_2SO_4 $<$ Na_2SO_3.

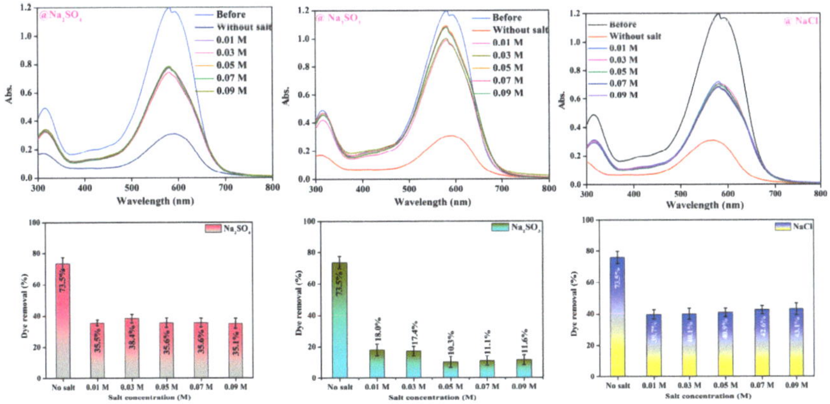

Figure 12. The UV-Vis absorbance spectra of the impact of salts at varying concentrations and their dye removal efficiency in the removal of TB by NZB LTH (Dye: TB; dye concentration: 20 mg·L^{-1}; Dosage: 80 mg; pH: Natural (~4.5–5.0); salts: Na_2SO_4, Na_2SO_3, and NaCl; salt concentration: 0.0–0.09 M; time: 1 h; volume: 25 mL; temperature: Room temperature).

4.5. Reusability

Since it saves the energy and cost of the process, the key stage for the adsorption process is the reuse of the adsorbent after adsorption. After TB adsorption by NZB LTH, desorption was obtained by individual contact with 25 mL of methanol, 0.1 M NaOH, and ethanol. A 12 h drying procedure at 60 °C was conducted on the NZB LTH in an oven to prepare it for use in the subsequent dye adsorption process after desorption. Similarly, the TB removal treatment was performed five times. As seen in Figure 13d, the NZB removed 82.5% of the dye during the zeroth cycle. The adsorbent then removed a TB percentage ranging from 59.0 to 15.2%, 69.7 to 9.6%, and 56.8 to 25.7%, respectively, using the reagents methanol, ethanol, and 0.1 M NaOH during the first to fifth cycles. The decline in dye removal percentage after each cycle may be attributed to dye molecules taking up permanently in the pores of NZB LTH after the zeroth cycle, as well as a minor loss of adsorbent quantity due to washing after each cycle; thus, increasing the number of cycles corresponds to a decrease in the dye removal percentage. However, when compared to methanol and ethanol, NaOH effectively desorbed TB from the surface of NZB LTH, even after five cycles. Therefore, 0.1 M NaOH proved to be an effective reagent for desorbing TB, allowing for up to three reuses of the NZB LTH. Zubair et al. found similar results when they used various LDHs, such as 0.5 M NaOH, ethanol, and 0.5 M NaOH + NaCl, to desorb EBT dye for up to three cycles [89]. Accordingly, the adsorption capacity of the CoFe, ZnFe, and NiFe LDHs drastically decreased from the first to the third cycles. However, 0.5 M NaOH showed better reusability compared to other agents. The dye molecules' strong adsorption on the adsorbent surfaces through hydrogen bonding or electrostatic interaction may be the cause of their decreasing adsorption capacity or removal efficiency as the number of cycles increases. Another study used water to desorb Coomassie brilliant blue (CBB) from the surfaces of Mg-Cu LDHs; however, after three cycles, this LDH was able to remove 50% of the CBB [90]. Therefore, it might be hard to remove the dye molecules from the LDHs or LTHs using a solvent or water process. To make the process more reusable, it is suggested that other methods be found to remove the dye from the adsorbent surfaces. The UV-Vis absorbance spectra of TB-adsorbed NZB before and after dye adsorption for 0–5 cycles are shown in Figure 13a–c. The absorbance peak becomes higher with each cycle, but the absorbance peaks for 0.1 M NaOH were lower than those for methanol and ethanol when they are used for the recycles.

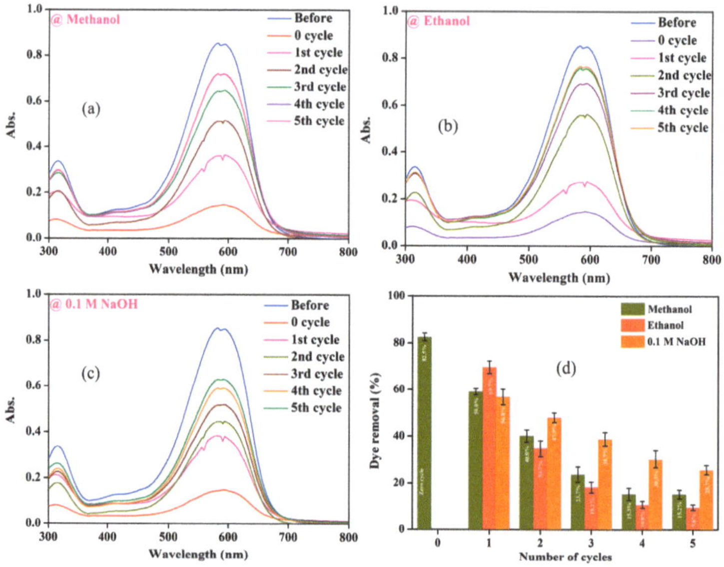

Figure 13. (**a–c**) UV-Vis absorbance spectra regarding the reusability of NZB LTH before and after TB adsorption (0 to 5 cycles); (**d**) the calculated dye removal percentage corresponds to the number of cycles. (Dosage: 80 mg, dye concentration: 15 mg·L^{-1}, time: 1 h, pH: Natural (~4.5–5.0), volume: 25 mL, temperature: RT, regenerating agents: methanol, ethanol, and 0.1 M NaOH).

4.6. Comparison

A comparison of the maximum adsorption capacity (q_{max}) of the adsorbent for dye removal with other reported adsorbents is provided in Table 4. However, comparing the findings of adsorbents with one another may be contrary as various pieces of research utilized different synthesis techniques, adsorbent types, dosages, dye concentrations, pH, volume of the dye solution, etc. In general, the q_{max} of the adsorbents is contingent upon the concentration of the dye, the volume of the solution, and the dosage. A higher q_{max} may result in a minimal adsorbent dosage, which is accompanied by a higher solution volume and dye concentration. A lowered q_{max} may result in a maximal dosage with a lower volume and dye concentration. The q_{max} of the NZB LTH (5.3 mg·g^{-1}) was lower than that of other adsorbents, as indicated in Table 4. The low value attained in this investigation may be attributed to the use of a high dosage (0.08 g) in a low volume (25 mL) and a low-concentrated dye (5–30 mg·L^{-1}). Typically, the adsorbents, including montmorillonite-modified hexadecyltrimethylammonium bromide and polyethylene glycol (MMT@HTAB@PEG), hydroxyapatite-modified reduced graphene oxide (HA@rGO), zinc nanoparticle-modified luffa sponge (LS-Zn), and magnesium oxide nanoparticles (MgO), were able to achieve a high q_{max} of 267.3, 146.5, 129.8, and 93.5 mg g^{-1}, respectively, as a result of the higher volume of solution (50 mL). Even so, the adsorbents exhibited a difference in high and low q_{max} based on the dosage and dye concentrations. Furthermore, the pH 2 used by MMT@HTAB@PEG to remove TB may not be environmentally friendly. Adsorbents like pyrolyzed maple bark (PMB), poly(diallyldimethylammonium chloride)/graphene oxide hydrogel (PDDA/GO), and GO-doped nano-hydroxyapatite (GO-nHAp) had moderate q_{max} values of 72.5, 50.0, and 41.0 mg g^{-1}, which could be attributed to the low solution volume or concentration of the dye used. NZB LTH has poorer efficacy than the others; however, it has a natural pH (~4.5–5.0), is simple to synthesize, environmentally friendly, and inexpensive. Future research might be conducted to improve its dye removal performance or alternative effective LTHs could be synthesized for the efficient removal of organic dyes.

Table 4. Comparing Ni-Zn-Bi LTH (NZB LTH) to other adsorbents for the removal of Trypan blue (TB).

Adsorbents	Dye	Dye Concentration	pH	Volume (mL)	Dosage (g)	q_{max} ($mg{\cdot}g^{-1}$)	Ref.
MMT@HTAB@PEG	Trypan blue	20–120 $mg{\cdot}L^{-1}$	2.0	50	0.03	267.3	[91]
HA@rGO	Trypan blue	50 $mg{\cdot}L^{-1}$	Natural	50	0.1	146.5	[92]
LS-Zn	Trypan blue	10–200 $mg{\cdot}L^{-1}$	7.0	50	1.0	129.8	[93]
MgO	Trypan blue	15–45 $mg{\cdot}L^{-1}$	5.4	50	0.02	93.5	[94]
PMB	Trypan blue	1000 $mg{\cdot}L^{-1}$	7.5	10	0.1	72.5	[95]
PDDA/GO	Trypan blue	0.001–1 mM	6.0	0.5	0.001	50.0	[96]
GO-nHAp	Trypan blue	50–200 $mg{\cdot}L^{-1}$	6.7	25	0.05	41.0	[97]
NZB LTH	Trypan blue	5–30 $mg{\cdot}L^{-1}$	~4.5–5.0	25	0.08	5.3	This work

5. Materials and Methods

Zinc (II) nitrate hexahydrate ($Zn(NO_3)_2{\cdot}6H_2O$, 95.0%), sodium chloride (NaCl, 99.0%), sodium hydroxide (NaOH, 93.0%), hydrochloric acid (HCl, 35–37%), and sodium sulfate anhydrous (Na_2SO_4, 99.0%) were obtained from Duksan pure chemicals Co., Ltd., Ansan-si, Republic of Korea. Nickel (II) nitrate hexahydrate ($Ni(NO_3)_2{\cdot}6H_2O$, 97.0%), Bismuth (III) nitrate pentahydrate ($Bi(NO_3)_3{\cdot}5H_2O$, 98.0%), and TB dye (Figure S1) were obtained from Daejung chemicals & metals Co., Ltd., Siheung-si, Republic of Korea. Sodium sulfite anhydrous (Na_2SO_3, 95.0%) was received from DC Chemicals Co., Ltd., Shanghai, China. The chemicals used in this study were utilized without any additional purification steps.

5.1. Synthesis of NiZnBi LTH (NZB LTH or NZB)

NZB was produced via the co-precipitation process. Initially, 0.1 M $Ni(NO_3)_2{\cdot}6H_2O$, 0.05 M $Zn(NO_3)_2{\cdot}6H_2O$, and 0.025 M $Bi(NO_3)_3{\cdot}5H_2O$ were added to 100 mL of purified water, which was then stirred for 30 min to ensure that all of the substances were completely dissolved. However, due to the insoluble nature of $Bi(NO_3)_3{\cdot}5H_2O$ in water, a series of drops of concentrated HCl was introduced into the mixture solution to solubilize it. Then, to achieve a precipitate, 1.0 M NaOH was added slowly to the above solutions to bring them to a pH of 10. The solution, which had been adjusted for pH, was constantly stirred at 60 °C for 12 h. Subsequently, the particulates were immersed in the solution using purified water until the pH leveled at 7.0. This process aided in the purification of the samples by removing any excess NaOH. Subsequently, the purified particles were dried in an oven at 60 °C to acquire the NZB. Furthermore, batch adsorption and characterization investigations were performed utilizing the acquired NZB.

5.2. Batch Adsorption Study

An experiment on batch adsorption was conducted using concentrations of Trypan blue (TB) ranging from 5 to 30 $mg{\cdot}L^{-1}$. The batch adsorption at room temperature included testing different parameters, including pH, adsorbent dosage, initial dye concentration, and time. The pH of the dye solution was varied using 1 M NaOH and 1 M HCl to explore the effect of pH. To study TB adsorption, 80 mg of NZB was added to 25 mL of various TB solution concentrations in a 100 mL conical flask. The flasks were then placed in a shaking bath and left to shake for 1 h. After that, the NZB that had adsorbed TB was isolated from the dye solution using 10 min centrifugation at 3000 rpm. The supernatant collected thereafter was used to assess the efficiency of NZB in removing TB by a calibration plot at λ_{max} 590 nm. The following Equations (1) and (2) were employed to calculate the removal percentage (R%) and adsorption capacity (q_e ($mg{\cdot}g^{-1}$)) of TB. The TB removal performance was analyzed in triplicate for each parameter influence to calculate standard errors.

$$R\% = \frac{C_i - C_f}{C_i} \times 100 \quad (1)$$

where C_i and C_f are the dye concentrations before and after adsorption, respectively.

$$q_e\left(mg{\cdot}g^{-1}\right) = \frac{C_i - C_f}{M} \times V \quad (2)$$

where V is the dye solution in liters (L) and M is the adsorbent mass in grams (g).

5.3. Characterization Techniques

X-ray diffraction (XRD) with an X′Pert PRO from PANalytical, Almelo, Netherlands, was used to analyze the structural stability of the adsorbents before and after NZB dye adsorption. The scan category applied was 'Continuous', with a scan rate of 5° min^{-1} and scan range and step size of analysis of 10–80° and 0.026, respectively. The functional groups of adsorbents between wavenumbers of 650 and 4000 cm^{-1} were analyzed by Fourier transform-infrared spectroscopy-attenuated total reflection (FTIR-ATR) using QATR-S, Shimadzu, Kyoto, Japan. The adsorbent surface area was studied at 77 K using Micromeritics, 3-FLEX, Norcross, GA, USA, and porosity was measured using the Barrett–Joyner–Halenda (BJH) method. The morphology and elemental composition and mapping of NZB were analyzed by field-emission scanning electron microscope (FESEM, S-4800, Hitachi, Tokyo, Japan) and energy dispersive X-ray spectrometer (EDX, Horiba EMAX, Horiba Co., Ltd., Kyoto, Japan), respectively. X-ray photoelectron spectroscopy (XPS, AXIS ULTRA DLD, Kratos Analytical instrument, Manchester, UK) using a monochromatic light source (Al K alpha) was utilized to study the oxidation states of elements in the NZB, and The XPS-PEAK41, Version 4.1, a free software developed by Raymund W.M. Kwok at The Chinese University of Hong Kong in Shatin, Hong Kong, was employed to conduct XPS peak fitting. The initial and final dye concentrations were analyzed using a UV-Vis spectrophotometer (Ubi-1800, MicroDigital Co., Ltd., Seongnam-si, Republic of Korea). This study used the shaking bath model JSSI-070T, 63L capacity, JS Research Co., Ltd., Gongju-si, Republic of Korea, for adsorption studies.

6. Conclusions

This study used co-precipitation to synthesize Ni-Zn-Bi LTH (NZB) for the removal of Trypan blue (TB) from water. The physiochemical characteristics of NZB were investigated both before and following dye adsorption using a variety of techniques. The dye solution's natural pH (~4.5–5.0) was used for all adsorption experiments, and the q_{max} of NZB was calculated to be 5.3 mg g^{-1} using the Langmuir model. Regarding selectivity, NZB is very effective at removing anionic dyes and completely removes cationic dyes when combined with TB. Studies found that increasing the salt concentration (0.01–0.09 M) of Na_2SO_3 reduced TB removal compared to NaCl and Na_2SO_4. The removal efficiency of TB is 15.2%, 9.6%, and 25.7%, respectively, as a result of the reusability of NZB in methanol, ethanol, and NaOH reagents for five cycles. Based on the experiments, NZB LTH may be able to remove cationic or anionic dyes via surface modification of incompatible materials. The findings show that NZB LTH might be a promising material for removing TB and a variety of anionic and cationic dyes, with probable applications in the treatment of dye-based industrial effluent.

Supplementary Materials: The following supporting information can be downloaded at: https://www.mdpi.com/article/10.3390/inorganics12110296/s1, Figure S1: Chemical structure of Trypan blue (TB) dye; Figure S2: UV-Vis absorption spectra of the TB's solution at different pHs and effect of pHs such as 2, 4, natural (~4.5–5.0), 6, 8, and 10 for before and after dye adsorption by NZB LTH (Dye: Trypan Blue (TB), Concentration: 15 mg·L^{-1}, Dosage: 80 mg, Volume: 25 mL, Time: 1 h, Temperature: Room temperature (RT)); Figure S3: UV-Vis absorption spectra of the adsorbent dosage (10–80 mg) for before and after TB adsorption by NZB (Dye: TB, pH: Natural (~4.5–5.0), Concentration: 10 mg·L^{-1}, Volume: 25 mL, Time: 1 h, Temperature: RT); Figure S4: UV-Vis absorption spectra of the effect of initial dye concentration (5–30 mg L^{-1}) for before and after TB adsorption by NZB (Dye: TB, Dosage: 80 mg, pH: Natural (~4.5–5.0), Time: 1 h, Volume: 25 mL, Temperature: RT); Figure S5: Characterizing NZB after TB adsorption: (a) N_2 adsorption–desorption isotherm and BJH pore size distribution; (b) XRD pattern; (c) Raman spectrum; (d) FTIR-ATR spectrum; and (e) FESEM images at 5 and 1 μm scales; Figure S6: Characterizing NZB after TB adsorption: XPS full survey spectrum and high-resolution spectra of Bi 4f, C 1s, O 1s, Ni 2p, Zn 2p, and N 1s; Adsorption Isotherms and Kinetic Studies; Figure S7: Chemical structure of organic dyes.

Author Contributions: Conceptualization, G.S. and K.D.; methodology, G.S.; software, M.A.; validation, N.B. and M.A.; formal analysis, K.S. and T.S.; investigation, G.S.; resources, N.B.; data curation, N.B. and K.S.; writing—original draft preparation, G.S.; writing—review and editing, K.D. and T.H.O.; visualization, M.A.; supervision, T.H.O.; project administration, T.H.O.; funding acquisition, T.H.O. All authors have read and agreed to the published version of the manuscript.

Funding: This research was funded by the National Research Foundation of Korea (NRF) by the Korean Government (MSIT), grant no: 2022R1A2C100428.

Data Availability Statement: The original contributions presented in the study are included in the article/Supplementary Materials, further inquiries can be directed to the corresponding authors.

Conflicts of Interest: The authors declare no conflicts of interest.

References

1. Abu Elella, M.H.; Aamer, N.; Abdallah, H.M.; López-Maldonado, E.A.; Mohamed, Y.M.A.; El Nazer, H.A.; Mohamed, R.R. Novel high-efficient adsorbent based on modified gelatin/montmorillonite nanocomposite for removal of malachite green dye. *Sci. Rep.* **2024**, *14*, 1228. [CrossRef] [PubMed]
2. Zulkiflee, A.; Mansoob Khan, M.; Yusuf Khan, M.; Khan, A.; Hilni Harunsani, M. Nb_2O_5/BiOCl composite as a visible-light-active photocatalyst for the removal of RhB dye and photoelectrochemical studies. *J. Photochem. Photobiol. A* **2024**, *446*, 115177. [CrossRef]
3. Sriram, G.; Thangarasu, S.; Selvakumar, K.; Kurkuri, M.; Dhineshbabu, N.R.; Oh, T.H. Effective removal of Rose Bengal using Ni-Co-Zn layered triple hydroxide: Studies on batch adsorption, mechanism, selectivity, co-ions, and reusability. *Colloids Surf. A Physicochem. Eng. Asp.* **2024**, *685*, 133199. [CrossRef]
4. Varjani, S.; Rakholiya, P.; Shindhal, T.; Shah, A.V.; Ngo, H.H. Trends in dye industry effluent treatment and recovery of value added products. *J. Water Process Eng.* **2021**, *39*, 101734. [CrossRef]
5. Slama, H.B.; Chenari Bouket, A.; Pourhassan, Z.; Alenezi, F.N.; Silini, A.; Cherif-Silini, H.; Oszako, T.; Luptakova, L.; Golińska, P.; Belbahri, L. Diversity of Synthetic Dyes from Textile Industries, Discharge Impacts and Treatment Methods. *Appl. Sci.* **2021**, *11*, 6255. [CrossRef]
6. Louis, K.S.; Siegel, A.C. Cell Viability Analysis Using Trypan Blue: Manual and Automated Methods. In *Mammalian Cell Viability: Methods and Protocols*; Stoddart, M.J., Ed.; Humana Press: Totowa, NJ, USA, 2011; pp. 7–12. [CrossRef]
7. Abdelkader, E.A.; McBain, V.A.; Anand, M.; Scott, N.W.; Rehman Siddiqui, M.A.; Lois, N. In vivo safety of trypan blue use in vitreoretinal surgery. *Retina* **2011**, *31*, 1122–1127. [CrossRef] [PubMed]
8. Dutta, S.; Gupta, B.; Srivastava, S.K.; Gupta, A.K. Recent advances on the removal of dyes from wastewater using various adsorbents: A critical review. *Mater. Adv.* **2021**, *2*, 4497–4531. [CrossRef]
9. Hussain, B.; Sajad, M.; Usman, H.; Al-Ghanim, K.A.; Riaz, M.N.; Berenjian, A.; Mahboob, S.; Show, P.L. Assessment of hepatotoxicity and nephrotoxicity in Cirrhinus mrigala induced by trypan blue—An azo dye. *Environ. Res.* **2022**, *215*, 114120. [CrossRef]
10. Ledakowicz, S.; Paździor, K. Recent Achievements in Dyes Removal Focused on Advanced Oxidation Processes Integrated with Biological Methods. *Molecules* **2021**, *26*, 870. [CrossRef]
11. Katheresan, V.; Kansedo, J.; Lau, S.Y. Efficiency of various recent wastewater dye removal methods: A review. *J. Environ. Chem. Eng.* **2018**, *6*, 4676–4697. [CrossRef]
12. Sriram, G.; Bendre, A.; Mariappan, E.; Altalhi, T.; Kigga, M.; Ching, Y.C.; Jung, H.-Y.; Bhaduri, B.; Kurkuri, M. Recent trends in the application of metal-organic frameworks (MOFs) for the removal of toxic dyes and their removal mechanism—A review. *Sustain. Mater. Technol.* **2022**, *31*, e00378. [CrossRef]
13. Bharathiraja, B.; Jayamuthunagai, J.; Praveenkumar, R.; Iyyappan, J. Phytoremediation Techniques for the Removal of Dye in Wastewater. In *Bioremediation: Applications for Environmental Protection and Management*; Varjani, S.J., Agarwal, A.K., Gnansounou, E., Gurunathan, B., Eds.; Springer: Singapore, 2018; pp. 243–252. [CrossRef]
14. Zafar, A.M.; Naeem, A.; Minhas, M.A.; Hasan, M.J.; Rafique, S.; Ikhlaq, A. Removal of reactive dyes from textile industrial effluent using electrocoagulation in different parametric conditions of aluminum electrodes. *Total Environ. Adv.* **2024**, *9*, 200087. [CrossRef]
15. Madhurima, V.P.; Kumari, K.; Jain, P.K. Synthesis and study of carbon nanomaterials through arc discharge technique for efficient adsorption of organic dyes. *Diam. Relat. Mater.* **2024**, *141*, 110538. [CrossRef]
16. Uğan, M.; Onac, C.; Kaya, A.; Köseoğlu, D.; Akdoğan, A. Removal of Reactive Red 195 dye from textile industry wastewater with Deep Eutectic Solvent-based green extraction. *J. Mol. Liq.* **2024**, *398*, 124249. [CrossRef]
17. Le Nam Vo, V.; Chung, Y.-M. Catalytic wet peroxide oxidation of organic dye with in-situ generated H_2O_2 over bifunctional Fe-Pt@Pd/SiO_2 catalyst prepared by double-metal complex salt approach. *Appl. Catal. A* **2024**, *676*, 119640. [CrossRef]

18. Ciğeroğlu, Z.; El Messaoudi, N.; Şenol, Z.M.; Başkan, G.; Georgin, J.; Gubernat, S. Clay-based nanomaterials and their adsorptive removal efficiency for dyes and antibiotics: A review. *Mater. Today Sustain.* **2024**, *26*, 100735. [CrossRef]
19. Modwi, A.; Elamin, M.R.; Abdulkhair, B.Y.; Elamin, N.Y.; Ali Ben Aissa, M.; Ben Said, R. Synthesis and characterization of Ti-doped$Y_2O_3@C_3N_4$ nanocomposite for the removal of dyes from aqueous solution. *Inorg. Chem. Commun.* **2023**, *158*, 111594. [CrossRef]
20. Rimzim; Singh, J.; Mittal, S.; Singh, H. Robust removal of cationic dyes by zinc ferrite composites in single and ternary dye systems. *Inorg. Chem. Commun.* **2023**, *153*, 110756. [CrossRef]
21. Rathi, B.S.; Kumar, P.S. Application of adsorption process for effective removal of emerging contaminants from water and wastewater. *Environ. Pollut.* **2021**, *280*, 116995. [CrossRef]
22. Velicu, M.; Fu, H.; Suri, R.P.S.; Woods, K. Use of adsorption process to remove organic mercury thimerosal from industrial process wastewater. *J. Hazard. Mater.* **2007**, *148*, 599–605. [CrossRef]
23. Papić, S.; Koprivanac, N.; Lončarić Božić, A.; Meteš, A. Removal of some reactive dyes from synthetic wastewater by combined Al(III) coagulation/carbon adsorption process. *Dye. Pigment.* **2004**, *62*, 291–298. [CrossRef]
24. Gonçalves, J.O.; Strieder, M.M.; Silva, L.F.O.; dos Reis, G.S.; Dotto, G.L. Advanced technologies in water treatment: Chitosan and its modifications as effective agents in the adsorption of contaminants. *Int. J. Biol. Macromol.* **2024**, *270*, 132307. [CrossRef] [PubMed]
25. Liu, P.; Wu, L.; Guo, Y.; Huang, X.; Guo, Z. High crystalline LDHs with strong adsorption properties effectively remove oil and micro-nano plastics. *J. Clean. Prod.* **2024**, *437*, 140628. [CrossRef]
26. Sarkar, A.; Mushahary, N.; Basumatary, F.; Das, B.; Basumatary, S.F.; Venkatesan, K.; Selvaraj, M.; Rokhum, S.L.; Basumatary, S. Efficiency of montmorillonite-based materials as adsorbents in dye removal for wastewater treatment. *J. Environ. Chem. Eng.* **2024**, *12*, 112519. [CrossRef]
27. Tsoutsa, E.K.; Tolkou, A.K.; Kyzas, G.Z.; Katsoyiannis, I.A. An Update on Agricultural Wastes Used as Natural Adsorbents or Coagulants in Single or Combined Systems for the Removal of Dyes from Wastewater. *Water Air Soil Pollut.* **2024**, *235*, 178. [CrossRef]
28. Akash, S.; Rameshwar, S.S.; Rajamohan, N.; Rajasimman, M.; Vo, D.-V.N. Metal oxide nanobiochar materials to remediate heavy metal and dye pollution: A review. *Environ. Chem. Lett.* **2024**, *22*, 2091–2112. [CrossRef]
29. Njuguna, D.; Schönherr, H. Tunable Gellan Gum Hydrogels as High Capacity Adsorbents for Rapid Dye Removal. *ACS Appl. Polym. Mater.* **2024**, *6*, 1528–1539. [CrossRef]
30. Yuan, Z.; Li, F.; Zhang, X.; Li, M.-C.; Chen, Y.; Hoop, C.F.d.; Qi, J.; Huang, X. Bio-based adsorption foam composed of MOF and polyethyleneimine-modified cellulose for selective anionic dye removal. *Environ. Res.* **2024**, *248*, 118263. [CrossRef]
31. Tamer, T.M.; Abbas, R.; Sadik, W.A.; Omer, A.M.; Abd-Ellatif, M.M.; Mohy-Eldin, M.S. Development of novel amino-ethyl chitosan hydrogel for the removal of methyl orange azo dye model. *Sci. Rep.* **2024**, *14*, 1284. [CrossRef]
32. Koli, A.; Kumar, A.; Pattanshetti, A.; Supale, A.; Garadkar, K.; Shen, J.; Shaikh, J.; Praserthdam, S.; Motkuri, R.K.; Sabale, S. Hierarchical Porous Activated Carbon from Wheat Bran Agro-Waste: Applications in Carbon Dioxide Capture, Dye Removal, Oxygen and Hydrogen Evolution Reactions. *ChemPlusChem* **2024**, *89*, e202300373. [CrossRef]
33. Koli, A.; Battu, A.K.; Motkuri, R.K.; Sabale, S. Hierarchical porous activated carbon derived from agro-waste for potential CO_2 capture and efficient dye removal applications. *Biomass Convers. Biorefinery* **2024**, *14*, 10177–10188. [CrossRef]
34. Bagherzadeh, M.; Salehi, G.; Rabiee, N. Rapid and efficient removal of methylene blue dye from aqueous solutions using extract-modified Zn–Al LDH. *Chemosphere* **2024**, *350*, 141011. [CrossRef] [PubMed]
35. Karim, A.R.; Danish, M.; Alam, M.G.; Majeed, S.; Alanazi, A.M. A review of pre- and post-surface-modified neem (Azadirachta indica) biomass adsorbent: Surface functionalization mechanism and application. *Chemosphere* **2024**, *351*, 141180. [CrossRef] [PubMed]
36. Wu, X.; Wu, D.; Fu, R.; Zeng, W. Preparation of carbon aerogels with different pore structures and their fixed bed adsorption properties for dye removal. *Dye. Pigment.* **2012**, *95*, 689–694. [CrossRef]
37. Wu, C.; Li, K.; Ni, X.; He, Y.; Wang, Y.; Ju, A. Efficient Template-Catalyzed In Situ Polymerization for Carbon Xerogels with Large Specific Surface Area and High Adsorption. *Langmuir* **2024**, *40*, 9985–9992. [CrossRef]
38. Kazemi, S.Y.; Pourfaraj, R.; Biparva, P. Synthesis of MgZnAl Layered Triple Hydroxide Nanoplates as an Efficient Adsorbent for Removing the Acid Yellow 76 Azo Dye. *ChemistrySelect* **2024**, *9*, e202400789. [CrossRef]
39. Singh, V.K.; Sett, A.; Karmakar, S. Waste to wealth: Facile activation of red mud waste and insights into industrial reactive dye removal from wastewater. *Chem. Eng. J.* **2024**, *481*, 148373. [CrossRef]
40. Noormohammadi, M.; Zabihi, M.; Faghihi, M. Novel chitosan–clay–iron nanocomposites supported on anodic aluminum as an efficient plate-shaped adsorbent for the removal of arsenic and dye from aqueous solutions. *J. Phys. Chem. Solids* **2024**, *187*, 111874. [CrossRef]
41. Wei, S.; Tan, Z.; Liu, Z.; Zuo, H.; Xia, Y.; Zhang, Y. Removal of methyl orange dye by high surface area biomass activated carbon prepared from bamboo fibers. *Ind. Crops Prod.* **2024**, *218*, 118991. [CrossRef]
42. Wang, A.-L.; Xu, H.; Li, G.-R. NiCoFe Layered Triple Hydroxides with Porous Structures as High-Performance Electrocatalysts for Overall Water Splitting. *ACS Energy Lett.* **2016**, *1*, 445–453. [CrossRef]

43. Bahadi, S.A.; Drmosh, Q.A.; Onaizi, S.A. Synergism between polyethyleneimine, graphene oxide, and MgFeAl-layered triple hydroxide in removing acid red 1 dye and bisphenol A from contaminated water samples. *Colloids Surf. A Physicochem. Eng. Asp.* **2024**, *688*, 133579. [CrossRef]
44. Mahmoud, R.; Kotb, N.M.; GadelHak, Y.; El-Ela, F.I.A.; Shehata, A.Z.; Othman, S.I.; Allam, A.A.; Rudayni, H.A.; Zaher, A. Investigation of ternary Zn–Co–Fe layered double hydroxide as a multifunctional 2D layered adsorbent for moxifloxacin and antifungal disinfection. *Sci. Rep.* **2024**, *14*, 806. [CrossRef] [PubMed]
45. Mittal, J. Recent progress in the synthesis of Layered Double Hydroxides and their application for the adsorptive removal of dyes: A review. *J. Environ. Manag.* **2021**, *295*, 113017. [CrossRef]
46. Daud, M.; Hai, A.; Banat, F.; Wazir, M.B.; Habib, M.; Bharath, G.; Al-Harthi, M.A. A review on the recent advances, challenges and future aspect of layered double hydroxides (LDH)—Containing hybrids as promising adsorbents for dyes removal. *J. Mol. Liq.* **2019**, *288*, 110989. [CrossRef]
47. Shan, R.; Yan, L.; Yang, K.; Yu, S.; Hao, Y.; Yu, H.; Du, B. Magnetic Fe_3O_4/MgAl-LDH composite for effective removal of three red dyes from aqueous solution. *Chem. Eng. J.* **2014**, *252*, 38–46. [CrossRef]
48. Asadi, E.; Haroonian, P.; Ghaedi, M.; Asfaram, A. Removal of tartrazine and indigo carmine with layered double hydroxide-modified graphite nanocomposite: Isotherm, kinetics and response surface methodology (RSM) optimization. *J. Mol. Liq.* **2024**, *402*, 124769. [CrossRef]
49. Gupta, V.; Gupta, S.; Miura, N. Electrochemically synthesized large area network of CoxNiyAlz layered triple hydroxides nanosheets: A high performance supercapacitor. *J. Power Sources* **2009**, *189*, 1292–1295. [CrossRef]
50. Patil, K.; Babar, P.; Bae, H.; Jo, E.; Jang, J.S.; Bhoite, P.; Kolekar, S.; Kim, J.H. Enhanced electrocatalytic activity of a layered triple hydroxide (LTH) by modulating the electronic structure and active sites for efficient and stable urea electrolysis. *Sustain. Energy Fuels* **2022**, *6*, 474–483. [CrossRef]
51. Hamad, H.A.; Nageh, H.; El-Bery, H.M.; Kasry, A.; Carrasco-Marín, F.; Elhady, O.M.; Soliman, A.M.M.; El-Remaily, M.A.E.A.A.A. Unveiling the exceptional synergism-induced design of Co-Mg-Al layered triple hydroxides (LTHs) for boosting catalytic activity toward the green synthesis of indol-3-yl derivatives under mild conditions. *J. Colloid Interface Sci.* **2021**, *599*, 227–244. [CrossRef]
52. Khatun, S.; Roy, P. Cobalt chromium vanadium layered triple hydroxides as an efficient oxygen electrocatalyst for alkaline seawater splitting. *Chem. Commun.* **2022**, *58*, 1104–1107. [CrossRef]
53. Kandasamy, B.; Govindasamy, P.; Thangavelu, P.; Theerthagiri, J.; Min, A.; Choi, M.Y. Improved visible light photocatalytic degradation of yttrium doped NiMgAl layered triple hydroxides for the effective removal of methylene blue dye. *Chemosphere* **2022**, *290*, 133299. [CrossRef] [PubMed]
54. Ismail, U.M.; Onaizi, S.A.; Vohra, M.S. Novel MgCuAl-layered triple hydroxide for aqueous selenite and selenate treatment. *Emergent Mater.* **2024**, *7*, 521–532. [CrossRef]
55. Chowdhury, M.F.; Kim, C.-M.; Jang, A. High-efficient and rapid removal of anionic and cationic dyes using a facile synthesized sole adsorbent NiAlFe-layered triple hydroxide (LTH). *Chemosphere* **2023**, *332*, 138878. [CrossRef] [PubMed]
56. Gao, Y.; Zhang, T.; Mao, Y.; Wang, J.; Sun, C. Highly efficient bifunctional layered triple Co, Fe, Ru hydroxides and oxides composite electrocatalysts for Zinc-Air batteries. *J. Electroanal. Chem.* **2023**, *935*, 117315. [CrossRef]
57. Sriram, G.; Dhanabalan, K.; Ajeya, K.V.; Aruchamy, K.; Ching, Y.C.; Oh, T.H.; Jung, H.-Y.; Kurkuri, M. Recent progress in anion exchange membranes (AEMs) in water electrolysis: Synthesis, physio-chemical analysis, properties, and applications. *J. Mater. Chem. A* **2023**, *11*, 20886–21008. [CrossRef]
58. Wang, G.; He, L.; Guo, Z.; Li, M. Ternary metal layered hydroxides: As promising electrode materials for supercapacitors. *J. Energy Storage* **2023**, *72*, 108544. [CrossRef]
59. Alnasrawi, F.A.; Mohammed, A.A.; Al-Musawi, T.J. Synthesis, characterization and adsorptive performance of CuMgAl-layered double hydroxides/montmorillonite nanocomposite for the removal of Zn(II) ions. *Environ. Nanotechnol. Monit. Manag.* **2023**, *19*, 100771. [CrossRef]
60. Xiong, C.; Cao, W.; Long, Q.; Chen, J.; Yu, Y.; Lian, X.; Huang, J.; Du, G.; Chen, N. Etching-induced ion exchange engineering of two-dimensional layered NiFeCo-based hydroxides for high energy charge storage. *Dalton Trans.* **2024**, *53*, 1295–1306. [CrossRef]
61. Al-Jamimi, H.A.; Bahadi, S.A.; BinMakhashen, G.M.; Onaizi, S.A. Optimal hybrid artificial intelligence models for predicting the adsorptive removal of dyes and phenols from aqueous solutions using an amine-functionalized graphene oxide/layered triple hydroxide nanocomposite. *J. Mol. Liq.* **2023**, *391*, 123374. [CrossRef]
62. Abdel-Hady, E.E.; Mohamed, H.F.M.; Hafez, S.H.M.; Fahmy, A.M.M.; Magdy, A.; Mohamed, A.S.; Ali, E.O.; Abdelhamed, H.R.; Mahmoud, O.M. Textural properties and adsorption behavior of Zn–Mg–Al layered double hydroxide upon crystal violet dye removal as a low cost, effective, and recyclable adsorbent. *Sci. Rep.* **2023**, *13*, 6435. [CrossRef]
63. Zaghouane-Boudiaf, H.; Boutahala, M.; Arab, L. Removal of methyl orange from aqueous solution by uncalcined and calcined MgNiAl layered double hydroxides (LDHs). *Chem. Eng. J.* **2012**, *187*, 142–149. [CrossRef]
64. Waheed, T.; Din, S.u.; Ming, L.; Ahmad, P.; Min, P.; Haq, S.; Khandaker, M.U.; Boukhris, I.; Faruque, M.R.I.; Rehman, F.U.; et al. Porous Hierarchical Ni/Mg/Al Layered Double Hydroxide for Adsorption of Methyl Orange from Aqueous Solution. *Nanomaterials* **2023**, *13*, 1943. [CrossRef] [PubMed]
65. Kim, M.; Kim, S.; Ahn, H.; Koh, Y.; Kim, K.; Lee, M.K.; Lee, J.W.; Kang, Y.T. Investigation on sacrificial hydrolysis reaction of octadecyltrimethoxysilane for moisture resistance enhancement of metal–organic framework. *Sep. Purif. Technol.* **2024**, *350*, 127957. [CrossRef]

66. Yuan, C.; Xiong, S.; Zhang, X.; Shen, L.; Zhang, F.; Gao, B.; Su, L. Template-free synthesis of ordered mesoporous NiO/poly(sodium-4-styrene sulfonate) functionalized carbon nanotubes composite for electrochemical capacitors. *Nano Res.* **2009**, *2*, 722–732. [CrossRef]
67. Wang, D.; Xu, R.; Wang, X.; Li, Y. NiO nanorings and their unexpected catalytic property for CO oxidation. *Nanotechnology* **2006**, *17*, 979. [CrossRef] [PubMed]
68. Zhang, J.T.; Liu, S.; Pan, G.L.; Li, G.R.; Gao, X.P. A 3D hierarchical porous α-$Ni(OH)_2$/graphite nanosheet composite as an electrode material for supercapacitors. *J. Mater. Chem. A* **2014**, *2*, 1524–1529. [CrossRef]
69. Winiarski, J.; Tylus, W.; Winiarska, K.; Szczygieł, I.; Szczygieł, B. XPS and FT-IR Characterization of Selected Synthetic Corrosion Products of Zinc Expected in Neutral Environment Containing Chloride Ions. *J. Spectrosc.* **2018**, *2018*, 2079278. [CrossRef]
70. Wang, M.; Jiang, L.; Kim, E.J.; Hahn, S.H. Electronic structure and optical properties of $Zn(OH)_2$: LDA + U calculations and intense yellow luminescence. *RSC Adv.* **2015**, *5*, 87496–87503. [CrossRef]
71. Ghaedi, M.; Khafri, H.Z.; Asfaram, A.; Goudarzi, A. Response surface methodology approach for optimization of adsorption of Janus Green B from aqueous solution onto ZnO/$Zn(OH)_2$-NP-AC: Kinetic and isotherm study. *Spectrochim. Acta Part A Mol. Biomol. Spectrosc.* **2016**, *152*, 233–240. [CrossRef]
72. Hobosyan, M.A.; Yolchinyan, S.A.; Martirosyan, K.S. A novel nano-energetic system based on bismuth hydroxide. *RSC Adv.* **2016**, *6*, 66564–66570. [CrossRef]
73. Liu, S.; Ren, G.; Gao, X.; Li, Z.; Wang, L.; Meng, X. A novel bismuth hydroxide ($Bi(OH)_3$) semiconductor with highly-efficient photocatalytic activity. *Chem. Commun.* **2022**, *58*, 8198–8201. [CrossRef] [PubMed]
74. Zhang, W.; Huang, X.; Tan, Y.; Gao, Y.; Wu, J.; Hu, J.; Stein, A.; Tang, B. A facile approach to prepare $Bi(OH)_3$ nanoflakes as high-performance pseudocapacitor materials. *New J. Chem.* **2015**, *39*, 5927–5930. [CrossRef]
75. Huang, H.; Wang, H.-L.; Jiang, W.-F. Solar-driven $Bi_6O_5(OH)_3(NO_3)_5(H_2O)_3$/$Bi_2WO_6$ heterojunction for efficient degradation of organic pollutants: Insights into adsorption mechanism, charge transfer and degradation pathway. *Sep. Purif. Technol.* **2024**, *349*, 127747. [CrossRef]
76. Zeng, X.; Cai, Z.; Zhang, C.; Wang, D.; Xu, J.; Wang, X. Novel NiFe-LDH@Ni-MOF/NF heterostructured electrocatalysts for efficient oxygen evolution. *Mater. Res. Lett.* **2022**, *10*, 88–96. [CrossRef]
77. Jian, W.; Jin, Z.; Yang, J.; Meng, G.; Liu, H.; Liu, H. Anticorrosion and antibiofouling performance of in-situ prepared layered double hydroxide coating modified by sodium pyrithione on aluminum alloy 7075. *J. Ind. Eng. Chem.* **2022**, *113*, 419–430. [CrossRef]
78. Manríquez, M.E.; Hernández-Cortez, J.G.; Wang, J.A.; Chen, L.F.; Zuñiga-Moreno, A.; Gómez, R. Synthesis of transition metal doped lamellar double hydroxides as base catalysts for acetone aldol condensation. *Appl. Clay Sci.* **2015**, *118*, 188–194. [CrossRef]
79. Xu, J.; Song, Y.; Tan, Q.; Jiang, L. Chloride absorption by nitrate, nitrite and aminobenzoate intercalated layered double hydroxides. *J. Mater. Sci.* **2017**, *52*, 5908–5916. [CrossRef]
80. Abo El-Reesh, G.Y.; Farghali, A.A.; Taha, M.; Mahmoud, R.K. Novel synthesis of Ni/Fe layered double hydroxides using urea and glycerol and their enhanced adsorption behavior for Cr(VI) removal. *Sci. Rep.* **2020**, *10*, 587. [CrossRef]
81. Li, J.; Zhang, S.; Chen, Y.; Liu, T.; Liu, C.; Zhang, X.; Yi, M.; Chu, Z.; Han, X. A novel three-dimensional hierarchical CuAl layered double hydroxide with excellent catalytic activity for degradation of methyl orange. *RSC Adv.* **2017**, *7*, 29051–29057. [CrossRef]
82. Spectroscopy of Carboxylic Acids and Nitriles. Available online: https://chem.libretexts.org/@go/page/448775 (accessed on 8 November 2024).
83. Kudo, S.; Nakashima, S. Changes in IR band areas and band shifts during water adsorption to lecithin and ceramide. *Spectrochim. Acta Part A: Mol. Biomol. Spectrosc.* **2020**, *228*, 117779. [CrossRef]
84. Nait-Merzoug, A.; Guellati, O.; Djaber, S.; Habib, N.; Harat, A.; El-Haskouri, J.; Begin, D.; Guerioune, M. Ni/Zn Layered Double Hydroxide (LDH) Micro/Nanosystems and Their Azorubine Adsorption Performance. *Appl. Sci.* **2021**, *11*, 8899. [CrossRef]
85. Ouyang, Y.; Xu, Y.; Zhao, L.; Deng, M.; Yang, P.; Peng, G.; Ke, G. Preparation of ZnNiAl-LDHs microspheres and their adsorption behavior and mechanism on U(VI). *Sci. Rep.* **2021**, *11*, 21625. [CrossRef] [PubMed]
86. Rego, R.M.; Sriram, G.; Ajeya, K.V.; Jung, H.-Y.; Kurkuri, M.D.; Kigga, M. Cerium based UiO-66 MOF as a multipollutant adsorbent for universal water purification. *J. Hazard. Mater.* **2021**, *416*, 125941. [CrossRef]
87. Lade, H.; Kadam, A.; Paul, D.; Govindwar, S. A Low-Cost Wheat Bran Medium for Biodegradation of the Benzidine-Based Carcinogenic Dye Trypan Blue Using a Microbial Consortium. *Int. J. Environ. Res. Public Health* **2015**, *12*, 3480–3505. [CrossRef] [PubMed]
88. Malik, P.K. Dye removal from wastewater using activated carbon developed from sawdust: Adsorption equilibrium and kinetics. *J. Hazard. Mater.* **2004**, *113*, 81–88. [CrossRef]
89. Zubair, M.; Aziz, H.A.; Ahmad, M.A.; Ihsanullah, I.; Al-Harthi, M.A. Adsorption and reusability performance of M-Fe (M = Co, Cu, Zn and Ni) layered double hydroxides for the removal of hazardous Eriochrome Black T dye from different water streams. *J. Water Process Eng.* **2021**, *42*, 102060. [CrossRef]
90. Tabti, H.A.; Ammam, A.; Guezzen, B.; Boudinar, M.; Kadeche, A.; Ramdani, A.; Doumi, B.; Ech-Chergui, A.N.; Boudia, R.A.; Adjdir, M. Exploring the potential of Cu-LDHs composite for efficient Coomassie brilliant blue dye removal: Evaluating adsorption isotherms and antibacterial activity. *Res. Chem. Intermed.* **2024**, *50*, 2793–2825. [CrossRef]

91. Sardi, A.; Bounaceur, B.; Mokhtar, A.; Boukoussa, B.; Abbes, M.T.; Chaibi, W.; Nacer, A.; Khadidja, K.B.; Issam, I.; Iqbal, J.; et al. Kinetics and Thermodynamic Studies for Removal of Trypan Blue and Methylene Blue from Water Using Nano Clay Filled Composite of HTAB and PEG and its Antibacterial Activity. *J. Polym. Environ.* **2023**, *31*, 5065–5088. [CrossRef]
92. Karthikeyan, P.; Nikitha, M.; Pandi, K.; Meenakshi, S.; Park, C.M. Effective and selective removal of organic pollutants from aqueous solutions using 1D hydroxyapatite-decorated 2D reduced graphene oxide nanocomposite. *J. Mol. Liq.* **2021**, *331*, 115795. [CrossRef]
93. Nadaroglu, H.; Cicek, S.; Gungor, A.A. Removing Trypan blue dye using nano-Zn modified Luffa sponge. *Spectrochim. Acta Part A Mol. Biomol. Spectrosc.* **2017**, *172*, 2–8. [CrossRef]
94. Priyadarshini, B.; Patra, T.; Sahoo, T.R. An efficient and comparative adsorption of Congo red and Trypan blue dyes on MgO nanoparticles: Kinetics, thermodynamics and isotherm studies. *J. Magnes. Alloys* **2021**, *9*, 478–488. [CrossRef]
95. Sismanoglu, T.; Aroguz, A.Z. Adsorption kinetics of diazo-dye from aqueous solutions by using natural origin low-cost biosorbents. *Desalination Water Treat.* **2015**, *54*, 736–743. [CrossRef]
96. Wang, X.; Liu, Z.; Ye, X.; Hu, K.; Zhong, H.; Yu, J.; Jin, M.; Guo, Z. A facile one-step approach to functionalized graphene oxide-based hydrogels used as effective adsorbents toward anionic dyes. *Appl. Surf. Sci.* **2014**, *308*, 82–90. [CrossRef]
97. Prabhu, S.M.; Khan, A.; Hasmath Farzana, M.; Hwang, G.C.; Lee, W.; Lee, G. Synthesis and characterization of graphene oxide-doped nano-hydroxyapatite and its adsorption performance of toxic diazo dyes from aqueous solution. *J. Mol. Liq.* **2018**, *269*, 746–754. [CrossRef]

inorganics

MDPI

Article

Synthesis of TiO_2/SBA-15 Nanocomposites by Hydrolysis of Organometallic Ti Precursors for Photocatalytic NO Abatement

Ons El Atti [1,2], Julie Hot [2,*], Katia Fajerwerg [1,*], Christian Lorber [1], Bénédicte Lebeau [3,4], Andrey Ryzhikov [3,4], Myrtil Kahn [1], Vincent Collière [1], Yannick Coppel [1], Nicolas Ratel-Ramond [5], Philippe Ménini [6] and Pierre Fau [5,*]

1 Laboratoire de Chimie de Coordination, CNRS, UPR 8241, 205 Route de Narbonne, 31077 Toulouse, CEDEX 4, France; ons.elatti@lcc-toulouse.fr (O.E.A.); christian.lorber@lcc-toulouse.fr (C.L.); myrtil.kahn@lcc-toulouse.fr (M.K.); vincent.colliere@lcc-toulouse.fr (V.C.); yannick.coppel@lcc-toulouse.fr (Y.C.)
2 Laboratoire Matériaux et Durabilité des Constructions, INSA/UPS Génie Civil, 135 Avenue de Rangueil, 31077 Toulouse, CEDEX 4, France
3 Université de Haute-Alsace, CNRS, Institut de Science des Matériaux de Mulhouse (IS2M), UMR 7361, Axe Matériaux à Porosité Contrôlée (MPC), 68093 Mulhouse, CEDEX, France; benedicte.lebeau@uha.fr (B.L.); andrey.ryzhikov@uha.fr (A.R.)
4 Université de Strasbourg, 67000 Strasbourg, France
5 Laboratoire de Physique et Chimie des Nano-Objets, UMR 5215, INSA/CNRS/Université de Toulouse, 135 Avenue de Rangueil, 31077 Toulouse, CEDEX 4, France; nicolas.ratel-ramond@insa-toulouse.fr
6 Laboratoire d'Analyse et d'Architecture des Systèmes, CNRS, UPR 8001, 7, Avenue du Colonel Roche, 31031 Toulouse, CEDEX 4, France; menini@laas.fr
* Correspondence: hot@insa-toulouse.fr (J.H.); katia.fajerwerg@lcc-toulouse.fr (K.F.); pfau@insa-toulouse.fr (P.F.)

Abstract: The development of advanced photocatalysts for air pollution removal is essential to improve indoor air quality. TiO_2/mesoporous silica SBA-15 nanocomposites were synthesized using an organometallic decoration method, which leverages the high reactivity of Ti precursors to be hydrolyzed on the surface water groups of silica supports. Both lab-made Ti(III) amidinate and commercial Ti(IV) amino precursors were utilized to react with water-rich SBA-15, obtained through a hydration process. The hydrated SBA-15 and the TiO_2/SBA-15 nanocomposites were characterized using TGA, FTIR, ^1H and ^{29}Si NMR, TEM, SEM, N_2 physisorption, XRD, and WAXS. This one-step TiO_2 decoration method achieved a loading of up to 51.5 wt.% of approximately 9 nm anatase particles on the SBA-15 surface. This structuring provided excellent accessibility of TiO_2 particles for photocatalytic applications under pollutant gas and UV-A light exposure. The combination with the high specific surface area of SBA-15 resulted in the efficient degradation of 400 ppb of NO pollutant gas. Due to synergistic effects, the best nanocomposite in this study demonstrated a NO abatement performance of 4.0% per used mg of TiO_2, which is 40% more efficient than the reference photocatalytic material TiO_2 P-25.

Keywords: mesoporous silica; titanium(III) amidinate precursor; TiO_2 nanoparticles; photocatalysis; NO degradation; air quality

Citation: El Atti, O.; Hot, J.; Fajerwerg, K.; Lorber, C.; Lebeau, B.; Ryzhikov, A.; Kahn, M.; Collière, V.; Coppel, Y.; Ratel-Ramond, N.; et al. Synthesis of TiO_2/SBA-15 Nanocomposites by Hydrolysis of Organometallic Ti Precursors for Photocatalytic NO Abatement. *Inorganics* **2024**, *12*, 183. https://doi.org/10.3390/inorganics12070183

Academic Editors: Torben R. Jensen and Roberto Nisticò

Received: 23 May 2024
Revised: 26 June 2024
Accepted: 26 June 2024
Published: 29 June 2024

1. Introduction

Indoor air pollution is a major societal concern in our industrial societies. Indeed, the toxic NO_x ($NO + NO_2$) compounds present in ambient air caused by human activities (e.g., road traffic, industrial plants, and fossil energy combustion) [1] are an environmental threat to human health. In particular, the WHO (World Health Organization) recently revised the air quality guidelines of NO_2 to lower thresholds (from 40 $\mu g/m^3$ in 2005 to 10 $\mu g/m^3$ in 2021 for long-term exposure and from 200 to 25 $\mu g/m^3$ for short-term exposure) [2].

Photocatalysis driven by metal oxide semiconductors has been employed for decades to oxidize and mineralize atmospheric pollutants and reduce the overall pollution levels of NO_x, either in outdoor or indoor conditions [3]. This method relies on the ability of a photocatalytic semiconducting oxide (SCO), such as zinc oxide (ZnO) or titanium dioxide (TiO_2), to promote, under ambient air and solar light exposure, electron/hole pairs that conduce to the formation of free radicals involved in pollutant degradation reactions [4,5]. One of the most studied materials applied in this field is TiO_2, particularly the now archetypical Evonik P-25, which contains a mixture of both anatase (around 80 wt.%) and the high-temperature rutile phase [6,7]. Although the presence of the most photocatalytically active anatase phase is a mandatory requirement, some other physico-chemical parameters can be tuned to enhance the photocatalytic properties of this material. Two parameters are of particular importance to achieve improved photocatalyst properties. First, the specific surface of the SCO has to be increased to allow higher adsorption of pollutant species on its surface. This is generally achieved through the use of nanosized metal oxides, where the surface-to-volume ratio is maximized [8,9]. However, downsizing the particle size in excess can be counterproductive for catalysis if the crystalline domains of the SCO become too small [10]. Indeed, the recombination rate of electron/hole pairs in the SCO is largely driven by the density of point defect in the structure of the material, and highly disordered systems, or quasi-amorphous structures, exhibit low electron/hole pair lifetime due to fast recombination effects [11].

The optimized material for photocatalysis should, therefore, be the result of a compromise between low dimensionality, associated with high specific surface levels, and large crystalline domains where a low recombination rate of excitons is achieved. This delicate balance is difficult to reach with the bare TiO_2 compound only, but one of the possible strategies to overcome this problem is to associate a high specific surface material directly in strong interaction with a well-crystallized TiO_2 anatase phase [12]. A close interface between materials involved in photocatalytic nanocomposites is also a key parameter in reducing the charge carriers' recombination due to surface defects [13,14]. Indeed, SBA-15 is a well-known mesoporous silica material (mesopore size from 3 to 6 nm large) exhibiting a very high specific surface (up to 1000 m^2/g) due to its structuration [15]. Thanks to its porous structure, SBA-15 has been used as an efficient NO_x physisorption material in surface photovoltage (SPV) devices for NO_x gas sensing [16]. SBA-15 is also a material of choice that has already been studied as catalyst support for selective catalytic reduction (SCR) of NO_x found in oxygen-rich exhaust gases at high temperatures [17,18]. The combination of porous SBA-15, which has efficient adsorption properties for gaseous or liquid species, with the photocatalytically active TiO_2 semiconductor has been proposed due to the interesting expected synergism within this hybrid material [8,19–23]. Different chemical protocols are suggested for synthesizing this composite according to the interest of having or not the SCO located inside, outside (or both locations) the pores of the silica matrix. In a one-step protocol, colloidal TiO_2 nanocrystals (prepared from Ti(IV) isopropoxide) are mixed together with the molecular precursors used to build the porous silica matrix (tetraethyl orthosilicate and triblock copolymer Pluronic P123) and followed by an hydrothermal annealing at around 373 K [24]. This technique leads to the efficient incorporation of atomic titanium into the silica matrix and very small TiO_2 particles (mean size 7 nm) within the pores of the SBA-15. From the photocatalysis point of view, these low-crystallinity TiO_2 particles may not be suitable for enabling a low recombination rate of photo-excited charge carriers. Another method that is often employed is the SBA-15, which is a hard template that can be decorated in a second step with TiO_2 prepared from the solution impregnation of a Ti precursor. Landau [25,26] has described a method called CSD (chemical solution decomposition) where a Ti(IV) alkoxide precursor [$Ti(OnBu)_4$)] is maintained inside the pores of the SAB-15 matrix in an autoclave under pressure. After thermal decomposition, the TiO_2 loading is up to 30 wt.% for one run and can be doubled if the procedure is repeated. In that case, the TiO_2 particles are exclusively located in the mesopores of the silica matrix. Several other examples in the literature describe the use of Ti(IV) precursor for the decoration of mesoporous silica matrixes [19,21,22,27–29]. One important parameter for mastering the size of the TiO_2 nanoparticles is the use of a high-reactivity precursor in order to facilitate the nucleation and growth of nanocrystals under mild conditions [22,30]. In this study, we have prepared a rather unusual precursor based on a Ti(III) amidinate composition [31] that should offer higher reactivity than the usual Ti(IV) ones. This precursor has been used for the first time to decorate hydrated SBA-15 to yield TiO_x nanoparticles located within and outside the mesoporous silica grains. Our chemical process exploits the high reactivity of this type of organometallic precursor toward hydroxyl groups and water molecules adsorbed on silica surface, to directly generate either a crystalline or amorphous metal oxide phase [32–34]. The same procedure was used with a less reactive Ti(IV) precursor (tetrakis(ethylmethylamido)titanium). The two different hybrid nanocomposites have been characterized and their photocatalytic performances were compared for 400 ppb NO abatement under UV-A exposure in humidified air. The use of Ti(III) precursor with hydrated SBA-15 allowed to prepare a photocatalytic material showing a 40% improvement in NO degradation compared to bare P-25 photocatalyst. This result highlights the importance of synergetic mechanisms (specific surface of the matrix and crystalline structure of TiO_2) for photocatalysis reactions and the key role of the metal precursor to yield nanomaterials with optimized performances.

2. Results and Discussion

2.1. SBA-15 Hydration Methods and Characterization

Our team has previously developed a technique that takes advantage of the presence of molecular water on nanostructured supports on which metalorganic precursors react locally to give rise to the growth of nanosized metal oxide particles [32–34]. It is, therefore, important to master and maximize the amount of water in SBA-15 to allow the reaction with the metalorganic precursor to occur. Two methods have been set up in order to control the hydration of the SBA-15 surface. The first method (method A), described by Y. Belmoujahid [35], consists of placing the silica powder in a closed glass vessel containing water saturated with NaCl salt at room temperature (RT) and allowing a constant relative humidity of 75% RH. The water intake saturation in SBA-15 is obtained after around 4 h.

The second method (method B) is rather straightforward and consists of dispersing the silica powder into distilled water placed in a closed glass vessel and letting it boil under reflux at 100 °C for 2 h [36].

The amount of water (wt.%) in the SBA-15 samples was measured with TGA analysis after they were dried for 1 h under a primary vacuum at RT (Figure 1). The water present in the "as-received" SBA-15 powder was evacuated at around 80 °C and corresponded to 8.5% of the total weight. The sample treated with hydration method A (saturated NaCl solution) contained 23.8 wt.% of water, whereas the sample obtained by method B (boiling water at reflux) contained 55 wt.% of water at the same temperature.

The FTIR analyses of the different silica powders confirm this evolution (Figure S1). The siloxane (–Si–O–Si–) band appears as a broad strong peak centered at 1082 cm^{-1}. The bands at 3426 and 1610 cm^{-1} are attributed to the stretching and bending vibrations of the hydroxyl groups and the adsorbed water molecules, respectively. The structural bands of the mesoporous silica are not modified by the adsorption of water.

There are only a few differences between the FTIR spectra of the SBA-15 as-received and that hydrated by method A. Conversely, the spectrum of the sample treated with method B exhibits a higher intensity of the band at 1610 cm^{-1}. The higher loading of water in the sample treated with hydration method B should be more adapted to the next step, where the hydrolysis of the organometallic precursor may lead to the generation of TiO_x nanoparticles on SBA-15.

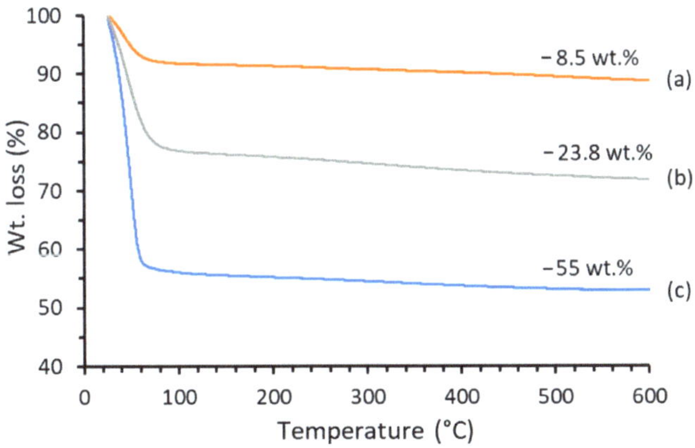

Figure 1. TGA analysis of SBA-15 powders according to hydration methods. (**a**) As-received; (**b**) exposed to air atmosphere with 75% RH for 4 h (method A); and (**c**) soaked in boiling water for 2 h (method B).

Figure 2 shows the ^{1}H NMR MAS (magic angle spinning) spectra of mesoporous SBA-15 silica in the solid state according to the different hydration methods. This figure shows a predominant signal in the region around 4.8 ppm (4.92, 4.99, and 4.71 ppm for the SBA-15 as-received, in method A, and in method B, respectively). This resonance is typical of water molecules in weak interaction with the surface of silica. Part of this signal may also be associated with OH surface groups where their H are rapidly exchanged with those of water. A higher-frequency chemical shift indicates a greater proportion of physisorbed water molecules and/or of hydroxyl OH surface groups, whereas a lower-frequency shift indicates the formation of larger free-water clusters. Their relative populations also have an impact on the linewidth of the resonances because the water in the clusters has greater local mobility than physisorbed molecules and OH groups, leading to sharper resonances for the former. The ^{1}H spectra show that the initial silica already contains a significant amount of water. Hydration method A leads to a slightly higher proportion of physisorbed water and/or OH surface groups. The use of boiling water (method B) leads to the incorporation of a large quantity of free-water clusters and possibly a reduction in the amount of OH surface groups.

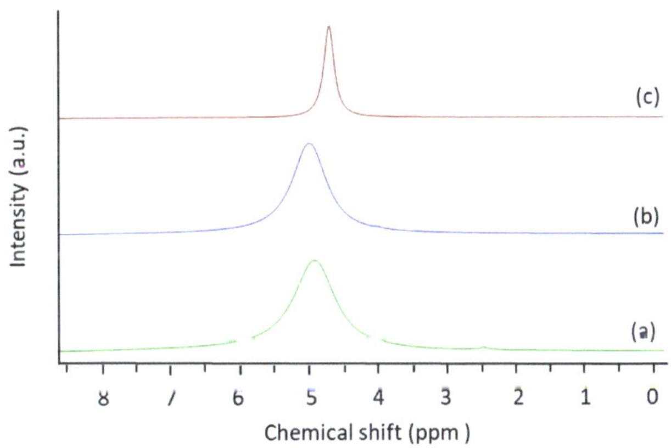

Figure 2. ^{1}H NMR MAS spectra of SBA-15 powders according to hydration methods. (**a**) As-received; (**b**) exposed to air with 75% RH for 4 h (method A); and (**c**) soaked in boiling water for 2 h (method B).

The ^{29}Si MAS NMR spectra (Figure S2) are characterized by three signals typical of Q^4 framework silica sites ($(SiO)_4Si$) (at ~−111 ppm), Q^3 silanol sites ($(SiO)_3$–Si-OH) (at ~−101 ppm) and Q^2 ($(SiO)_2$–(Si-$(OH)_2$)) species (at ~−92 ppm). The relative intensity of Q^3 and Q^2 resonances increases in the ^{29}Si CPMAS NMR (Figure S3) compared to MAS experiments as a result of the greater transfer of magnetization from neighboring H than for Q^4. As with ^{1}H NMR, the ^{29}Si NMR spectra of the as-received and hydrated SBA-15 are very close (Figures S2 and S3 green and blue spectra). On the other hand, the spectra of silica hydrated in boiling water (method B) show some differences (Figures S2 and S3 red spectra), with a higher relative proportion of Q^4. The population decrease of Q^2 for the sample prepared with method B is slightly lower than for Q^3 (Figure S2). Low-frequency displacement is also observed in the sample treated with method B and is greater for Q^4 (1.6 ppm in CPMAS and 1.0 ppm in MAS) than for Q^3 (0.6 ppm in CPMAS) and Q^2 (~0.2 ppm in CPMAS). These results indicate a slight but undoubtful modification of the SBA-15 surface in the presence of boiling water, with a lower proportion of OH on the surface. Therefore, the exposition of silica to boiling water leads to the formation of Si–O–Si bonds by dehydration condensation between the hydroxyl groups themselves.

Small-angle XRD pattern and N_2 sorption isotherms of the SBA-15-based materials according to the different hydration methods are reported in Figures S4 and S5, respectively. The corresponding textural properties are presented in Table 1.

Table 1. Textural properties of the different SBA-15 samples.

SBA-15 Sample	S_{BET} (m^2/g)	V_{micro} (cm^3/g)	V_{meso} (cm^3/g)	D_p (nm)
As-received SBA-15	1041	0.16	0.99	6.2
(SBA-15)/A	721	0.08	0.77	6.2
(SBA-15)/B	470	0.04	0.81	7.2

S_{BET} is the specific surface area obtained using the BET model, V_{micro} is the microporous volume extrapolated from the t-plots, $V_{meso} = V_{tot} - V_{micro}$ is the mesopore volume where V_{tot} is the total pore volume obtained at $P/P_0 = 0.9$, and D_p is the mean pore diameter obtained using the BdB model on the desorption branch.

Small-angle XRD patterns of SBA-15 samples show three diffraction peaks (100), (110), and (200) at the same position, which are characteristic of the 2D hexagonal pore (with P6mm hexagonal symmetry typical for SBA-15 mesoporous materials) with the same unit cell a_{hex} of 10.6 nm. No significant evolution of the structure of the silica was noticed, regardless of the hydration treatments used. The same observation was also confirmed by the absence of any change in the microstructure of the different SBA-15 revealed by SEM observations (Figure S6).

Whatever the hydration method, the SBA-15 isotherms are all of type I and IV according to the IUPAC classification [37]. As a result, the as-received SBA-15 material has a high specific surface area (S_{BET} = 1041 m^2/g) and large pore volume (V = 1.15 cm^3/g, V meso + V micro) of 6.2 nm diameter. The SBA-15 synthesized in these conditions simultaneously exhibits mesopores and connections between the main mesopores walls through micropores [38]. It is noteworthy that the hydration method particularly influences the microporous volume (pore size < 2 nm). Indeed, the latter decreases when the amount of water increases, particularly with hydration method B (Table 1, Figure 1). In this case, V_{micro} has decreased by a factor of four compared to the as-received SBA-15 sample. The other textural parameters (V_{meso} and D_p) were less influenced by the hydration method used. The results strongly suggest that hydration treatments induce a phenomenon of silica dissolution/redeposition condensation within micropores, as reported by Galarneau et al. [39]. For the sample hydrated with method B, D_p increased from 6.2 to 7.2 nm, which was due to a decrease in the wall thickness because the unit cell a_{hex} is the same, indicating the same pore-to-pore distance. This evolution is in good agreement with ^{29}Si NMR, which indicates the condensation of hydroxyl groups upon hydration with method B and the formation of more Si–O–Si bonds.

2.2. Bare TiO_2 Powder Syntheses

The synthesis of TiO_x-yH_2O nanopowders was obtained by the controlled hydrolysis under an argon atmosphere of a metalorganic Ti precursor dissolved in an organic solvent (Figure S7). In this procedure, moisture slowly diffuses from the argon gas phase into the organic solvent containing the metalorganic precursor and induces the breaking of nitrogen to metal bonds to eventually produce metal oxide nanoparticles [40]. For the purpose of this study, we tested two different precursors with distinct titanium oxidation numbers. The tetrakis(ethylmethylamido)titanium (TEMAT; see **1** in Scheme 1) is commonly used in the atomic layer deposition (ALD) technique for TiO_2 films [28,41] and demonstrates a Ti center with an oxidation state of IV. In addition, an even more reactive metalorganic precursor for water has been envisaged. We synthesized and studied the hydrolysis in an organic solution of the (N-N′ diisopropylacetamidinato) Ti (III) precursor (Ti-Amd; see **2** in Scheme 1). The presence of a reduced oxidation number for the Ti center is expected to bring a better reactivity of the compound to water. The hydrolysis process, set up under an argon atmosphere at RT, is achieved through the slow diffusion of water from the gas phase into the organic solvent reacting with the Ti precursor, as described in Scheme 1. According to the reaction, 2 molar equivalents of water are necessary for a stoichiometric reaction. In the experimental conditions, an amount of 4 molar equivalents of water is employed in order to saturate the gas phase with water vapor and ensure the full hydrolysis of the dissolved precursors. In these conditions, the hydrolysis reaction takes place over two days.

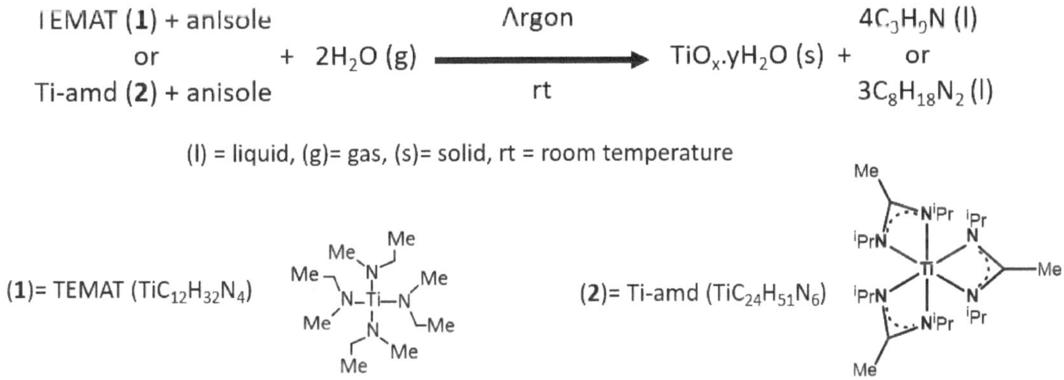

Scheme 1. Hydrolysis reaction of precursor **1** or **2** with a stoichiometric amount of water.

The samples are characterized by wide-angle X-ray scattering (WAXS), a very sensitive technique to determine the local structure of materials with very small crystalline domains (Figure 3). Whatever the precursor composition **1** or **2**, after calcination at 150 °C, the samples have no discernible crystalline domains and correspond to barely amorphous structures (Figure 3a). The pair distribution function (PDF) of the sample obtained from precursor **2** and calcined at 350 °C shows the very first coherence domains corresponding to a crystalline phase (Figure 3b). The structures of TiO_2 anatase (COD 7206075) and TiO_2 brookite (COD 9004140) were refined against the PDF obtained from the WAXS measurement of this sample (Figure 3d,e). Refinement results show that the local structure ($r < 10$ Å) is comparable to that of brookite (Figure 3d), while the average structure ($r > 10$ Å) is close to anatase (Figure 3e). The size of the coherent crystal domains calculated for spherical grains corresponds to a diameter of ca. 3.6 nm. Reversely, the oxide sample obtained from **1** still remains amorphous at this temperature.

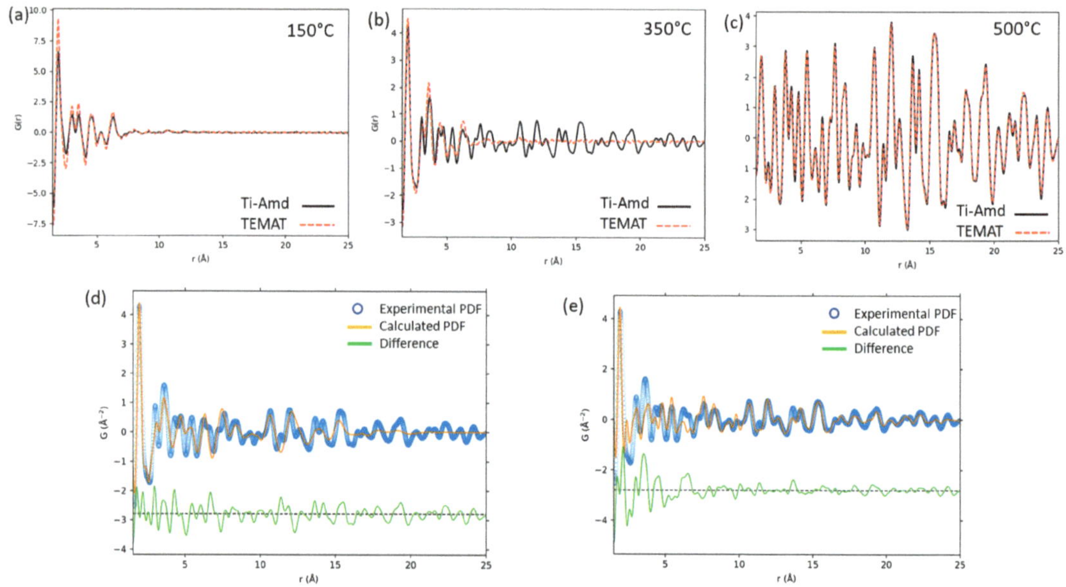

Figure 3. Comparison of PDF obtained from WAXS measurement of TiOx powders from TEMAT (**1**, red dotted line) and Ti-Amd (**2**, plain black line) precursors according to the calcination temperature at: (**a**) 150 °C, (**b**) 350 °C, and (**c**) 500 °C. Refinement of the pair distribution function obtained on TiO_2-amd calcinated at 350 °C with (**d**) brookite and (**e**) anatase structures.

This result clearly shows the higher reactivity of **2** toward hydrolysis, giving rise to an easier growth of crystalline domains at the moderate temperature of 350 °C.

As the calcination temperature increases (500 °C), both samples prepared with TiO_2/Ti-amd or TiO_2/TEMAT unambiguously show the anatase structure of TiO_2, as shown in Figures 3c and S8. According to PDF analysis, TiO_2/TEMAT exhibits coherent domains of 17.5 nm (±2.3 nm), whereas TiO_2/Ti-Amd has a slightly higher crystallite size of 19.8 nm (±2.8 nm), which is consistent with the earlier presence of the very first crystalline domains obtained at 350 °C with this precursor. The presence of the anatase phase after calcination at 500 °C is also confirmed by X-ray diffraction analysis of the different powders (Figure S9). According to the Scherrer equation applied on the (101) peak measured at 2 theta 25.5°, the crystallite size is around 18.4 nm and 21.0 nm for **1** and **2**, respectively.

Figure S10a,c shows the TEM images of the TiO_x-yH_2O powders obtained after hydrolysis of precursors **1** and **2** at RT. The powders have a rather ill-defined microstructure that may be formed by the agglomeration of amorphous particles. Calcination of the as-prepared samples has been performed to remove the water molecules and allow the crystallization of the TiO_2 phase. After calcination at 350 °C, there is no change in the structure of TiO_x grains obtained from **1** (Figure S10b), whereas a nano-structuration (ca. 4 nm grain size, determined on around 100 particles) already appears for the sample prepared with **2** (Figure S10d). After calcination at 500 °C under air, the structuration of the powder is clearly nanocrystalline for the two samples where similar nanosized grain size of ca. 13.5 nm (σ = 3.6 nm) and 13.1 nm (σ = 3.9 nm) are measured for TiO_2 obtained from **1** and **2**, respectively (Figure 4).

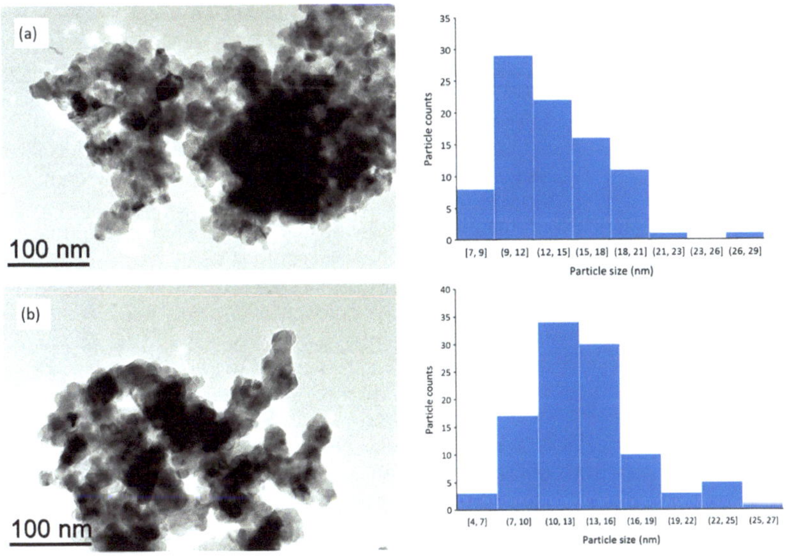

Figure 4. TEM images (left) and corresponding histograms of particle size distribution (right) for TiO_2 powders (determined by ImageJ software (v1.54j) on around 100 particles) after calcination at 500 °C from the hydrolysis of (**a**) TEMAT (**1**) and (**b**) Ti-Amd (**2**).

2.3. *Synthesis of TiO_2/SBA-15 Nanocomposites*

The decoration of SBA-15 with TiO_2 is performed only on the silica samples prepared with the hydration method B because it leads to a higher amount of water in the silica structure. Indeed, our strategy is to maximize the loading of TiO_2 on SBA-15 by using the hydrated sample that contains a higher amount of adsorbed water.

The mass of the hydrated SBA-15 in our experimental protocol is 200 mg and, since it contains 55 wt.% water (Figure 1), this corresponds to 110 mg of water (i.e., 6 mmol) and 90 mg of SiO_2 per sample. We use an amount of 1.3 mmol of Ti precursor **1** or **2** to react with the hydrated SBA-15, which corresponds to 4.6 molar equivalent of water relative to titanium. Therefore there is a large excess of water compared to the employed Ti precursor and similar to the ratio used for the synthesis of bare TiO_x powders. This excess of water (a stoichiometric amount of 2 molar equivalents according to Scheme 1) is necessary to ensure the complete hydrolysis of the titanium precursor on the silica surface. Assuming the hydrolysis reaction is complete, 1.3 mmol of TiO_2 (62 mg of Ti, 103 mg of TiO_2) will form on the silica, which corresponds to a theoretical mass ratio of Ti equal to 32% (or a mass ratio of 51.5% for TiO_2).

The ICP-MS analyses of the two nanocomposite samples prepared with precursors **1** and **2** are presented in Table 2.

Table 2. ICP-MS analyses of Ti content present inside the TiO_2/SBA-15 nanocomposites obtained from **1** and **2** and calcined at 500 °C.

Precursor	Theoretical Ti Content (wt.%)	ICP Measured Ti Content (wt.%)
TEMAT (**1**)	32	30.09 30.27 Mean: 30.18
Ti-Amd (**2**)	32	31.94 31.17 Mean: 31.55

Interestingly, the experimental amounts of Ti in TiO_2/SBA-15 samples obtained from precursors **1** and **2** (30.2 and 31.5 wt.%, respectively) are very close to the theoretical value previously calculated. This result indicates that, whatever the composition of the Ti precursor, the hydrolysis process is very efficient and leads to the total transformation of the precursor on the silica matrix surface.

Low magnification (×500) SEM images of the nanocomposites calcined at 500 °C are shown in Figure S11. They reveal that the TiO_2 aggregates are exclusively linked to the silica blocks and are distributed on the silica matrix surface.

Figure 5 shows the high magnification SEM image (×5000) of the nanocomposites acquired in the chemical contrast mode (back-scattered electrons). The images confirm that the TiO_2 nanoparticles form aggregated structures over the silica matrix. However, although it appears homogeneous at low magnification (Figure S11), the structuration at high magnification reveals a rather erratic dispersion of TiO_2 aggregated nanoparticles on the porous silica. Some areas of SBA-15 bundles appear to be devoid of TiO_2 particles, while in other locations, TiO_2 nanoparticles are clustered together. This TiO_2 dispersion over the silica matrix is the same for the two Ti precursors studied. In addition, no free TiO_2 aggregate is located apart from the SBA-15 matrix. A SEM-EDS analysis (Figure 6) of the different components of the nanocomposites confirms this particular repartition of TiO_2 nanoparticles over the SBA-15 support.

Figure 5. SEM images of TiO_2/SBA-15 calcined at 500 °C and prepared with (**a**) TEMAT (**1**) and (**b**) Ti-Amd (**2**). The images are in chemical contrast mode (back-scattered electrons). The brighter spots on the image correspond to an element with a higher atomic number and indicate the presence of Ti. Magnification is ×5000.

TiO_2 nanoparticles are concentrated in specific zones of the silica matrix. There are locations where the silica matrix is not decorated with TiO_2 whereas the photocatalytic oxide is concentrated into agglomerated structures. This peculiar repartition is very different from what we have previously observed on the decoration of WO_3-$2H_2O$ nanoplatelets with ZnO nanoparticles by using a similar hydrolysis method. In the former case, ZnO appears as a very homogeneous nanostructured layer distributed all over the surface of the WO_3-$2H_2O$ support [33]. This result was achieved thanks to the availability of one of the structural water molecules present in the support to hydrolyze the Zn precursor. In the case of the SBA-15 support, it seems that the water added by the boiling process is not homogeneously distributed over the silica. Instead, water-rich areas, such as clusters or droplets of water, could be randomly distributed over the SBA-15 surface. This assumption is also supported by the ^1H MAS NMR analyses that reveal the presence of water clusters in the sample hydrated by method B (boiling water). The Ti precursor in contact with these water reservoirs reacts to locally produce a high density of TiO_x-yH_2O structures. The interesting point is that these water reservoirs remain tightly linked to the substrate because all the TiO_x structures are attached to the silica support in a similar way. This mechanism is supported by TEM imaging of the nanocomposites (Figure 7). In these images, the TiO_2 nanoparticles appear irregularly dispersed on the silica surface and do not form a continuous layer. Similar TiO_2 grapes are observed for the nanocomposites prepared with precursor **1** or **2**. This analysis confirms that despite the aggregated structure of TiO_2 nanoparticles, they are mainly present on the surface of SBA-15 grains and not dispersed homogeneously into the pores (Figure 7c,d).

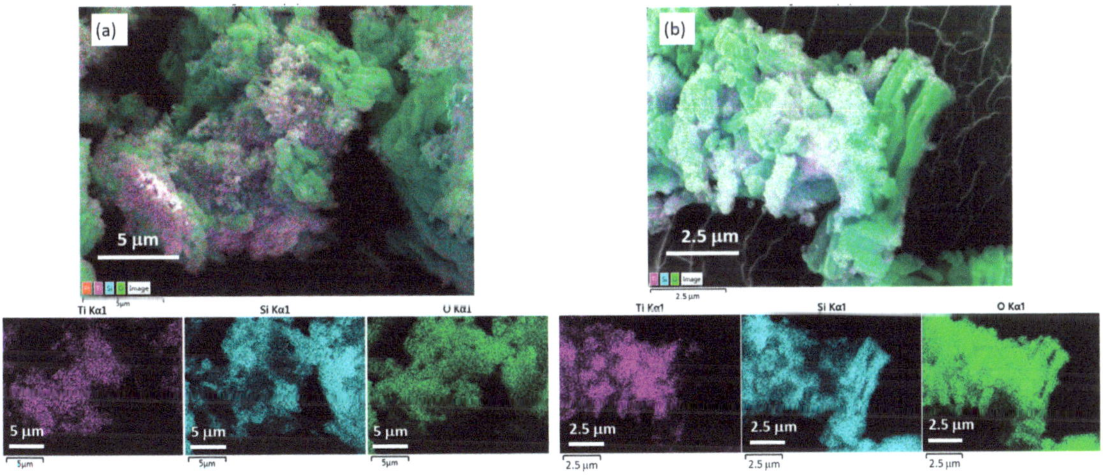

Figure 6. SEM images of TiO_2/SBA-15 prepared with (**a**) TEMAT (**1**) (top) and images in artificial color corresponding to the EDS chemical mapping of elements Ti, Si, and O from the sample (bottom); and (**b**) Ti-Amd (**2**) (top) and EDS chemical mapping for Ti, Si, and O (bottom). Ti atoms are shown in pink, Si atoms are shown in blue, and O atoms are shown in green. A flash platinum layer is deposited on the sample in order to enhance its electrical conductivity and improve the image quality (Pt, red dots displayed uniformly on the image). Magnification is ×5000 for (**a**) and ×10,000 for (**b**).

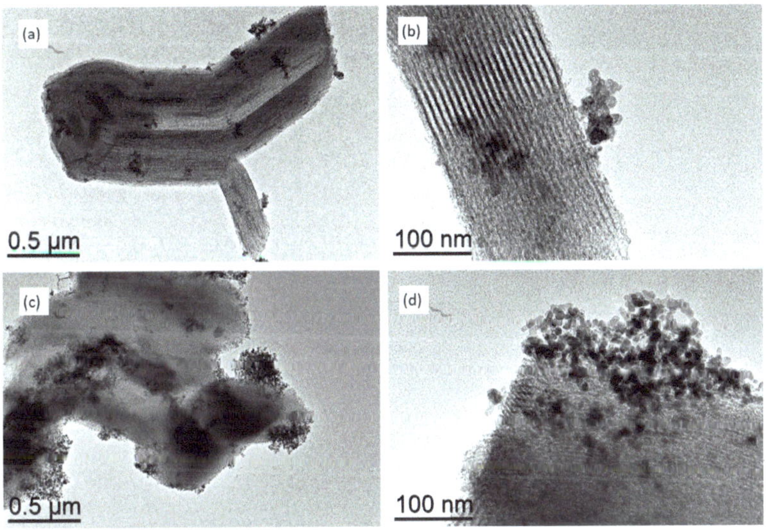

Figure 7. TEM images of TiO_2/SBA-15 nanocomposites obtained from hydrolysis of TEMAT (**1**) (**a**) magnification ×10,000, (**b**) magnification ×100,000, and from hydrolysis of Ti-Amd (**2**) (**c**) magnification ×10,000, (**d**) magnification ×100,000.

The mean grain size of TiO_2 particles formed on SBA-15 supports is close to 9 nm whatever the employed Ti precursor (9.8 and 9.1 nm for precursor **1** and **2**, respectively). This value is clearly lower than that of bare TiO_2 nanoparticles obtained by hydrolysis in solution, which is close to 13 nm (13.5 and 13.1 nm for precursor **1** and **2**, respectively). This result underlines the important role played by the support in the hydrolysis and growth mechanism of the oxide particles. The growth of TiO_2 crystals occurs mainly at the SBA-15 surface, particularly on the micropores where water clusters are present and not inside the mesopores. The interaction between the silica support and TiO_2 induces the particles' stabilization and their limited size. This low dimensionality of the anatase phase is also evidenced by the XRD diffractograms of the nanocomposites, which show all the peaks relating to the crystalline anatase structure but with rather large half-height peaks that reflect the limited size of the crystalline domains (Figure S12 vs. Figure S9).

The main physico-chemical characteristics of the SBA-15 as-received, TiO_2 oxides, and TiO_2/SBA-15 nanocomposites are summarized in Table 3.

Table 3. Main physico-chemical characteristics of the samples tested for photocatalysis.

Sample	TiO_2 Content (wt.%)	S_{BET} (m^2/g)	TiO_2 Crystal Size (nm)	
			TEM	XRD
TiO_2 P-25	100	50	21.0	31.0
$TiO_{2\ TEMAT}$	100	51	13.5	18.4
$TiO_{2\ Ti\text{-}Amd}$	100	21	13.1	21
As-received SBA-15	-	1041	-	-
SBA-15 $_{hydrated\ B}$	-	470	-	-
$TiO_{2\ TEMAT}$/SBA-15	50.4	297	9.8	11.7
$TiO_{2\ Ti\text{-}Amd}$/SBA-15	52.6	251	9.3	11.9

Based on the previous findings, the decoration of hydrated SB1-15 with TiO_x NPs, prepared from the hydrolysis of Ti-Amd precursor, is schematically depicted in Figure 8.

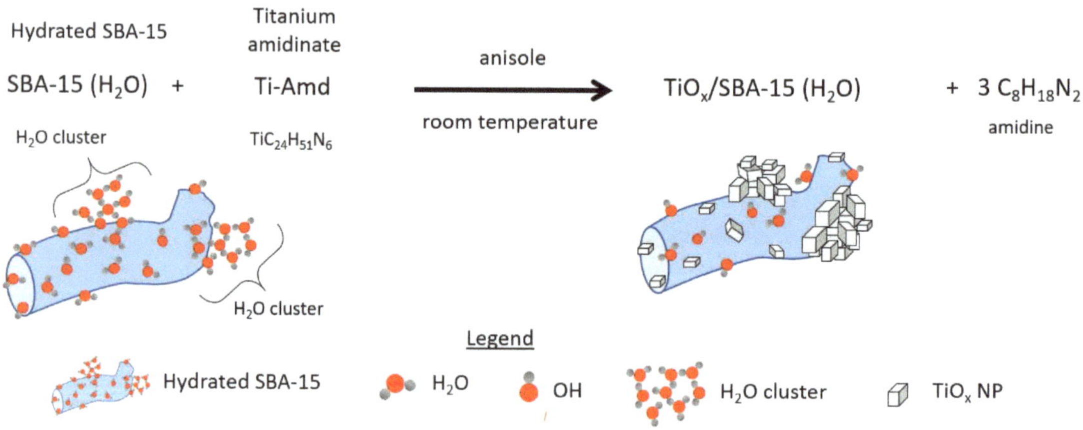

Figure 8. Schematic description of the reaction hydrolysis of Ti-Amd precursor on hydrated SBA-15.

2.4. NO Abatement Tests

NO degradation percentages were calculated from Equation (1) (see Section 3.6 Evaluation of the Photocatalytic Activity) for 20 mg of photocatalytic dispersion applied to the surface of a glass substrate (length × width: 10 × 5 cm^2). The photocatalytic runs were performed on three main types of samples: nanocomposites (TiO_2/SBA-15 obtained from **1** and **2**), bare TiO_2 powders (P-25, TiO_2 powder calcined at 500 °C and obtained from **1** and **2**) and physical mixtures of SBA-15 and TiO_2 powders to assess the role of the silica matrix. However, the TiO_2/SBA-15 nanocomposites and the SBA-15 mixed with TiO_2 powders are not 100% TiO_2 by weight, unlike the bare oxides (TiO_2 synthesized from **1** and **2**, and TiO_2 P-25). The effective quantity of TiO_2 deposited on the glass surface was therefore considered according to ICP-MS quantification (Table 2) in order to obtain the NO degradation results for the same surface density of TiO_2, i.e., 1 mg of TiO_2 per 50 cm^2 of the substrate surface (0.02 mg/cm^2). The NO degradation results obtained under artificial UV-A light for bare TiO_2, TiO_2/SBA-15 nanocomposites, and SBA-15 physically mixed with bare TiO_2 are shown in Figure 9. The photocatalytic activity of SBA-15 alone was also assessed using the same protocol and found to be equal to zero.

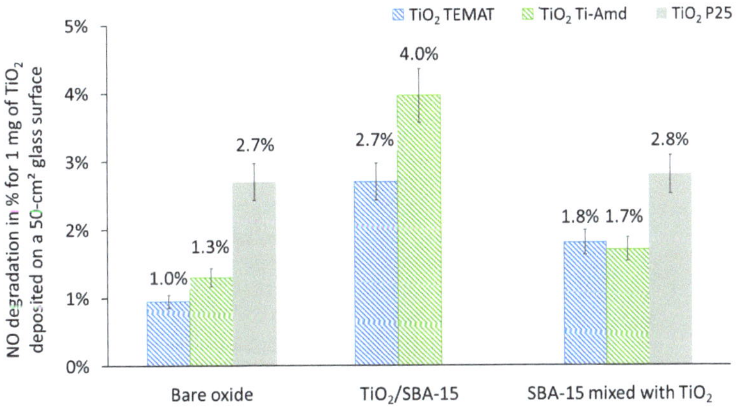

Figure 9. Degradation of NO under UV-A artificial light for the same surface density of TiO_2 (0.02 mg of TiO_2 per cm^2 of glass surface) obtained with bare TiO_2 (TiO_2 synthesized from **1** (TEMAT) and **2** (Ti-Amd) calcined at 500 °C, and TiO_2 P-25), TiO_2/SBA-15 nanocomposites (calcined at 500 °C) and SBA-15 physically mixed with bare TiO_2 oxides (calcined at 500 °C and obtained from **1** and **2**).

For bare oxides, the highest percentage of NO degradation was obtained with TiO_2 P-25 (2.7%). The efficiency of bare oxides obtained from precursors **1** (1%) and **2** (1.3%) was around twice as low. The photocatalytic efficiency of TiO_2 depends on several factors, including crystal structure, crystalline phase, crystallinity, particle size, and surface area, and the balance between these factors determines the overall activity of a given photocatalyst [42]. The air depollution performance of TiO_2 P-25 has been widely reported in the literature and is mainly attributed to the synergistic effect between the anatase and rutile crystalline phases, which reduces photogenerated charge recombination [43,44]. The physico-chemical characterization presented above showed that crystalline anatase was the only phase detected for TiO_2 nanoparticles obtained from **1** and **2** (after calcination at 500 °C, Figure S9), which could explain the lowest activity compared to TiO_2 P-25. Moreover, TiO_2 from **1** had a mean particle size of 13.5 nm and a specific surface area of 51 m^2/g compared with values of 13.1 nm and 21 m^2/g, respectively, for TiO_2 nanoparticles from **2** (Table 3). The two TiO_2 anatase powders prepared in this study demonstrated the same performance for NO abatement (around 1%), which is expected for metal oxides that have similar structural properties.

For the same TiO_2 surface density (1 mg of TiO_2 per 50 cm^2 of substrate surface), the TiO_2/SBA-15 nanocomposites showed similar and even superior performance than TiO_2 P-25. Note that photocatalytic activity is highly dependent on experimental conditions [45], making it difficult to compare the performance of the nanocomposites tested with other materials in the literature. For this reason, we compared the NO degradation percentages obtained with the TiO_2/SBA-15 nanocomposites with those of the reference photocatalyst TiO_2 P-25 tested under the same experimental conditions. The best NO degradation (4.0%) was obtained with $TiO_{2\ Ti\text{-}Amd}$/SBA-15 (around 1.5 times higher than that of TiO_2

P-25), while TiO_2 TEMAT/SBA-15 had the same NO degradation percentage than TiO_2 P-25. These results clearly highlighted the following: (1) the advantage of decorating mesoporous SBA-15 silica with TiO_2 nanoparticles compared to the corresponding bare oxides as the performance is increasing by a factor of 3 (1% vs. 2.7% and 1.3% vs. 4.0%), and (2) the higher efficiency of TiO_2 Ti-Amd/SBA-15 compared to bare TiO_2 P-25. The higher photoactivity of TiO_2/SBA-15 nanocomposites compared to bare TiO_2 may be related to three synergistic factors that made NO pollutant molecules more accessible to the active sites and thus improve the photocatalytic reaction: (1) smaller TiO_2 particle size (~12 nm for nanocomposites and ~20 nm for bare oxides, from XRD analysis; Table 3), (2) larger surface area (around 250–300 m^2/g for nanocomposites and around 20–50 m^2/g for bare oxide, Table 3), and (3) well dispersed TiO_2 nanoparticles over the surface of the silica matrix. The positive effect of these factors and the higher photocatalytic activity of mesoporous silica–titania compared to bare TiO_2 were notably highlighted by other authors [46–48].

The last series of results allows us to compare the photocatalytic activity of the physical mixing of TiO_2 powders and SBA-15 to the chemically synthesized nanocomposites (with identical TiO_2/SBA-15 mass ratio). Interestingly, the physical mixing also showed an improvement in NO degradation compared to the corresponding bare TiO_2 powders. However, degradation percentages were lower than those of TiO_2/SBA-15 nanocomposites. A higher UV absorption intensity in the mixed oxides due to the presence of SBA-15 could be responsible for this increase. A stronger UV absorption intensity implies that more TiO_2 particles can be activated by light and, therefore, more photogenerated charges can be promoted, leading to enhanced photocatalytic activity. The role of the received light irradiation on photocatalytic activity was highlighted by several authors in the literature. Notably, Li and Kim [47] partially attributed the higher photo-oxidation of benzene of TiO_2-xSiO_2 composites compared to bare titania to greater UV absorption intensity. Alonso-Tellez et al. [49] demonstrated that the higher irradiance received by the UV100 TiO_2 coating and the deeper penetration of light within this coating than in the case of the TiO_2 P-25 coating led to increased conversion and mineralization of the methylethylketone pollutant. This effect is even more pronounced in the case of the chemically prepared nanocomposites of this study, especially for the one prepared with the precursor Ti-Amd.

This suggests a more efficient UV-A sensitization of TiO_2 particles when they are well dispersed and closely attached to the hydrated SBA-15 support, thanks to the use of a highly reactive Ti precursor.

3. Materials and Methods

3.1. Synthesis of SBA-15 Powder

The ordered mesoporous silica SBA-15 was prepared at IS2M (Mulhouse, France) according to the protocol described by Zhao et al. [15]. In a 250 mL polypropylene flask, 4 g of triblock copolymer P123 (Aldrich, Saint Louis, MO, USA) were dissolved in HCl aqueous solution (19.5 mL of 37 wt.% HCl and 127 mL of distilled H_2O). The flask was placed in a water bath at 40 °C with magnetic stirring (500 rpm) for about 3 h to allow a complete dissolution of P123. Then, 8.62 g of Tetraethylorthosilicate TEOS (Aldrich, Saint Louis, MO, USA) were added by maintaining the stirring and the temperature conditions. The molar composition of the gel was 1 TEOS: 0.017 P123: 5.68 HCl: 197 H_2O. The solution was stirred at 40 °C for 24 h and then the bottle was placed without any stirring in an oven for 24 h at 90 °C. The precipitated solid (as-made SBA-15) was recovered by filtration on a Büchner funnel, washed with 200 mL of distilled water, and dried for 48 h at 70 °C. To eliminate the porogen agent (P123) and thus release the porosity of the material, the as-made SBA-15 was calcined under air in a muffle furnace at 300 or 500 °C for 4 h (the heating time from ambient to final temperature was 6 h). The mesoporous silica was hydrated after synthesis in two different ways. The first method (method A) consisted of hydrating the mesoporous silica in a controlled relative humidity atmosphere for 4 h. A 75% relative humidity is created in a closed glass enclosure using an aqueous saturated NaCl (Aldrich) in Milli-Q water solution. The second method (method B) consisted in boiling the mesoporous silica directly in Milli-Q water (purification system) for 2 h at 100 °C. Typically, around 500 mg of SBA-15 mesoporous silica was brought to reflux in 100 mL of Milli-Q water [36]. Afterward, the water was removed using centrifugation, and the silica paste was dried for 1 h under a vacuum at RT.

3.2. Synthesis of Titanium Tris-Amidinate Precursor

The titanium amidinate (Ti-Amd) precursor was synthesized according to an adaptation of the protocol described by Gordon et al. [31]. All reactions and manipulations were conducted under a nitrogen atmosphere using either a glovebox or a standard Schlenk line technique. The used solvents were dried and collected using a solvent purification system. For the synthesis of the lithium amidinate, a solution of methyllithium (1.6 M in Et_2O, 40 mL, 0.063 mol) in Et_2O was added dropwise using a bromine bulb to a solution of 1,3-diisopropylcarbodiimide (6.9 g, 0.055 mol) in 80 mL of Et_2O at −30 °C. The mixture was warmed to RT and stirred for 4 h. The solvent was then removed under reduced pressure. The resulting white solid was transferred to the glovebox and extracted with pentane (washing three times with 30 mL), then dried under reduced pressure. A solution of the obtained solid in ether was added dropwise to a solution of titanium chloride tetrahydrofuran $TiCl_3.(THF)_3$ (7.8 g, 0.054 mol) in 50 mL of Et_2O. The reaction mixture was stirred for 12 h in the glovebox. The resulting solution was filtered through a pad of Celite® on a glass frit to give a brown solution. The concentration of the filtrate and its cooling to −33 °C in the glovebox afforded a brown solid.

3.3. Synthesis of Bare TiO_2 Powder

TEMAT (tetrakis(ethylmethylamido)titanium IV) was received from Sigma-Aldrich supplier. Ti-Amd (N-N′ diisopropylacetamidinato)titanium III) precursor was prepared in the laboratory according to the protocol described in Section 3.2. The hydrolysis of the organometallic precursors was carried out in a proprietary designed glass reactor (Figure S7), comprising two separated and concentric reservoirs: one containing the dissolved precursor in anisole, and the second placed in the center of the vessel in order to receive the water (4 molar equivalent compared to titanium precursor) introduced from the septum placed above and in the center of the airtight vessel lid. This reactor allowed the slow hydrolysis of the titanium precursor in organic solvent by diffusion of the water from the gas phase to the solvent one, the reaction occurring at the gas-solvent interface. A white TiO_x-yH_2O powder was collected after 2 days, and the powder was purified by washing it in toluene and centrifuged before drying it under a vacuum at RT. The samples were then calcined at various temperatures from 150 to 500 °C for 2 h in air.

3.4. Synthesis of TiO_2/SBA-15 Nanocomposites

The nanocomposites of TiO_2/SBA-15 were synthesized as follows: a volume of 400 µL of Tetrakis(ethylmethylamido)titanium (TEMAT) (99%, STREM, Newburyport, MA, USA), or 613 mg of Ti amidinate precursor, was added to a suspension of 200 mg of hydrated SBA-15 within 12 mL of anisole. Hence, the amount of water introduced in the reaction medium through the hydrated silica corresponds to around 4.6 molar equivalent (i.e., very similar to the amount used to prepare bareTiO_2 by water moisture presented in Section 3.3). The mixture was stirred for 24 h at RT. At the end of the reaction, the resulting solution was centrifuged (15 min, 3000 rpm) and then transferred to the glovebox. The precipitate was repeatedly washed three times with anisole and centrifuged. At the end of the washing procedure, the TiO_x/SBA-15 was dried under a vacuum for 2 h in order to fully remove the anisole solvent. Thermal annealing was performed under air at various temperatures from 150 to 500 °C to achieve fully oxidized and crystallized TiO_2/SBA-15 nanocomposites.

3.5. Characterization Techniques

Thermogravimetric analysis (TGA) was performed using a Setaram thermobalance (Setaram Engineering, Caluire-et-Cuire, France) with a ramp of 10 °C/min in the 30–600 °C range under ambient air.

NMR experiments were recorded on a Bruker Avance 400 III HD spectrometer (Bruker Corp., Billerica, MA, USA) operating at magnetic fields of 9.4 T. Samples were packed into 4 mm zirconia rotors and were rotated at a frequency of 10 kHz at 295 K. ^{1}H MAS were performed with the DEPTH pulse sequence and a recycle delay of 5 s. ^{29}Si MAS were acquired with the UDEFT pulse sequence [50] with a recycle delay of 80 s. Ramped cross-polarization (CP) ^{1}H $\rightarrow$ ^{29}Si MAS spectra were recorded with a recycle delay of 2 s and a contact time of 2 ms. Chemical shifts were referenced to TMS. Fourier transform infra-red spectroscopy (FTIR) spectra were obtained by analyses of SBA-15 powders mixed in a KBr pellet using a Perkin–Elmer 100 spectrometer (Perkin–Elmer, Waltham, MA, USA). The spectra were registered between 4000 and 350 cm^{-1} wavenumbers.

Total scattering (WAXS) measurements were performed on a Malvern Panalytical Empyrean III diffractometer (Malvern, Worcestershire, UK) in transmission geometry, and equipped with a Mo anode (λ = 0.71 Å) and a Galipix3D detector. The maximum reached q value is q_{max} = 16.6 Å^{-1}, thus corresponding to a maximum angle of $2\theta_{max}$ = 143°. Powder specimens were placed in 1 mm diameter capillaries. For each specimen measurement, an empty capillary contribution was subtracted. Pair distribution functions (PDF) were extracted using PDFgetX3 software v2.1.2. Least square refinements of experimental PDF were performed using WAXS_toolbox software v1.0 suite [51], based on the diffpy-cmi package [52]. Crystalline structural models of TiO_2 phases were taken from the Crystal Open Database [53]. Powder X-ray diffraction patterns were obtained on a Seifert XRD 3000 TT X-ray diffractometer (Seifert X-ray, Deutschland, Germany), with Cu-Kα radiation, fitted with a diffracted-beam graphite monochromator. Data were collected in the 2θ configuration between an angle of 10 and 70°.

Physisorption measurements were made with the Micromeritics ASAP 2020 analyzer (Micromeritics, Norcross, GA, USA). The samples were degassed at 200 °C under high vacuum for 12 h before analyses. A quantity of 200 mg of powder was used to carry out the measurement. The specific surface area of the samples was determined using the linear part of the Brunauer, Emmet, and Teller (BET) plot, and the pore size was estimated using the desorption branch of the isotherms and the Broekhoff–De Boer (BdB) model. The microporous volume was extrapolated from the linear part of the t-plots.

Transmission electron microscopy (TEM) images were obtained with a Jeol 1400 transmission electron microscope at 120 kV (Jeol Ltd., Akashima, Tokyo, Japan). Field emission scanning electron microscopy (FESEM) images were obtained using a Jeol JSM-6700F microscope operating at 10 kV.

3.6. Evaluation of the Photocatalytic Activity

TiO_2-decorated mesoporous silica powders hydrated using method B with each precursor (**1** for TEMAT and **2** for Ti-Amd at 500 °C) were dispersed in water (10 mg/mL) with sonication for 30 min. Each TiO_2/SBA-15 nanocomposite dispersion was then applied to the surface of the glass substrate (10 cm × 5 cm = 50 cm^2) using a glass pipette. The quantity of the dispersion deposited on the surface was 20 mg for each sample. The same dispersion preparation and deposit protocols were followed for bare oxides (TiO_2 obtained from **1** and **2** at 500 °C as described in Section 3.3, and TiO_2 P-25). Dispersions of SBA-15 mixed (physical mixing) with bare oxides (TiO_2 obtained from **1** and **2**, and TiO_2 P-25) were also prepared and applied to the surface of the glass substrate. The weight percentage of TiO_2 added to these mixtures was chosen to be the same as the one found in TiO_2/SBA-15 nanocomposites (ICP-MS results, Table 2).

Photocatalytic degradation of NO was evaluated using the experimental set-up described by Hot et al. [54] and Castelló Lux et al. [4], to which the reader is kindly referred for a more detailed description of the methodology. It is adapted from standard ISO 22197-

1 [55]. The main points of the protocol to keep in mind as as follows: (1) continuous polluted air (at a flow rate of 1.5 L/min) was injected through the by-pass to check that the desired concentration was reached (400 ppb NO), (2) the polluted air was injected into a borosilicate reactor containing the photocatalytic sample to be tested in the dark for 10 min, (3) the photocatalysis was activated by switching on a UV-A light (Narva LT-T8 Blacklight blue 18 W 073 fluorescent tube) placed above the surface of the sample for 20 min (1 W/m^2 on the sample surface measured with a Gigahertz–Optik radiometer (UV-A detector: UV-3717 model, 315–400 nm; Gigahertz–Optik GmbH, Türkenfeld, Germany), and (4) before the end of the test, the polluted air was returned through the by-pass to check the concentration. Relative humidity and temperature were kept constant at 50% and 20 °C, respectively. The reactor was placed in a light-tight box to isolate the sample from room light and to receive only artificial illumination when the light was switched on. The NO concentration was measured by a chemiluminescence analyzer (model AC32M, Envea SA, Poissy, France).

The photo-oxidation of NO was assessed with Equation (1):

$$\text{NO degradation } (\%) = 100 \times \frac{[\text{NO}]_{\text{initial}} - [\text{NO}]_{\text{final}}}{[\text{NO}]_{\text{initial}}} \quad (1)$$

where $[NO]_{initial}$ was the concentration in ppb measured by the analyzer at the exit of the reactor before light activation once the steady state was established, and $[NO]_{final}$ was the concentration in ppb measured at the exit of the reactor after the light activation and averaged over the last 10 min.

Photocatalytic tests were conducted on glass samples coated with TiO_2 P-25 oxide, TiO_2/SBA-15 nanocomposite, and SBA-15 mixed with TiO_2 P-25 oxide. Each test was repeated twice.

4. Conclusions

In this paper, we have investigated a novel metalorganic method for the preparation of bare TiO_2 and TiO_2/SBA-15 nanocomposite powders aimed at the photocatalytic abatement of NO pollutant gas (400 ppb). The photocatalytic activity of the different materials prepared in this study is compared, at a laboratory reactor scale, with that of commercial TiO_2 P-25 powder. The mechanical combination of TiO_2 powders with mesoporous SBA-15 silica results in improved photocatalytic abatement of NO compared to bare TiO_2 powders. This improvement is likely due to a more efficient UV diffusion pathway toward the TiO_2 grains. However, chemically prepared TiO_2/SBA-15 nanocomposites demonstrate even higher photocatalytic activity compared with the simple physical mixing of TiO_2 powders with SBA-15 and bare TiO_2 oxides. Specifically, the nanocomposite prepared from Ti-Amd precursor, which is more reactive than TEMAT under hydrolysis conditions, exhibits the best activity for NO abatement with a value of 4.0% (NO abatement/mg of TiO_2 deposited on the glass substrate surface), which represents a 40% increase compared to TiO_2 P-25. Thus, maximizing the interaction and relationship between porous silica and TiO_2 nanoparticles is helpful to achieve a higher synergistic effect for the photocatalytic degradation of NO pollutants. Future work will focus on incorporating Au nanoparticles onto chemically prepared TiO_2/SBA-15 nanocomposites to further enhance the catalytic activity thanks to the plasmonic properties of the noble metal nanoparticles in the visible domain of the spectrum.

Supplementary Materials: The following supporting information can be downloaded at: https://www.mdpi.com/article/10.3390/inorganics12070183/s1, Figure S1: FTIR spectra for SBA-15 according to the hydration pre-treatments; Figure S2: ^{29}Si MAS NMR spectra of SBA-15 powders according to the hydration pre-treatments; Figure S3: ^{29}Si CP-MAS NMR spectra of SBA-15 powders according to the hydration pre-treatments; Figure S4: Small-angle XRD diagrams of SBA-15 powders according to the hydration pre-treatments; Figure S5: N_2 adsorption isotherms of SBA-15 powders according to the hydration pre-treatments; Figure S6: SEM images of SBA-15 powders according to the hydration pre-treatments; Figure S7: Scheme of the glass vessel used for the controlled hydrolysis of metalorganic precursor in organic solvent under argon; Figure S8: Refinement of anatase structures against PDF obtained on (a) TiO_2-amd and (b) TiO_2-TEMAT after calcination at 500 °C; Figure S9: X-ray diffraction diagrams of TiO_2 powders obtained after annealing at 500 °C under air of precursor (a) Ti-Amd and (b) TEMAT; Figure S10: TEM images of TiO_x powders obtained from hydrolysis of TEMAT (a) as-prepared, (b) calcined at 350 °C under air, and from hydrolysis of Ti-Amd (c) as-prepared, (d) calcined at 350 °C under air. Magnification is ×50,000 for (a,c), and ×100,000 for (b,d); Figure S11: SEM image of TiO_2/SBA15 powders calcined at 500 °C showing the homogeneous repartition of the TiO_2 over the silica matrix in the sample (a,c) TiO_2 obtained from TEMAT hydrolysis (magnification ×500 and ×2000 respectively), (b,d) TiO_2 obtained from Ti-Amd hydrolysis (magnification ×500 and ×5000 respectively); Figure S12: XRD analyses of nanocomposites TiO_2/SBA-15 calcined at 500 °C and (a) prepared from precursor TEMAT, (b) prepared from precursor Ti-Amd.

Author Contributions: O.E.A. carried out the synthesis of oxides and nanocomposites, performed the photocatalytic tests and data analyses, prepared the figures, and wrote the experimental section of the manuscript; J.H. coordinated the joint research, supervised the photocatalytic experiments, wrote the catalysis section, and participated in reviewing the manuscript; K.F. coordinated the joint research, supervised the hydrolysis experiments, conducted data analyses, and participated in reviewing the manuscript; C.L. supervised the metalorganic preparation of Ti-Amd precursor and its characterization; B.L. and A.R. supervised the preparation of SBA-15 material employed for the study, performed the N_2 adsorption experiments and analyses, participated to scientific working meetings; M.K. participated to scientific working meetings; V.C. was in charge of the SEM, SEM-EDS and TEM analyses; Y.C. carried out the NMR experiment and NMR analysis section; N.R.-R. performed the WAXS experiments and analysis section; P.M. participated in scientific working meetings; P.F. coordinated the joint research, supervised the study, the hydrolysis method, and data analyses, wrote the draft, and finalized the manuscript. All authors have read and agreed to the published version of the manuscript.

Funding: The research project DPOLAIR was funded by the Allocation Doctorale Interdisciplinaire (ADI 21) program of the Université Fédérale de Toulouse Midi-Pyrénées (UFTMP) and the Région Occitanie.

Data Availability Statement: The original contributions presented in the study are included in the article/Supplementary Material, further inquiries can be directed to the corresponding author/s.

Acknowledgments: Simon Cayez (LPCNO-INSA) for XRD measures, Stéphane Leblond du Pouy (Centre de micro-caractérisation Castaing, Toulouse, France) for SEM analyses, and Frederic Averseng (Laboratoire de Réactivité de Surface, Sorbonne Université, Paris, France) for EPR discussion.

Conflicts of Interest: The authors declare no conflicts of interest.

References

1. Lorente, A.; Boersma, K.F.; Eskes, H.J.; Veefkind, J.P.; van Geffen, J.H.G.M.; de Zeeuw, M.B.; Denier van der Gon, H.A.C.; Beirle, S.; Krol, M.C. Quantification of nitrogen oxides emissions from build-up of pollution over Paris with TROPOMI. *Sci. Rep.* **2019**, *9*, 20033. [CrossRef] [PubMed]
2. WHO Global Air Quality Guidelines. Available online: https://www.who.int/news-room/questions-and-answers/item/who-global-air-quality-guidelines (accessed on 29 March 2024).
3. Ângelo, J.; Andrade, L.; Madeira, L.M.; Mendes, A. An overview of photocatalysis phenomena applied to NO_x abatement. *J. Environ. Manag.* **2013**, *129*, 522–539. [CrossRef]
4. Castelló Lux, K.; Hot, J.; Fau, P.; Bertron, A.; Kahn, M.L.; Ringot, E.; Fajerwerg, K. Nano-gold decorated ZnO: An alternative photocatalyst promising for NO_x degradation. *Chem. Eng. Sci.* **2023**, *267*, 118377. [CrossRef]
5. Seiss, V.; Thiel, S.; Eichelbaum, M. Preparation and Real World Applications of Titania Composite Materials for Photocatalytic Surface, Air, and Water Purification: State of the Art. *Inorganics* **2022**, *10*, 139. [CrossRef]
6. Uddin, M.J.; Cesano, F.; Chowdhury, A.R.; Trad, T.; Cravanzola, S.; Martra, G.; Mino, L.; Zecchina, A.; Scarano, D. Surface Structure and Phase Composition of TiO_2 P25 Particles After Thermal Treatments and HF Etching. *Front. Mater.* **2020**, *7*, 192. [CrossRef]
7. Qian, R.; Zong, H.; Schneider, J.; Zhou, G.; Zhao, T.; Li, Y.; Yang, J.; Bahnemann, D.W.; Pan, J.H. Charge carrier trapping, recombination and transfer during TiO_2 photocatalysis: An overview. *Catal. Today* **2019**, *335*, 78–90. [CrossRef]
8. Noureen, L.; Wang, Q.; Humayun, M.; Shah, W.A.; Xu, Q.; Wang, X. Recent advances in structural engineering of photocatalysts for environmental remediation. *Environ. Res.* **2023**, *219*, 115084. [CrossRef]

9. Toma, F.L.; Bertrand, G.; Klein, D.; Coddet, C. Photocatalytic removal of nitrogen oxides via titanium dioxide. *Environ. Chem. Lett.* **2004**, *2*, 117–121. [CrossRef]
10. Du, L.; Furube, A.; Yamamoto, K.; Hara, K.; Katoh, R.; Tachiya, M. Plasmon-Induced Charge Separation and Recombination Dynamics in Gold−TiO_2 Nanoparticle Systems: Dependence on TiO_2 Particle Size. *J. Phys. Chem. C* **2009**, *113*, 6454–6462. [CrossRef]
11. Sun, S.; Song, P.; Cui, J.; Liang, S. Amorphous TiO_2 nanostructures: Synthesis, fundamental properties and photocatalytic applications. *Catal. Sci. Technol.* **2019**, *9*, 4198–4215. [CrossRef]
12. MiarAlipour, S.; Friedmann, D.; Scott, J.; Amal, R. TiO_2/porous adsorbents: Recent advances and novel applications. *J. Hazard. Mater.* **2018**, *341*, 404–423. [CrossRef]
13. Jiang, W.; Ji, W.; Au, C.-T. Surface/Interfacial Catalysis of (Metal)/Oxide System: Structure and Performance Control. *ChemCatChem* **2018**, *10*, 2125–2163. [CrossRef]
14. Kumar, S.G.; Rao, K.S.R.K. Comparison of modification strategies towards enhanced charge carrier separation and photocatalytic degradation activity of metal oxide semiconductors (TiO_2, WO_3 and ZnO). *Appl. Surf. Sci.* **2017**, *391*, 124–148. [CrossRef]
15. Zhao, D.; Huo, Q.; Feng, J.; Chmelka, B.F.; Stucky, G.D. Nonionic Triblock and Star Diblock Copolymer and Oligomeric Surfactant Syntheses of Highly Ordered, Hydrothermally Stable, Mesoporous Silica Structures. *J. Am. Chem. Soc.* **1998**, *120*, 6024–6036. [CrossRef]
16. Yamada, T.; Zhou, H.; Uchida, H.; Honma, I.; Katsube, T. Experimental and Theoretical NO_x Physisorption Analyses of Mesoporous Film (SBA-15 and SBA-16) Constructed Surface Photo Voltage (SPV) Sensor. *J. Phys. Chem. B* **2004**, *108*, 13341–13346. [CrossRef]
17. Boutros, M.; Onfroy, T.; Da Costa, P. Mesostructured or Alumina-mesostructured Silica SBA-16 as Potential Support for NO_x Reduction and Ethanol Oxidation. *Catal. Lett.* **2010**, *139*, 50–55. [CrossRef]
18. Bin, F.; Song, C.; Lv, G.; Song, J.; Cao, X.; Pang, H.; Wang, K. Structural Characterization and Selective Catalytic Reduction of Nitrogen Oxides with Ammonia: A Comparison between Co/ZSM-5 and Co/SBA-15. *J. Phys. Chem. C* **2012**, *116*, 26262–26274. [CrossRef]
19. Sharma, M.V.P.; Kumari, V.D.; Subrahmanyam, M. TiO_2 supported over SBA-15: An efficient photocatalyst for the pesticide degradation using solar light. *Chemosphere* **2008**, *73*, 1562–1569. [CrossRef]
20. Acosta-Silva, Y.J.; Nava, R.; Hernandez-Morales, V.; Macias-Sanchez, S.A.; Gomez-Herrera, M.L.; Pawelec, B. Methylene Blue photodegradation over titania-decorated SBA-15. *Appl. Catal. B* **2011**, *110*, 108–117. [CrossRef]
21. Wang, X.-j.; Li, F.-t.; Hao, Y.-j.; Liu, S.-j.; Yang, M.-l. TiO_2/SBA-15 composites prepared using H2TiO3 by hydrothermal method and its photocatalytic activity. *Mater. Lett.* **2013**, *99*, 38–41. [CrossRef]
22. Besancon, M.; Michelin, L.; Josien, L.; Vidal, L.; Assaker, K.; Bonne, M.; Lebeau, B.; Blin, J.-L. Influence of the porous texture of SBA-15 mesoporous silica on the anatase formation in TiO_2-SiO_2 nanocomposites. *New J. Chem.* **2016**, *40*, 4386–4397. [CrossRef]
23. Conceicao, D.S.; Graca, C.A.L.; Ferreira, D.P.; Ferraria, A.M.; Fonseca, I.M.; Botelho do Rego, A.M.; Teixeira, A.C.S.C.; Vieira Ferreira, L.F. Photochemical insights of TiO_2 decorated mesoporous SBA-15 materials and their influence on the photodegradation of organic contaminants. *Microporous Mesoporous Mater.* **2017**, *253*, 203–214. [CrossRef]
24. Araujo, M.M.; Silva, L.K.R.; Sczancoski, J.C.; Orlandi, M.O.; Longo, E.; Santos, A.G.D.; Sa, J.L.S.; Santos, R.S.; Luz, G.E.; Cavalcante, L.S. Anatase TiO_2 nanocrystals anchored at inside of SBA-15 mesopores and their optical behavior. *Appl. Surf. Sci.* **2016**, *389*, 1137–1147. [CrossRef]
25. Vradman, L.; Peer, Y.; Mann-Kiperman, A.; Landau, M.V. Thermal decomposition-precipitation inside the nanoreactors. High loading of W-oxide nanoparticles into the nanotubes of SBA-15. In *Studies in Surface Science and Catalysis*; Park, S.-E., Ryoo, R., Ahn, W.-S., Lee, C.W., Chang, J.-S., Eds.; Elsevier: Amsterdam, The Netherlands, 2003; Volume 146, pp. 121–124.
26. Landau, M.V.; Vradman, L.; Wang, X.; Titelman, L. High loading TiO_2 and ZrO_2 nanocrystals ensembles inside the mesopores of SBA-15: Preparation, texture and stability. *Microporous Mesoporous Mater.* **2005**, *78*, 117–129. [CrossRef]
27. Peza-Ledesma, C.L.; Escamilla-Perea, L.; Nava, R.; Pawelec, B.; Fierro, J.L.G. Supported gold catalysts in SBA-15 modified with TiO_2 for oxidation of carbon monoxide. *Appl. Catal. A Gen.* **2010**, *375*, 37–48. [CrossRef]
28. Ke, W.; Liu, Y.; Wang, X.; Qin, X.; Chen, L.; Palomino, R.M.; Simonovis, J.P.; Lee, I.; Waluyo, I.; Rodriguez, J.A.; et al. Nucleation and Initial Stages of Growth during the Atomic Layer Deposition of Titanium Oxide on Mesoporous Silica. *Nano Lett.* **2020**, *20*, 6884–6890. [CrossRef]
29. Wang, C.-Y.; Shen, K.; Gorte, R.J.; Vohs, J.M. Preparation of SBA-15-Supported Metals by Vapor-Phase Infiltration. *Inorganics* **2022**, *10*, 215. [CrossRef]
30. Gardon, J.A.; Pike, S.D. Hydrolysis of organometallic and metal–amide precursors: Synthesis routes to oxo-bridged heterometallic complexes, metal-oxo clusters and metal oxide nanoparticles. *Dalton Trans.* **2018**, *47*, 3638–3662. [CrossRef]
31. Lim, B.S.; Rahtu, A.; Park, J.-S.; Gordon, R.G. Synthesis and Characterization of Volatile, Thermally Stable, Reactive Transition Metal Amidinates. *Inorg. Chem.* **2003**, *42*, 7951–7958. [CrossRef]
32. Cure, J.; Piettre, K.; Coppel, Y.; Beche, E.; Esvan, J.; Colliere, V.; Chaudret, B.; Fau, P. Solution Layer Deposition: A Technique for the Growth of Ultra-Pure Manganese Oxides on Silica at Room Temperature. *Angew. Chem. Int. Ed.* **2016**, *55*, 3027–3030. [CrossRef]

33. Castello Lux, K.; Fajerwerg, K.; Hot, J.; Ringot, E.; Bertron, A.; Collière, V.; Kahn, M.L.; Loridant, S.; Coppel, Y.; Fau, P. Nano-Structuration of WO_3 Nanoleaves by Localized Hydrolysis of an Organometallic Zn Precursor: Application to Photocatalytic NO2 Abatement. *Nanomaterials* **2022**, *12*, 4360. [CrossRef] [PubMed]
34. Jonca, J.; Castello-Lux, K.; Fajerwerg, K.; Kahn, M.L.; Colliere, V.; Menini, P.; Sowka, I.; Fau, P. Gas Sensing Properties of $CuWO_4$@WO_3 n-n Heterojunction Prepared by Direct Hydrolysis of Mesitylcopper (I) on $WO_3 \cdot 2H_2O$ Nanoleaves. *Chemosensors* **2023**, *11*, 495. [CrossRef]
35. Belmoujahid, Y.; Bonne, M.; Scudeller, Y.; Schleich, D.; Grohens, Y.; Lebeau, B. SBA-15 mesoporous silica as a super insulating material. *Eur. Phys. J. Spec. Top.* **2015**, *224*, 1775–1785. [CrossRef]
36. Wang, W.; Song, M. Photocatalytic activity of titania-containing mesoporous SBA-15 silica. *Microporous Mesoporous Mater.* **2006**, *96*, 255–261. [CrossRef]
37. Thommes, M. Physisorption of gases, with special reference to the evaluation of surface area and pore size distribution (IUPAC Technical Report). *Chem. Int.* **2016**, *38*, 25. [CrossRef]
38. Galarneau, A.; Cambon, H.; Di Renzo, F.; Ryoo, R.; Choi, M.; Fajula, F. Microporosity and connections between pores in SBA-15 mesostructured silicas as a function of the temperature of synthesis. *New J. Chem.* **2003**, *27*, 73–79. [CrossRef]
39. Galarneau, A.; Nader, M.; Guenneau, F.; Di Renzo, F.; Gedeon, A. Understanding the Stability in Water of Mesoporous SBA-15 and MCM-41. *J. Phys. Chem. C* **2007**, *111*, 8268–8277. [CrossRef]
40. Amiens, C.; Chaudret, B.; Ciuculescu-Pradines, D.; Collière, V.; Fajerwerg, K.; Fau, P.; Kahn, M.; Maisonnat, A.; Soulantica, K.; Philippot, K. Organometallic approach for the synthesis of nanostructures. *New J. Chem.* **2013**, *37*, 3374–3401. [CrossRef]
41. Niemelä, J.-P.; Marin, G.; Karppinen, M. Titanium dioxide thin films by atomic layer deposition: A review. *Semicond. Sci. Technol.* **2017**, *32*, 093005. [CrossRef]
42. Mamaghani, A.H.; Haghighat, F.; Lee, C.-S. Photocatalytic oxidation technology for indoor environment air purification: The state-of-the-art. *Appl. Catal. B Environ.* **2017**, *203*, 247–269. [CrossRef]
43. Bickley, R.I.; Gonzalez-Carreno, T.; Lees, J.S.; Palmisano, L.; Tilley, R.J.D. A structural investigation of titanium dioxide photocatalysts. *J. Solid State Chem.* **1991**, *92*, 178–190. [CrossRef]
44. Hendrix, A.L.Y.; Yu, Q.; Brouwers, J. Titania-Silica Composites: A Review on the Photocatalytic Activity and Synthesis Methods. *World J. Nano Sci. Eng.* **2015**, *5*, 161–177. [CrossRef]
45. Russell, H.S.; Frederickson, L.B.; Hertel, O.; Ellermann, T.; Jensen, S.S. A Review of Photocatalytic Materials for Urban NO_x Remediation. *Catalysts* **2021**, *11*, 675. [CrossRef]
46. Inada, M.; Enomoto, N.; Hojo, J. Fabrication and structural analysis of mesoporous silica–titania for environmental purification. *Microporous Mesoporous Mater.* **2013**, *182*, 173–177. [CrossRef]
47. Li, Y.; Kim, S.-J. Synthesis and Characterization of Nano titania Particles Embedded in Mesoporous Silica with Both High Photocatalytic Activity and Adsorption Capability. *J. Phys. Chem. B* **2005**, *109*, 12309–12315. [CrossRef] [PubMed]
48. Xu, Y.; Zheng, W.; Liu, W. Enhanced photocatalytic activity of supported TiO_2: Dispersing effect of SiO_2. *J. Photochem. Photobiol. A Chem.* **1999**, *122*, 57–60. [CrossRef]
49. Alonso-Tellez, A.; Masson, R.; Robert, D.; Keller, N.; Keller, V. Comparison of Hombikat UV100 and P25 TiO_2 performance in gas-phase photocatalytic oxidation reactions. *J. Photochem. Photobiol. A Chem.* **2012**, *250*, 58–65. [CrossRef]
50. Duong, N.T.; Trébosc, J.; Lafon, O.; Amoureux, J.-P. Improved sensitivity and quantification for 29Si NMR experiments on solids using UDEFT (Uniform Driven Equilibrium Fourier Transform). *Solid State Nucl. Magn. Reson.* **2019**, *100*, 52–62. [CrossRef] [PubMed]
51. Ratel-Ramond, N. Nicoratel/WAXS_toolbox: PDF-for-nano (v1.0.0-alpha). *Zenodo*, 2024. [CrossRef]
52. Juhás, P.; Farrow, C.L.; Yang, X.; Knox, K.R.; Billinge, S.J. Complex modeling: A strategy and software program for combining multiple information sources to solve ill posed structure and nanostructure inverse problems. *Acta Crystallogr. Sect. A Found. Adv.* **2015**, *71*, 562–568. [CrossRef]
53. Gražulis, S.; Chateigner, D.; Downs, R.T.; Yokochi, A.F.T.; Quirós, M.; Lutterotti, L.; Manakova, E.; Butkus, J.; Moeck, P.; Le Bail, A. Crystallography Open Database—An open-access collection of crystal structures. *J. Appl. Crystallogr.* **2009**, *42*, 726–729. [CrossRef]
54. Hot, J.; Topalov, J.; Ringot, E.; Bertron, A. Investigation on Parameters Affecting the Effectiveness of Photocatalytic Functional Coatings to Degrade NO: TiO_2 Amount on Surface, Illumination, and Substrate Roughness. *Int. J. Photoenergy* **2017**, *2017*, 6241615. [CrossRef]
55. *ISO 22197-1:2016*; Fine Ceramics (Advanced Ceramics, Advanced Technical Ceramics)—Test Method for Air-Purification Performance of Semiconducting Photocatalytic Materials—Part 1: Removal of Nitric Oxide. ISO: Geneva, Switzerland, 2016. Available online: https://www.iso.org/standard/65416.html (accessed on 20 November 2023).

inorganics

Article

Thermoelectric Characteristics of β-Ag_2Se_{1+x} Prepared via a Combined Rapid Mechano-Thermal Approach

Katarína Gáborová [1,*], Michal Hegedüs [2], Petr Levinský [3], František Mihok [1,4], Miloš Matvija [1], Karel Knížek [3], Ondrej Milkovič [1,5], Dagmara Vatraľová [1,4], Jiří Hejtmánek [3] and Karel Saksl [1,4]

1 Faculty of Materials, Metallurgy and Recycling, Technical University of Košice, Letná 1/9, 042 00 Košice, Slovakia; milos.matvija@tuke.sk (M.M.); karel.saksl@tuke.sk (K.S.)
2 Synthon, s.r.o., 67801 Blansko, Czech Republic
3 FZU—Institute of Physics of the Czech Academy of Sciences, Cukrovarnická 10/112, 162 00 Prague, Czech Republic; knizek@fzu.cz (K.K.)
4 Institute of Materials Research, Slovak Academy of Sciences, Watsonova 47, 040 01 Košice, Slovakia
5 Institute of Experimental Physics, Slovak Academy of Sciences, Watsonova 47, 040 01 Košice, Slovakia
* Correspondence: katarina.gaborova@tuke.sk

Abstract: This study investigates the thermoelectric properties of Se-rich β-Ag_2Se synthesized via a mechanochemical method followed by spark plasma sintering (SPS) in less than 30 min of the total reaction time. Importantly, only a short 10 min milling process followed by appropriate SPS was enough to produce single-phase Ag_2Se_{1+x} samples with varying selenium content (where x = 0, 0.01, 0.02, 0.04). The introduction of excess selenium significantly influenced the thermoelectric performance, optimizing the carrier concentration during synthesis and resulting in substantial thermoelectric improvements. The sample with nominal composition $Ag_2Se_{1.01}$ exhibited a high dimensionless figure-of-merit (ZT) >0.9 at 385 K, which is nearly six times higher than the reference sample (β-Ag_2Se). Our findings bring valuable insight into the technology of optimization of thermoelectric characteristics of Se-rich β-Ag_2Se, highlighting its potential for applications in thermoelectric devices. The study demonstrates the energetically efficient and environmental advantage of our mechanochemical route to produce Se-rich β-Ag_2Se, providing a solvent-free and commercially viable alternative synthesis for energy (thermoelectric and solar energy).

Keywords: thermoelectric properties; β-Ag_2Se_{1+x}; mechanochemistry; planetary ball milling

Citation: Gáborová, K.; Hegedüs, M.; Levinský, P.; Mihok, F.; Matvija, M.; Knížek, K.; Milkovič, O.; Vatraľová, D.; Hejtmánek, J.; Saksl, K. Thermoelectric Characteristics of β-Ag_2Se_{1+x} Prepared via a Combined Rapid Mechano-Thermal Approach. *Inorganics* **2024**, *12*, 334. https://doi.org/10.3390/inorganics12120334

Academic Editors: Torben R. Jensen and Roberto Nisticò

Received: 7 November 2024
Revised: 4 December 2024
Accepted: 19 December 2024
Published: 21 December 2024

1. Introduction

In the quest for environmentally friendly and efficient thermoelectric (TE) materials capable of harnessing waste heat for clean electricity generation, the attention has shifted towards materials facilitating the direct conversion of heat to electricity and vice versa without generating excessive noise and emissions [1].

To achieve widespread utilization of TE materials, it is crucial that they exhibit high performance and cost-effectiveness. The energy conversion efficiency of TE materials is quantified by the dimensionless figure-of-merit ZT:

$$ZT = \frac{S^2}{\rho\lambda}, \tag{1}$$

where S is the Seebeck coefficient, ρ is the electrical resistivity, λ is the thermal conductivity and T is the absolute temperature [2]. The figure-of-merit ZT directly influences the efficiency of TE devices like thermoelectric generators (TEGs) and heat pumps [3–6]. Higher ZT values are desired, and materials of interest for TE applications are expected to have a figure-of-merit ZT of at least one [7].

Ag_2Se exists in two main structural phases, each with distinct physical properties that significantly influence its applicability in energy conversion technologies. At low temperatures (LT) below 407 K, Ag_2Se crystallizes in the orthorhombic β-phase (space group $P2_12_12_1$), which is a semiconductor. This phase is characterized by a highly ordered crystal lattice, exhibiting low electrical resistivity, relatively high Seebeck coefficients, and low lattice thermal conductivity. These attributes make the β-phase highly suitable for thermoelectric applications, particularly in the conversion of waste heat into [8,9]. Above 407 K, Ag_2Se undergoes a first-order phase transition to the cubic α-phase (space group $Im\bar{3}m$) [10,11]. The α-phase is characterized as a superionic conductor, where silver ions exhibit high mobility within the lattice, leading to significantly increased ionic conductivity. While this property is advantageous for certain applications, such as solid-state electrolytes, it results in reduced thermoelectric performance due to higher electrical resistivity and altered thermal conductivity in the α-phase. Consequently, the β-phase of Ag_2Se is the primary focus for thermoelectric applications because its semiconducting behavior and stable structure support the optimized interplay of electrical and thermal properties necessary for achieving high ZT values [12,13]. The dominance of the β-phase at low temperatures aligns with the overarching goal of thermoelectric materials to efficiently convert waste heat into electricity in operational environments typically below the α-to-β phase transition temperature.

Noteworthy advancements in this domain include the work of Mi et al., which achieved the highest figure-of-merit ZT of 0.96 at T = 401 K through the implementation of excess selenium in Ag_2Se [14]. Mi et al. achieved this by preparing Se-rich β-Ag_2Se using high-temperature annealing followed by controlled cooling, emphasizing compositional tuning and phase stability to enhance thermoelectric properties. Chen et al. further contributed by preparing the LT orthorhombic phase β-Ag_2Se, characterized by high porosity and attained a maximum ZT of 0.9 at T = 390 K [15]. Their unique approach involved the fabrication of porous β-Ag_2Se through a melting and sintering process, leveraging enhanced phonon scattering properties of the porous structure to reduce thermal conductivity. Building on this, Jood et al. successfully suppressed the metastable structure of Ag_2Se using a slight excess of selenium, resulting in the highest ZT of 1.0 at T = 375 K [16]. This underscores the potential of selenium excess as a strategy for enhancing TE properties, warranting further in-depth studies for the development of an economical and effective preparation method.

Furthermore, the combination of silver (Ag) and selenium (Se) presents significant advantages regarding lower toxicity. This differs from the combinations of silver with tellurium (Te) or bismuth (Bi). Additionally, Ag_2Se is a promising and stable thermoelectric (TE) material, maintaining its stability even at 300 K [15]. Additionally, this material is cheaper to produce by omitting rare and expensive Te and replacing it with Se [17]. Ferhat et al. recently identified Ag_2Se as a promising candidate for TE applications in TE coolers and sensor devices [18]. This is attributed to the unique characteristics of the low-temperature modification of Ag_2Se, which exhibits unusually low lattice thermal conductivity (λ_L~0.5 $W \cdot m^{-1} \cdot K^{-1}$), coupled with low electrical resistivity (ρ~5.10^{-6} Ω·m) and a relatively high Seebeck coefficient (S~−150 $\mu V \cdot K^{-1}$ at 300 K) [19]. Given recent discoveries of significant positive volume magnetoresistance [20] and thin-film behavior [21] in Ag_2Se, it becomes imperative to undertake a detailed exploration of this system for TE applications at LTs as well [18].

In the broader context of the preparation of Ag_2Se-based compounds, the pursuit of enhanced performance and efficient synthesis methods remains a driving force. Conventional approaches to synthesizing Ag_2Se have drawbacks such as low yield, multiple steps, and long processing periods, contributing to elevated production costs and environmental impact [22]. Additionally, traditional melting–annealing–sintering processes for Ag_2Se are time-consuming and energy-intensive [23]. These limitations highlight the need for more efficient and environmentally friendly synthesis approaches for Ag_2Se and related materials. The present research challenges these conventional methods by introducing a novel approach to the preparation of Ag_2Se-based compounds. The data from various studies emphasize the importance of developing rapid, cost-effective, and less toxic synthesis methods to overcome the cons associated with the conventional synthesis routes of Ag_2Se [24].

A novel approach to the preparation of Ag_2Se-based compounds involves utilizing single-source molecular precursors for the synthesis of Ag_2Se nanocrystals, as demonstrated in one study [25]. Another study highlighted the successful synthesis of binary Ag_2Se and composite Ag^0:Ag_2Se through an ambient aqueous-solution-based approach without high-temperature heating or the use of organic solvents [26]. Additionally, a unique method involving the nanostructuring of g-Se with silver nanoparticles led to the formation of fine nanowires of trigonal selenium within the glass matrix, enhancing various physical properties effectively [27].

Drawing upon prior research, the transport properties of Ag_2Se demonstrate a significant sensitivity to synthetic methods, composition variations, and resulting defects [16]. For example, Huang et al. [28] employed multiple methods (melting, melting with spark plasma sintering (SPS), and mechanical alloying with SPS) to prepare Ag_2Se bulk materials, resulting in imperfections such as nanosized Ag- or Se-rich precipitates, micropores with Se-aggregated interfaces, and large voids. These imperfections have the potential to impact thermoelectric performance, uniformity, and reproducibility.

The thermoelectric performance of β-Ag_2Se prepared by conventional methods has been widely studied, with ZT values ranging from 0.8 to 1.0 in the 300–400 K range. For instance, Huang et al. reported ZT values of up to 0.9 using melting and spark plasma sintering [28], while Jood et al. achieved a ZT of 1.0 at 375 K by introducing selenium excess to stabilize the matrix [16].Other studies highlight how crystallographic orientation [13] and porosity control [29] significantly influence performance.

However, traditional synthesis methods face challenges such as high energy consumption, long processing times, and defect formation, which can degrade transport properties [18,28]. Newer approaches, such as nanostructuring and solution-based techniques, have shown promise in improving Seebeck coefficients and reducing thermal conductivity [26]. The rapid mechanochemical synthesis employed in this study stands out by enabling the precise control of selenium excess, resulting in a high ZT of 0.94 at 385 K. While the direct comparison with the conventional methods is difficult due to the novelty of this approach, our results align with general trends observed in the literature, demonstrating its potential for superior thermoelectric performance.

Jood et al.'s identification of a metastable phase in nominally stoichiometric Ag_2Se, which is believed to negatively impact transport properties, prompted successful interventions [16]. They prevented the formation of this metastable phase and stabilized the matrix by introducing a slight excess of anions (Se and S). This intervention resulted in achieving high mobility (2510 $cm^2 \cdot V^{-1} \cdot s^{-1}$ at 300 K, low lattice thermal conductivity (0.2 $W \cdot m^{-1} \cdot K^{-1}$ at RT), and a high ZT ($\approx$1.0 over 300–375 K) in anion-rich compositions. These findings underscore the complexity of Ag_2Se in comparison to previously described aspects and emphasize the need for a detailed exploration of this system for TE applications at LTs [18].

In this study, Se-doped polycrystalline Ag_2Se featuring fine grains was synthesized using mechanochemical synthesis followed by the SPS treatment under relatively mild conditions. This research focuses on an in-depth exploration of the thermoelectric properties, with particular attention given to elucidating the impact of the Se substitution on the enhancement of thermoelectric characteristics. The initial results of measuring the thermoelectric properties of the β-Ag_2Se material were recently published, and this extended study builds upon them by focusing on the preparation of Se-rich β-Ag_2Se [30]. The results from this study presented herein not only showcase the efficiency of the rapid synthesis approach but also underscore the superior thermoelectric properties exhibited by the Se-rich β-Ag_2Se, positioning it as a noteworthy candidate for applications in waste heat recovery and energy harvesting [31].

The purpose of this work is to develop and evaluate the aforementioned scalable, solvent-free, and energy-efficient approach for producing selenium-rich β-Ag_2Se by fine-tuning selenium content. Then, we analyze its effects on microstructure and carrier concentration. Overall, this study aims to optimize the material's thermoelectric properties and demonstrate its potential for achieving high ZT values.

2. Results and Discussion

2.1. Phase Composition and Microstructure Analysis

The structural aspects of milled samples and samples after SPS treatment with the chemical composition of Ag_2Se_{1+x} (x = 0, 0.01, 0.02, 0.04) were characterized using powder pXRD analysis. The pXRD analysis of milled samples confirmed the presence of the orthorhombic phase of β-Ag_2Se after 10 min of milling. In Figure 1a, the pXRD patterns of the powder samples after mechanochemical synthesis are depicted. The quantitative phase analysis confirmed that all milled Ag_2Se_{1+x} samples exhibited varying phase compositions, with the main phase being the orthorhombic β-Ag_2Se (space group $P2_12_12_1$, a = 4.333 Å, b = 7.062 Å, c = 7.761 Å), (see Table 1). The pXRD analysis of milled mixtures indicates that the slight excess of Se provides a material with a composition close to that of stoichiometric β-Ag_2Se. By altering the composition through milling, i.e., adding Se, the chemical composition improved, and in the $Ag_2Se_{1.01}$ sample, the β-Ag_2Se phase is the most dominant, approaching the stoichiometric composition of the pure β-Ag_2Se phase. This might be explained by the lower purity of the used Se compared to Ag or the volatility of some selenium chemical species. An additional amount of Se, however, hinders the formation of β-Ag_2Se during the 10 min milling period as more elemental Ag is found in the mixtures. Achieving proximity to the stoichiometric composition by adding Se to the matrix is therefore essential to optimize the final chemical composition, which leads to enhanced thermoelectric properties as discussed further.

The phase composition of the investigated Ag_2Se_{1+x} series after SPS was modified in favor of the pure orthorhombic β-Ag_2Se phase as confirmed by the Rietveld refinement (see the results for the selected sample with $x = 0.01$ in Figure 1b). This finding further supports the short milling time required to homogenize the input mixture, even if no full conversion is observed in all cases after the milling process. The total reaction time needed to obtain a phase of pure material was less than 30 min. After SPS treatment, the amount of elemental Ag in the stoichiometric mixture Ag_2Se_{1+x} ($x = 0$) was so small that it could not be practically quantified. As there were no visible reflections of Se after the SPS process ($x = 0.02, 0.04$), it was assumed all Se atoms had been incorporated into the crystal structure of the compound. All Se-rich samples comply with the phase pure orthorhombic phase β-Ag_2Se. This is in accordance with previously observed results of Mi et al. [14] and the phase diagram of the Ag-Se system [32]. HT-pXRD patterns were measured to detect a phase transition to the cubic phase α-Ag_2Se, which took place at 400 K (Figure 1c). A transitional state was captured at 403 K where both phases were present, suggesting that the phase transformation was not as fast and needed more energy. The results of the HT-pXRD measurement correspond well with the measured TE data.

This investigation, emphasizing the chemical composition and structure of the Ag_2Se_{1+x} series, harmonizes with and reinforces prior research delving into analogous materials. Several seminal studies [14,22,33,34] have investigated the crystal structures of Ag_2Se and its derivatives with varying selenium concentrations, shedding light on the intricate interplay between composition, structure, and TE properties. In line with these investigations, Mi et al. focused on the thermoelectric properties of the Ag_2Se_{1+x} series prepared via annealing at high temperatures, particularly highlighting the crystallographic characterization of the samples. Mi et al. observed that all Ag_2Se_{1+x} series samples crystallize with the orthorhombic structure further corroborating our findings from the pXRD analysis of the mechanochemically synthesized Ag_2Se_{1+x} series [14].

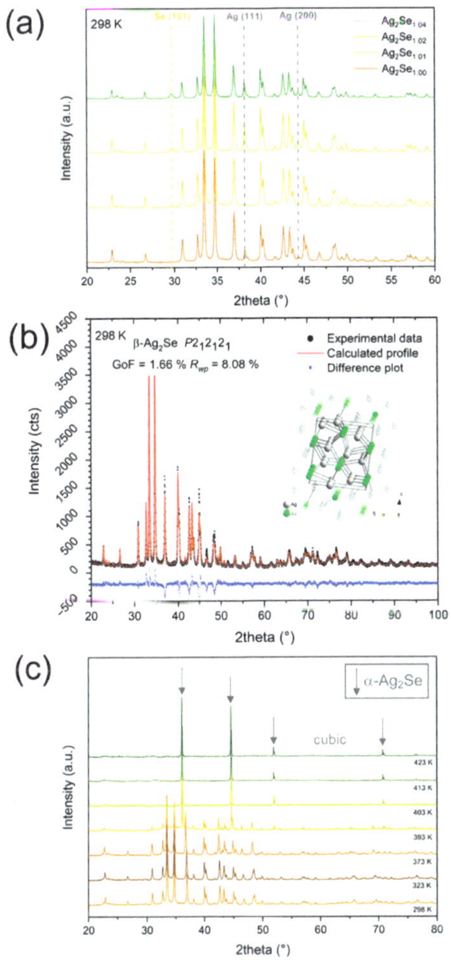

Figure 1. (**a**) pXRD patterns of Ag_2Se_{1+x} series after mechanochemical synthesis, (**b**) Rietveld refinement, and (**c**) HT-pXRD analysis of nominal $Ag_2Se_{1.01}$ sample after SPS.

Table 1. Representation of individual phases and lattice parameters in the Ag_2Se_{1+x} series before and after SPS as calculated from quantitative Rietveld analysis.

Sample	Individual Phases (mol.%)			Lattice Parameters for Ag_2Se_{1+x} Phase (Å)		
	Ag_2Se	Ag	Se	*a*	*b*	*c*
$Ag_2Se_{1.00}$	93.0	7.0	0.0	4.3352 ± 0.0002	7.0675 ± 0.0003	7.7723 ± 0.0003
$Ag_2Se_{1.01}$	99.1	0.6	0.3	4.3347 ± 0.0001	7.0673 ± 0.0002	7.7725 ± 0.0002
$Ag_2Se_{1.02}$	73.5	13.0	13.4	4.3350 ± 0.0001	7.0675 ± 0.0002	7.7726 ± 0.0002
$Ag_2Se_{1.04}$	73.2	10.4	16.4	4.3350 ± 0.0001	7.0674 ± 0.0002	7.7723 ± 0.0002
$Ag_2Se_{1.00}$ SPS	100	0.0	0.0	4.3367 ± 0.0003	7.0695 ± 0.0004	7.7742 ± 0.0004
$Ag_2Se_{1.01}$ SPS	100	0.0	0.0	4.3356 ± 0.0002	7.0676 ± 0.0003	7.7697 ± 0.0003
$Ag_2Se_{1.02}$ SPS	100	0.0	0.0	4.3356 ± 0.0002	7.0678 ± 0.0003	7.7704 ± 0.0003
$Ag_2Se_{1.04}$ SPS	100	0.0	0.0	4.3346 ± 0.0002	7.0663 ± 0.0003	7.7680 ± 0.0003

Figure 2 illustrates the microstructural evolution of β-Ag_2Se_{1+x} (x = 0, 0.01, 0.02, 0.04) captured via SEM. SEM images (a)–(d) show the samples after mechanochemical synthesis, highlighting the initial morphology. After mechanochemical synthesis, the samples exhibited similar morphologies, with the exception of x = 0, where two distinct morphologies were observed: rod-like and spherical, consistent with the prior findings detailed in previous study [30]. The morphologies of samples with x = 0.01, 0.02, and 0.04 are similar, characterized by the absence of rod-like particles.

Element	Atom (%)
Ag	68.4
Se	31.6
Total	100.0

Element	Atom (%)
Ag	67.3
Se	32.7
Total	100.0

Element	Atom (%)
Ag	68.4
Se	31.6
Total	100.0

Element	Atom (%)
Ag	68.2
Se	31.8
Total	100.0

Figure 2. SEM images of β-Ag_2Se_{1+x} (x = 0, 0.01, 0.02, 0.04 from left to right) (**a–d**) after mechanochemical synthesis and (**e–h**) after sintering including EDX with elemental mapping.

Subsequently, SEM images (e)–(h) show the samples after SPS with EDX values included. The sintering process induced significant changes in the microstructure. Observable changes include variations in grain boundaries and the appearance of porosity, which is indicative of structural reorganization during the sintering process. While the crystal structure remained unchanged, the process led to a tighter arrangement of individual powder particles, facilitating the formation of well-defined crystal grains. These observations align with the literature, which highlights that reorganization is inherent to the sintering process [35]. Additionally, the observed porosity and fracture surfaces were evident in the sample's morphology, and they are of critical importance for the thermoelectric (TE) properties of β-Ag_2Se. These features can significantly affect electrical transport properties by potentially suppressing activated charge carriers.

Furthermore, the EDX analysis after SPS confirms a close 2:1 ratio of Ag to Se, which is consistent with the expected stoichiometry of the main β-Ag_2Se phase in the system. EDX analysis reveals that elements Ag and Se are homogeneously distributed in all samples, further confirming findings from pXRD analysis for β-Ag_2Se preparation through the combined mechano-thermal method.

2.2. Thermoelectric Properties

The temperature dependence of electrical resistivity (ρ), Seebeck coefficient (S), thermal conductivity (λ), and the figure-of-merit ZT is illustrated for the Ag_2Se_{1+x} series in the temperature range of 5–440 K in Figure 3. All four parameters clearly indicate a distinct phase transition from the orthorhombic to the cubic phase at around 400 K, which is consistent with the literature [36,37]. Discrepancies in the curves measuring thermoelectric properties at low and high temperatures can be attributed to the use of different instruments.

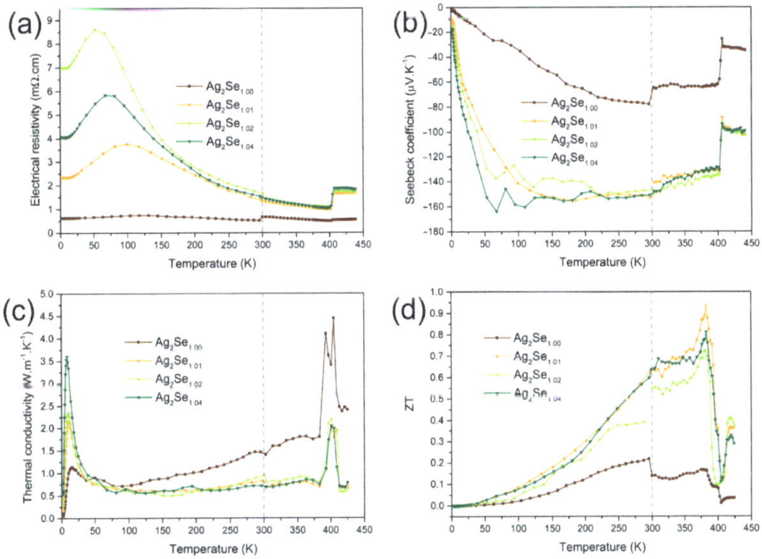

Figure 3. Thermoelectric properties of the Ag_2Se_{1+x} system (x = 0; 0.01; 0.02; 0.04) as a function of temperature in the range of 5–440 K: (**a**) electrical resistivity, ρ; (**b**) Seebeck coefficient, S; (**c**) thermal conductivity, λ; and (**d**) figure-of-merit ZT.

Electrical resistivity as a function of temperature is shown in Figure 3a. The $Ag_2Se_{1.00}$ sample consistently exhibits lower resistivity values compared to the other samples (x = 0.01, 0.02, 0.04). At low temperatures, the resistivity of all samples shows an initial increase as temperature decreases, reaching a maximum at 55–73 K. This behavior corresponds to the semiconductor nature of the materials, where carrier mobility decreases with decreasing temperature. Samples undergo a transition to a metallic character below this maximum with resistivity decreasing and eventually saturating at temperatures below 10 K. This temperature-dependent trend is characteristic of the Ag_2Se system and aligns with the previous study [38].

In the temperature range of 150–375 K, electrical resistivity (ρ) decreases with increasing temperature, which is typical for semiconductors, as the thermal excitation of carriers dominates over scattering effects [14,16,39]. The higher resistivity of samples observed is attributed to a lower concentration of free electrons caused by the changing Se content [40,41]. For example, Jood et al. reported that the electron concentration for stoichiometric Ag_2Se was approximately 6×10^{18} cm^{-3} [16], decreasing to 4×10^{18} cm^{-3} in $Ag_2Se_{1.01}$ and remaining stable for higher excess Se contents.

The temperature dependence of the Seebeck coefficient (S) is shown in Figure 3b. Negative values across the entire temperature range confirm the n-type semiconductor nature of all samples. The absolute values of S are higher for samples (x = 0.01, 0.02, 0.04) between 20 K and 300 K. The observed increase in the Seebeck coefficient for the selected samples correlates with the reduced carrier concentration due to selenium content variation. Concurrently, the reduction in thermal conductivity is attributed to enhanced phonon scattering caused by the changing selenium content. These effects synergistically contribute to the improved ZT value observed in the $Ag_2Se_{1.01}$ sample with peak ZT of 0.94 at 385 K.

The thermal conductivity (λ) data are presented in Figure 3c. Thermal conductivity is generally low across the whole temperature range, which is a key feature for thermoelectric materials. For stoichiometric $Ag_2Se_{1.00}$, λ is higher due to a significant electronic contribution to heat transport. At LTs, the lattice contribution dominates, and a pronounced phononic peak at around 10 K is observed, which is typical for crystalline materials. This peak correlates with the Se content, where a lower carrier concentration in Se- modified samples reduces phonon scattering and results in higher lattice thermal conductivity.

At HTs, thermal conductivity becomes weakly temperature-dependent, with the electronic contribution playing a more prominent role. In the range of 300–450 K, the λ values exhibit only minor variations around the phase transition. However, the slight increase in λ near 380–400 K should be regarded with caution, as it is challenging to measure thermal conductivity precisely in the vicinity of a phase transition.

The temperature dependence of the figure-of-merit (ZT) is shown in Figure 3d. For all samples, ZT increases with temperature, reaching a maximum just before the phase transition at approximately 385 K. Beyond this transition, ZT values drop sharply, decreasing by a factor of ~2. The maximum ZT of 0.94 is achieved for the $Ag_2Se_{1.01}$ sample at 385 K, highlighting its optimized balance of thermoelectric properties. Se-modified samples show consistent trends, with their ZT values peaking at the same temperature, further emphasizing the influence of phase transitions on thermoelectric performance. This behavior underscores the significance of fine-tuning the composition of Ag_2Se_{1+x} to optimize the thermoelectric properties, particularly around the phase transition temperature.

The observed ZT peak near 385 K in the samples can be attributed to a combination of factors that influence the material's thermal and electrical properties. As the temperature approaches the orthorhombic-to-cubic phase transition, several phenomena occur that enhance the thermoelectric performance. Firstly, the reduction in thermal conductivity is a critical factor contributing to enhanced ZT. As the temperature nears the phase transition, the lattice thermal conductivity continues to drop while the electronic contribution remains stable, thus reducing the total thermal conductivity [42]. Secondly, the optimized carrier concentration and mobility in the Ag_2Se_{1+x} system are pivotal in achieving high thermoelectric performance. Tuning the selenium amount in the Ag_2Se_{1+x} system leads to a reduction in the concentration of free electrons, which optimizes carrier mobility. This balance between carrier concentration and mobility results in an increased Seebeck coefficient, particularly near the phase transition, without a significant rise in electrical resistivity [43,44].

Furthermore, the temperature-induced enhancements associated with the orthorhombic phase contribute to improved thermoelectric performance. The orthorhombic phase has more inherent semiconducting properties than the cubic phase, making it better suited for thermoelectric applications. As the material approaches the phase transition, it exhibits enhanced electronic properties, including a higher Seebeck coefficient and more stable electrical resistivity [45,46]. This transition not only influences the electronic structure but also affects the overall thermoelectric efficiency, as the material's response to temperature changes becomes more favorable for thermoelectric generation.

These factors together contribute to the observed ZT peak near 385 K just before the phase transition to the cubic phase. Once the phase transition occurs, the material undergoes structural changes that disrupt the favorable balance of thermal and electrical properties, leading to a sharp decrease in ZT. This behavior underscores the importance of optimizing the composition of Ag_2Se_{1+x} around the phase transition to achieve the best thermoelectric performance.

The observed trends are consistent with previous research, confirming the potential of this system for thermoelectric applications [7,9,31,33].

3. Materials and Methods

3.1. Mechanochemical Synthesis

Se-rich samples of β-Ag_2Se_{1+x} (x = 0, 0.01, 0.02, 0.04) were prepared from the following elements: Ag (99.9%, 125 μm, Thermo Fisher Scientific Chemicals, Waltham, MA, USA) and Se (99.5%, 74 μm, Aldrich, Darmstadt, Germany). The simple mechanochemical synthesis of β-Ag_2Se_{1+x} was carried out by milling in a laboratory planetary mill PULVERISETTE 6 classic line (Fritsch, Idar-Oberstein, Germany), following Equations (2)–(5):

$$2\ Ag\ (s) + 1.00\ Se\ (s) \rightarrow Ag_2Se_{1.00}\ (s) \quad (2)$$

$$2\ Ag\ (s) + 1.01\ Se\ (s) \rightarrow Ag_2Se_{1.01}\ (s) \quad (3)$$

$$2\ Ag\ (s) + 1.02\ Se\ (s) \rightarrow Ag_2Se_{1.02}\ (s) \quad (4)$$

$$2\ Ag\ (s) + 1.04\ Se\ (s) \rightarrow Ag_2Se_{1.04}\ (s) \quad (5)$$

and under the milling conditions specified in Table 2.

Table 2. Conditions for the mechanochemical synthesis of β-Ag_2Se.

Volume of the Milling Chamber:	250 mL
Material of Milling Chamber and Medium:	WC
Loading of Milling Chamber:	50 balls (Ø10 mm)
Ball-to-Powder Ratio (BPR):	73:1
Total Mass of Input Reactants:	5 g
Milling Atmosphere:	Ar
Addition:	Methanol (<1 mL)
Rotation Speed:	550 rpm
Milling Time:	10 min

3.2. Pellets for the Measurements of Thermoelectric Properties

Ultimately, the powders from mechanochemical synthesis were subjected to the spark plasma sintering (SPS) process within a graphite die, with a sintering temperature of 723 K, a holding time of 15 min, and an applied pressure of 80 MPa, resulting in the formation of pellets (Ø 10 mm). To remove the excess graphite layer remaining on the surface after SPS, the pellets were polished. To facilitate further measurements, the polished pellet was divided into smaller geometrically appropriate parts using a diamond wire saw. All analyses were performed on the parts originating from a single pellet. The sample's mass was determined using analytical balances, and dimensions were obtained through multiple measurements using a calibrated digital caliper. The theoretical density, determined based on lattice parameters obtained from the ICSD database (Inorganic Crystal Structure Database), was calculated to be 8.23 g·cm^{-3}. The density of the prepared tablets was calculated from geometric dimensions and sample mass. However, the experimental density values of all samples were approximately 15% lower than the theoretical density. This discrepancy arose due to several factors, including residual porosity within the pellets caused by the SPS process, microstructural inhomogeneity, and surface effects or residual impurities. Despite this discrepancy, the experimental densities are sufficient to support the high thermoelectric performance demonstrated by the samples, indicating that the remaining porosity or inhomogeneity does not negatively impact the key transport properties. The studied series with the chemical composition Ag_2Se_{1+x} and calculated pellet densities are given in Table 3.

3.3. Characterization Techniques

The phase purity of β-Ag_2Se_{1+x} was checked by powder X-ray diffraction acquired on powder Bruker D8 Advance diffractometer with CuKα radiation equipped with a Lynxeye XE-T detector. Phase transition occurring around 404 K was further studied using powder X-ray diffraction (pXRD) measurements performed in Bragg–Brentano geometry using a

Cu lamp and strip D/teX detector by Rigaku Ultima IV type II (Tokyo, Japan). The sample was placed in a standard holder in the high-temperature chamber HTK 1200N (Anton Paar, Graz, Austria) in an oxygen atmosphere. The chamber was heated to 423 K and then cooled back to room temperature. During this process, the temperature change was controlled at a constant rate, and the goniometer was rotated at a speed of 0.02 degrees per minute (θ/min).

Table 3. Calculated densities of the pellets with the mixture composition β-Ag_2Se_{1+x}.

Sample Composition	Experimental Density of Pellet ($g{\cdot}cm^{-3}$)	Experimental Density of Pellet (%)
$Ag_2Se_{1.00}$	7.07	86.1
$Ag_2Se_{1.01}$	6.94	84.4
$Ag_2Se_{1.02}$	6.71	81.6
$Ag_2Se_{1.04}$	6.95	84.6

A scanning electron microscopy (SEM) investigation was conducted employing the MIRA3 FE-SEM microscope (TESCAN, Brno, Czech Republic), which was fitted with an energy-dispersive X-ray (EDX) detector (Oxford Instruments, Abingdon, UK) for EDX analysis.

To measure the TE properties below 300 K, a commercial apparatus called the Physical Property Measurement System (PPMS) from Quantum Design (San Diego, CA, USA) was utilized. For concurrent measurements of electrical resistivity, ρ, Seebeck coefficient, S, and thermal conductivity, λ, the Thermal Transport Option (TTO) sample holder was employed.

The electrical resistivity was measured simultaneously with the Seebeck coefficient via the four-terminal and static direct-current method using custom equipment. The sample together with the holder, was placed in a furnace controlled by the Lakeshore 331S temperature controller. A Keithley 2401 source meter served as the source for a gradient heater and a current source for measuring electrical resistivity. Meanwhile, an Agilent 34970A data acquisition unit equipped with a multiplexing card was responsible for scanning the voltages and temperatures across the sample. The thermal conductivity was measured by a laser flash method in a flowing N_2 atmosphere (LFA 467, Netsch, Selb, Germany) in the temperature range of 300–440 K and was obtained as a product of measured thermal diffusivity, heat capacity (using Pyroceram 9606 as a standard) and experimental density.

Low-temperature (LT) and high-temperature (HT) measurements allowed us to observe the temperature dependence of the thermoelectric properties and identify the maximum ZT values at required temperatures.

4. Conclusions

Mechanochemistry, a field gaining increasing prominence in materials science, involves chemical transformations induced by mechanical force. The planetary ball mill employed in this study served as a powerful tool to harness mechanochemical reactions, allowing for the expedited synthesis of Ag_2Se_{1+x} compounds. The distinctive advantage of this approach lies in its ability to achieve a high degree of mixing and reactivity over a short duration. This led to the formation of finely tuned thermoelectric materials with enhanced properties. It is noteworthy that the accelerated synthesis of Ag_2Se_{1+x} in a remarkably brief 10 min milling process sets a new benchmark for the synthesis of Ag_2Se-based compounds. The implications of this rapid mechanochemical synthesis are far-reaching, promising to streamline production processes, reduce energy consumption, and open new avenues for the exploration of thermoelectric materials.

This study successfully demonstrated the enhanced thermoelectric properties of β-Ag_2Se synthesized via a rapid mechanochemical approach followed by spark plasma sintering (SPS). The varying amounts of selenium (x = 0, 0.01, 0.02, 0.04) significantly affect the thermoelectric performance by optimizing carrier concentration and reducing the thermal conductivity of the material. The best-performing sample, with the composition of $Ag_2Se_{1.01}$, achieved a dimensionless figure-of-merit (ZT) of 0.94 at 385 K. While the ZT value of 0.94 for the $Ag_2Se_{1.01}$ sample is high, it does not surpass the best-known values in the literature, such as the ZT of 1.0 achieved by Jood et al. [16] or 0.96 reported by Mi et al. [14]. These higher values were achieved using more time-intensive or complex synthesis techniques, such as the annealing or suppression of metastable phases. Nonetheless, our rapid mechanochemical method combined with SPS demonstrates a competitive ZT value, offering a practical and scalable alternative for the synthesis of this thermoelectric material. The modification of selenium amount has been demonstrated as an effective strategy for tuning the thermoelectric performance of Ag_2Se, consistent with previous research findings.

Phase composition analysis confirms the predominance of the orthorhombic β-Ag_2Se phase, which highlights the effectiveness of the mechanochemical synthesis and subsequent SPS method. The SPS process plays a crucial role in achieving phase purity and transforming the sample into a phase-pure material despite the presence of significant amounts of unreacted precursors in the milled sample. This finding underscores the importance of SPS as a critical technological step in the preparation of high-purity β-Ag_2Se. Microstructural observations through SEM and EDX analyses reveal significant changes in grain boundaries and porosity post-sintering. This contributes to the material's enhanced thermoelectric characteristics.

The results indicate that the β-Ag_2Se place with the composition closest to the stoichiometric, and only minimal change in the Se content, leads to the best results mainly because of reduced lattice thermal conductivity and improvements in electrical conductivity, thereby achieving a high ZT, which is crucial for the efficient use of thermoelectric materials.

Overall, the results of this work not only confirm the effectiveness of mechanochemical and SPS methods for preparing high-performance β-Ag_2Se materials but also provide a foundation for future research aimed at further improving and applying these materials in the field of thermoelectrics. By demonstrating the benefits of selenium doping and the viability of the synthesis methods used, this study advances our understanding of how to enhance thermoelectric materials for practical and sustainable energy solutions.

Author Contributions: Conceptualization, K.G., M.H. and P.L.; Data curation, K.G., M.H., P.L., F.M., M.M. and K.K.; Formal analysis, K.G., M.H., F.M., M.M. and K.K.; Investigation, K.G., M.H., M.M., K.K. and O.M.; Methodology, K.G. and P.L.; Resources, K.G. and M.H.; Supervision, P.L. and J.H.; Validation, K.G., M.H., P.L., F.M., M.M., K.K. and J.H.; Visualization, K.G.; Writing—original draft, K.G. and M.H.; Writing—review and editing, K.G., M.H., P.L., F.M., M.M., K.K., D.V., J.H. and K.S. All authors have read and agreed to the published version of the manuscript.

Funding: This research was funded by the Slovak Research and Development Agency under the contract no. APPV no. 18-0357, providing the necessary resources for the successful execution of the research.

Data Availability Statement: The data are contained within the article.

Acknowledgments: The authors extend their sincere gratitude to the Institute of Physics of the Czech Academy of Sciences and the Institute of Geotechnics Slovak Academy of Sciences for their invaluable collaboration and support throughout the research conducted for this paper. Additionally, the authors acknowledge the support received from the National Scholarship Programme, which played a crucial role in facilitating the academic pursuits underlying this publication.

Conflicts of Interest: Author Michal Hegedüs was employed by Synthon, s.r.o. The remaining authors declare that the research was conducted in the absence of any commercial or financial relationships that could be construed as a potential conflict of interest.

References

1. Tan, G.; Zhao, L.-D.; Kanatzidis, M.G. Rationally Designing High-Performance Bulk Thermoelectric Materials. *Chem. Rev.* **2016**, *116*, 12123–12149. [CrossRef]
2. Liu, M.; Zhang, X.; Zhang, S.; Pei, Y. Ag_2Se as a Tougher Alternative to n-Type Bi_2Te_3 Thermoelectrics. *Nat. Commun.* **2024**, *15*, 6580. [CrossRef] [PubMed]
3. Lei, Y.; Qi, R.; Chen, M.; Chen, H.; Xing, C.; Sui, F.; Gu, L.; He, W.; Zhang, Y.; Baba, T.; et al. Microstructurally Tailored Thin β-Ag_2Se Films toward Commercial Flexible Thermoelectrics. *Adv. Mater.* **2022**, *34*, 2104786. [CrossRef] [PubMed]
4. He, J.; Tritt, T.M. Advances in Thermoelectric Materials Research: Looking Back and Moving Forward. *Science* **2017**, *357*, eaak9997. [CrossRef] [PubMed]
5. Shi, X.-L.; Zou, J.; Chen, Z.-G. Advanced Thermoelectric Design: From Materials and Structures to Devices. *Chem. Rev.* **2020**, *120*, 7399–7515. [CrossRef] [PubMed]
6. Qin, Y.; Yang, L.; Wei, J.; Yang, S.; Zhang, M.; Wang, X.; Yang, F. Doping Effect on Cu_2Se Thermoelectric Performance: A Review. *Materials* **2020**, *13*, 5704. [CrossRef] [PubMed]
7. Snyder, G.J.; Toberer, E.S. Complex Thermoelectric Materials. *Nat. Mater.* **2008**, *7*, 105–114. [CrossRef]
8. Xiao, C.; Xu, J.; Li, K.; Feng, J.; Yang, J.; Xie, Y. Superionic Phase Transition in Silver Chalcogenide Nanocrystals Realizing Optimized Thermoelectric Performance. *J. Am. Chem. Soc.* **2012**, *134*, 4287–4293. [CrossRef] [PubMed]
9. Perez-Taborda, J.A.; Caballero-Calero, O.; Vera-Londono, L.; Briones, F.; Martin-Gonzalez, M. High Thermoelectric zT in n-Type Silver Selenide Films at Room Temperature. *Adv. Energy Mater.* **2018**, *8*, 1702024. [CrossRef]
10. Lin, S.; Guo, L.; Wang, X.; Liu, Y.; Wu, Y.; Li, R.; Shao, H.; Jin, M. Revealing the Promising Near-Room-Temperature Thermoelectric Performance in Ag_2Se Single Crystals. *J. Materiomics* **2023**, *9*, 754–761. [CrossRef]
11. Hu, Q.-X.; Liu, W.-D.; Zhang, L.; Sun, W.; Gao, H.; Shi, X.-L.; Yang, Y.-L.; Liu, Q.; Chen, Z.-G. SWCNTs/Ag_2Se Film with Superior Bending Resistance and Enhanced Thermoelectric Performance via in Situ Compositing. *Chem. Eng. J.* **2023**, *457*, 141024. [CrossRef]
12. Byeon, D.; Sobota, R.; Delime-Codrin, K.; Choi, S.; Hirata, K.; Adachi, M.; Kiyama, M.; Matsuura, T.; Yamamoto, Y.; Matsunami, M.; et al. Discovery of Colossal Seebeck Effect in Metallic Cu_2Se. *Nat. Commun.* **2019**, *10*, 72. [CrossRef] [PubMed]
13. Ding, Y.; Qiu, Y.; Cai, K.; Yao, Q.; Chen, S.; Chen, L.; He, J. High Performance N-Type Ag_2Se Film on Nylon Membrane for Flexible Thermoelectric Power Generator. *Nat. Commun.* **2019**, *10*, 841. [CrossRef]
14. Mi, W.; Qiu, P.; Zhang, T.; Lv, Y.; Shi, X.; Chen, L. Thermoelectric Transport of Se-Rich Ag_2Se in Normal Phases and Phase Transitions. *Appl. Phys. Lett.* **2014**, *104*, 133903. [CrossRef]
15. Chen, J.; Sun, Q.; Bao, D.; Liu, T.; Liu, W.-D.; Liu, C.; Tang, J.; Zhou, D.; Yang, L.; Chen, Z.-G. Hierarchical Structures Advance Thermoelectric Properties of Porous N-Type β-Ag_2Se. *ACS Appl. Mater. Interfaces* **2020**, *12*, 51523–51529. [CrossRef]
16. Jood, P.; Chetty, R.; Ohta, M. Structural Stability Enables High Thermoelectric Performance in Room Temperature Ag_2Se. *J. Mater. Chem. A* **2020**, *8*, 13024–13037. [CrossRef]
17. Missen, O.P.; Ram, R.; Mills, S.J.; Etschmann, B.; Reith, F.; Shuster, J.; Smith, D.J.; Brugger, J. Love Is in the Earth: A Review of Tellurium (Bio)Geochemistry in Surface Environments. *Earth-Sci. Rev.* **2020**, *204*, 103150. [CrossRef]
18. Ferhat, M.; Nagao, J. Thermoelectric and Transport Properties of β-Ag_2Se Compounds. *J. Appl. Phys.* **2000**, *88*, 813–816. [CrossRef]
19. Conn, J.B.; Taylor, R.C. Thermoelectric and Crystallographic Properties of Ag_2Se. *J. Electrochem. Soc.* **1960**, *107*, 977. [CrossRef]
20. Yang, F.; Xiong, S.; Xia, Z.; Liu, F.; Han, C.; Zhang, D. Two-Step Synthesis of Silver Selenide Semiconductor with a Linear Magnetoresistance Effect. *Semicond. Sci. Technol.* **2012**, *27*, 125017. [CrossRef]
21. Ali, H.M.; Khudayer, I.H. Study Structure and Optical Properties of Ag_2Se, $Ag_2Se_{0.8}Te_{0.2}$ and $Ag_2Se_{0.8}S_{0.2}$ Thin Films. *J. Ovonic Res.* **2022**, *18*, 675–680. [CrossRef]
22. Wang, P.; Chen, J.-L.; Zhou, Q.; Liao, Y.T.; Peng, Y.; Liang, J.S.; Miao, L. Enhancing the Thermoelectric Performance of Ag_2Se by Non-Stoichiometric Defects. *Appl. Phys. Lett.* **2022**, *120*, 193902. [CrossRef]
23. Li, X.; Liu, Z.; Luo, K.; Yin, X.; Lin, X.; Zhu, C. Biomimetic Synthesis of Ag_2Se Quantum Dots with Enhanced Photothermal Properties and as "Gatekeepers" to Cap Mesoporous Silica Nanoparticles for Chemo–Photothermal Therapy. *Chem. Asian J.* **2019**, *14*, 155–161. [CrossRef]
24. Duan, H.Z.; Li, Y.L.; Zhao, K.P.; Qiu, P.F.; Shi, X.; Chen, L.D. Ultra-Fast Synthesis for Ag_2Se and CuAgSe Thermoelectric Materials. *JOM* **2016**, *68*, 2659–2665. [CrossRef]
25. Karmakar, G.; Tyagi, A.; Shah, A.Y.; Nigam, S.; Wadawale, A.P.; Kedarnath, G.; Vats, B.G.; Naveen Kumar, N.; Singh, V. Facile One Pot Synthesis of Highly Photoresponsive Coinage Metal Selenides ($Cu_{1.8}Se$ and Ag_2Se) Achieved through Novel Cu and Ag Pyridylselenolates as Molecular Precursors. *Dalton Trans.* **2022**, *51*, 12670–12685. [CrossRef]
26. Tee, S.Y.; Tan, X.Y.; Wang, X.; Lee, C.J.J.; Win, K.Y.; Ni, X.P.; Teo, S.L.; Seng, D.H.L.; Tanaka, Y.; Han, M.-Y. Aqueous Synthesis, Doping, and Processing of n-Type Ag_2Se for High Thermoelectric Performance at Near-Room-Temperature. *Inorg. Chem.* **2022**, *61*, 6451–6458. [CrossRef]
27. Kumar, A.; Mehta, N.; Dahshan, A. A New Approach for Nano-Structuring of Glassy Selenium (g-Se) Using Silver Nanoparticles (AgNPs) as Precursor. *Mater. Today Commun.* **2021**, *26*, 101719. [CrossRef]
28. Huang, S.; Wei, T.-R.; Chen, H.; Xiao, J.; Zhu, M.; Zhao, K.; Shi, X. Thermoelectric Ag_2Se: Imperfection, Homogeneity, and Reproducibility. *ACS Appl. Mater. Interfaces* **2021**, *13*, 60192–60199. [CrossRef]

29. Chen, J.; Yuan, H.; Zhu, Y.-K.; Zheng, K.; Ge, Z.-H.; Tang, J.; Zhou, D.; Yang, L.; Chen, Z.-G. Ternary $Ag_2Se_{1-x}Te_x$: A Near-Room-Temperature Thermoelectric Material with a Potentially High Figure of Merit. *Inorg. Chem.* **2021**, *60*, 14165–14173. [CrossRef]
30. Achimovičová, M.; Gáborová, K.; Girman, V.; Dutková, E.; Briančin, J.; Levinský, P.; Puchý, V. Simple Mechanochemical Synthesis, Characterization, Optical and Thermoelectric Properties of a Nanostructured Silver (I) Selenide Semiconductor. *Appl. Res.* **2024**, *3*, e202300076. [CrossRef]
31. Hu, Q.; Liu, W.; Zhang, L.; Gao, H.; Wang, D.; Wu, T.; Shi, X.; Li, M.; Liu, Q.; Yang, Y.; et al. Carrier Separation Boosts Thermoelectric Performance of Flexible *n*-Type Ag_2Se-Based Films. *Adv. Energy Mater.* **2024**, 2401890. [CrossRef]
32. Wu, H.; Shi, X.; Duan, J.; Liu, Q.; Chen, Z.-G. Advances in Ag_2Se-Based Thermoelectrics from Materials to Applications. *Energy Environ. Sci.* **2023**, *16*, 1870–1906. [CrossRef]
33. Wei, T.; Qiu, P.; Zhao, K.; Shi, X.; Chen, L. Ag_2Q-Based (Q = S, Se, Te) Silver Chalcogenide Thermoelectric Materials. *Adv. Mater.* **2023**, *35*, 2110236. [CrossRef]
34. Liang, J.; Qiu, P.; Zhu, Y.; Huang, H.; Gao, Z.; Zhang, Z.; Shi, X.; Chen, L. Crystalline Structure-Dependent Mechanical and Thermoelectric Performance in $Ag_2Se_{1-x}S_x$ System. *Research* **2020**, *2020*, 6591981. [CrossRef]
35. Palaporn, D.; Kurosaki, K.; Pinitsoontorn, S. Effect of Sintering Temperature on the Thermoelectric Properties of Ag_2Se Fabricated by Spark Plasma Sintering with High Compression. *Adv. Energy Sustain. Res.* **2023**, *4*, 2300082. [CrossRef]
36. Li, D.; Zhang, J.H.; Li, J.M.; Zhang, J.; Qin, X.Y. High Thermoelectric Performance for an Ag_2Se-Based Material Prepared by a Wet Chemical Method. *Mater. Chem. Front.* **2020**, *4*, 875–880. [CrossRef]
37. Yang, D.; Su, X.; Meng, F.; Wang, S.; Yan, Y.; Yang, J.; He, J.; Zhang, Q.; Uher, C.; Kanatzidis, M.G.; et al. Facile Room Temperature Solventless Synthesis of High Thermoelectric Performance Ag_2Se *via* a Dissociative Adsorption Reaction. *J. Mater. Chem. A* **2017**, *5*, 23243–23251. [CrossRef]
38. Jin, M.; Liang, J.; Qiu, P.; Huang, H.; Yue, Z.; Zhou, L.; Li, R.; Chen, L.; Shi, X. Investigation on Low-Temperature Thermoelectric Properties of Ag_2Se Polycrystal Fabricated by Using Zone-Melting Method. *J. Phys. Chem. Lett.* **2021**, *12*, 8246–8255. [CrossRef]
39. Lim, K.H.; Wong, K.W.; Liu, Y.; Zhang, Y.; Cadavid, D.; Cabot, A.; Ng, K.M. Critical Role of Nanoinclusions in Silver Selenide Nanocomposites as a Promising Room Temperature Thermoelectric Material. *J. Mater. Chem. C* **2019**, *7*, 2646–2652. [CrossRef]
40. Grientschnig, D.; Sitte, W. Interpretation of Ionic Transport Properties of Some Silver Chalcogenides. *J. Phys. Chem. Solids* **1991**, *52*, 805–820. [CrossRef]
41. Pasternak, M.; Benczer-Koller, N.; Yang, T.; Ruel, R.; Herber, R.H. Impurity-Induced Local Disorder in the Ordered State of the Superionic Conductor β-Ag_2Se. *Phys. Rev. B* **1983**, *27*, 2055–2058. [CrossRef]
42. Yu, T.; Ning, S.; Liu, Q.; Zhang, T.; Chen, X.; Qi, N.; Su, X.; Tang, X.; Chen, Z. Balanced High Thermoelectric Performance in N-Type and p-Type CuAgSe Realized through Vacancy Manipulation. *ACS Appl. Mater. Interfaces* **2023**, *15*, 40781–40791. [CrossRef] [PubMed]
43. Wu, Y.; Su, X.; Yang, D.; Zhang, Q.; Tang, X. Boosting Thermoelectric Properties of $AgBi_3(Se_y S_{1-y})_5$ Solid Solution via Entropy Engineering. *ACS Appl. Mater. Interfaces* **2021**, *13*, 4185–4191. [CrossRef]
44. Wang, W.; Shen, D.; Li, H.; Chen, C.; Chen, Y. Thermoelectric Optimization of N-Type $AgBiSe_2$ via Se Vacancy Control and Transition-Metal Doping. *ACS Appl. Energy Mater.* **2023**, *6*, 9709–9715. [CrossRef]
45. Wei, W.; Chang, C.; Yang, T.; Liu, J.; Tang, H.; Zhang, J.; Li, Y.; Xu, F.; Zhang, Z.; Li, J.-F.; et al. Achieving High Thermoelectric Figure of Merit in Polycrystalline SnSe via Introducing Sn Vacancies. *J. Am. Chem. Soc.* **2018**, *140*, 499–505. [CrossRef] [PubMed]
46. Lee, M.H.; Yun, J.H.; Kim, G.; Lee, J.E.; Park, S.-D.; Reith, H.; Schierning, G.; Nielsch, K.; Ko, W.; Li, A.-P.; et al. Synergetic Enhancement of Thermoelectric Performance by Selective Charge Anderson Localization–Delocalization Transition in n-Type Bi-Doped PbTe/Ag_2 Te Nanocomposite. *ACS Nano* **2019**, *13*, 3806–3815. [CrossRef] [PubMed]

inorganics

MDPI

Article

Cu-Doped TiO_2 Thin Films by Spin Coating: Investigation of Structural and Optical Properties

Syrine Sassi [1,2,*], Amal Bouich [1,*], Anouar Hajjaji [2], Lotfi Khezami [3], Brahim Bessais [2] and Bernabé Mari Soucase [1]

1 Departamento de Fisica Aplicada, School of Design Engineering, Universitat Politecnica de Valencia, Cami de Vera, 46022 València, Spain; bmari@fis.upv.es

2 Laboratoire de Photovoltaïque, Centre de Recherches et des Technologies de l'Energie, Technopole de Borj Cédria, BP 95, Hammam-Lif 2050, Tunisia; physicshajjaji@gmail.com (A.H.); brahim.bessais@crten.rnrt.tn (B.B.)

3 Department of Chemistry, College of Sciences, Imam Mohammad Ibn Saud Islamic University (IMSIU), Riyadh 13318, Saudi Arabia; lhmkhezami@imamu.edu.sa

* Correspondence: sassi.syrine13@gmail.com (S.S.); ambo1@doctor.upv.es (A.B.)

Abstract: Cu-doped TiO_2 films were synthesized directly on FTO glass with a spin coating method. With a variation in copper amount, samples were prepared with 0%, 1%, 2%, 4% and 8% of dopant concentrations. Morphological and structural characterization of undoped and Cu-doped TiO_2 samples were investigated and the obtained results showed the small, spherical shapes of the nanoparticles forming a thin film on top of FTO glass and their preferred orientation of TiO_2 anatase (101), which is the same for each sample. However, this peak exhibited a slight shift for the 2% sample, related to the inflation of the microstrain compared to the other samples. For the optical properties, the 4% sample displayed the highest transmittance whereas the 2% sample exhibited the lowest band gap energy of 2.96 eV. Moreover, the PL intensity seems to be at its highest for the 2% sample due to the present peaking defects in the structure, whereas the 8% sample shows a whole new signal that is related to copper oxide. These properties make this material a potential candidate to perform as an electron transport layer (ETL) in solar cells and enhance their power conversion efficiency.

Keywords: titanium dioxide TiO_2 nanoparticles; copper Cu doping; perovskite solar cells PSCs; electron transport layer ETL

Citation: Sassi, S.; Bouich, A.; Hajjaji, A.; Khezami, L.; Bessais, B.; Soucase, B.M. Cu-Doped TiO_2 Thin Films by Spin Coating: Investigation of Structural and Optical Properties. *Inorganics* **2024**, *12*, 188. https://doi.org/10.3390/inorganics12070188

Academic Editors: Torben R. Jensen and Roberto Nisticò

Received: 21 April 2024
Revised: 24 June 2024
Accepted: 3 July 2024
Published: 8 July 2024

1. Introduction

Renewable energy sources are witnessing an immense growth in industrial developments and technologies, as they started with conventional solar cells and biomass combustion and progressed to innovative hydrogen production systems that are still to this day under assessment [1]. Solar cells have their share in spreading and reinforcing renewable energy technologies into industry and their flourishing success [2]. Moreover, solar panels have been expanding and diverting over the years; they were pioneered with different materials and different systems, as you can find monocrystalline and polycrystalline as well as amorphous silicon solar cells, plasmonic, multi-junction, thin films, quantum dots, dye-sensitized, perovskite and organic–inorganic hybrid solar cells [3–5]. All of these types of solar panels have received great feedback on their efficiency and yielding [6]. Among the third generation, perovskite solar cells (PSCs) have gained a lot of recognition due to their high power conversion efficiency (PCE), which was recently reported to have increased from 3.8% in 2009 to 25.2% in 2020 [7], which is very close to the notional Shockley–Queisser efficiency limit (~33%), and also, perovskite materials possess a narrow direct band gap, leading to a high optical absorption coefficient and broadband absorption and thus a long charge carrier lifetime and high mobility, and they also offer a simple, low-cost and effective synthesis methods [8]; several types of perovskite have been developed to create a better absorber with excellent optoelectronic properties, like a slow recombination rate of e^-/h^+ pairs [9], and the ABX_3 formula is the most used one, where A is an organic ($CH_3NH_3^+$ or MA and FA) or alkali metal, B is a bivalent metal (Pb, Sn and Ge) and X is a halide anion, and this formula is called organic–inorganic metal halide perovskite [10], and we can also find other types like RP phase layered perovskite and ACI-type perovskite. From all of these types, it seems that the Pb^{2+}-based perovskite is needed for exhibiting excellent photoelectric properties [11]. After all of these successful achievements of perovskite solar cells, they are still deteriorated by the issue of charge carrier recombination [12]. So, engineering the lane of carriers between the perovskite active layer and the electron transport layer (ETL) is vital to enhance the PCE [13]. However, many factors affect the power conversion efficiency (PCE) of PSCs with an n-i-p structure or inverted p-i-n structure, and it seems that the most crucial is the charge carrier transfer procedure at the interfaces [14], such as the interface between electron transport layer (ETL) and perovskite and hole transport layer (HTL); it is better to construct a contact interface that reduces charge carrier trapping at the interface, which leads to their recombination. Some researchers have focused on engineering the ETL and HTL separately to optimize their electrical and optical properties, since the chemical stability of these layers as well as the thickness, absorption spectrum, doping and interfacial defects can badly effect the extraction and transportation of the photogenerated charge carriers from the perovskite absorber layer, leading to a reduction in the PCE [15–17]. The most commonly used metal oxides for an ETL are TiO_2, ZnO, SnO_2 and ZrO_2. But TiO_2 is frequently selected due to its low cost, great chemical stability, suitable conduction band and high electric conductivity [18]; even a solar cell with a TiO_2 ETL reported a 24.66% PCE. But the drawbacks presented by TiO_2 mainly consist of the large energy barrier at the perovskite/TiO_2 interface and its limited UV light absorption which hinders efficient charge transfer, causing researchers to turn to other metal oxides [19,20]. To overcome these drawbacks, tuning the energy levels is the optimal choice either via doping or/and inserting an interlayer between the ETL and perovskite layer to achieve a prime energy alignment. To create a smooth contact interface, some researchers have resorted to adding Cl ions to expand the electron diffusion length; S. D. Stranks et al. reported the aftermath of adding Cl ions to the perovskite solar cells, as the PCE went from 4.2% to 12.2% [21]. As for the alteration of TiO_2 ETL properties, for instance the Fermi level, band gap and electrical conductivity, the doping process has been highly suggested [22]; furthermore, many metals have been proposed as a supreme dopant candidate for a TiO_2 ETL in PSCs, namely Sn, Zn, Ag, Nb and W [23]. Shih-Husan Chen et al. reported that for PSCs with a *meso*-Sn-doped TiO_2 ETL, the PEC was improved from 16.86% to 20.55% due to the decrease in defect states and a slight upward shifting of the conduction band and valence, leading to improved carrier extraction and transport [24]. Also, Sadiq Shahriyar Nishat et al. reported a 16.44% PEC for a 4.17 mol% Zn-doped TiO_2 ETL in a PSC, which is considered slightly lower compared to the same amount of 4.17 mol% Sn-doped TiO_2 that showed a 0.63% higher PCE [22]. Chen et al. reported that a mesoporous Ag-doped TiO_2 ETL in PSCs exhibits a great PEC of 17.7% [25]. Among the transition metals, Cu has never been used in any research report for doping a TiO_2 ETL in a perovskite solar cell, even considering its high conductivity and abundance, but it has been widely studied in dye-sensitized solar cells (DSSCs) [26] and it shows great yielding in different doping concentrations of TiO_2 photoanodes and photocatalysis applications [27–30]. T. Raguram et al. revealed a maximum efficiency of 3.90% for 0.1 M Cu-TiO_2 for DSSC application and 97.12% for Rhodamine B photocatalytic degradation; the 0.1 M Cu-TiO_2 synthesized with the sol–gel method showed a great impact on the structural, optical, morphological and electrical properties of TiO_2, as if the incorporation of Cu concentrations perfectly tailored the optical band gap and crystallinity of TiO_2 nanoparticles, as it shifted from 3.2 eV to 2.3 eV with the addition of copper concentration [31]. A lot of methods exist for Cu-TiO_2 synthetization and deposition on FTO glass substrate that have accrued throughout the years, as each process gives the material a specific size, morphology and other typical properties. The electrodeposition [32], sol–gel [33], spin coating [34], hydrothermal [35] and spray pyrolysis [36] methods are the most commonly used, especially the spin coating process since it is a very simple, cost effective and easy to manipulate method without any dissipations of energy and resources [37]; its consistency of a homogeneous dispersion of nanoparticles on the surface of FTO glass is the main cause of its vast utilization in wide range of research reports [38]. Ying-Han Liao et al. fabricated a perovskite solar cell $CH_3NH_3PbI_{3-x}Cl_x/(Sn/TiO_2)/FTO$ with the spin coating method with a variation in the concentration of dopant Sn, with a PCE of 14.4% for the 1.0 mol% Sn/TiO_2 sample; all the samples showed a successful decrease in band gap and an increase in charge carrier mobility [39].

Herein, we will prepare an electron transport layer (ETL) that consists of Cu-doped TiO_2 for a perovskite solar cell with the spin coating method, while varying the dopant concentration; we will study the effect on the morphology, crystallinity and the optical properties of TiO_2, and thus the effect on e^-/h^+ mobility and recombination. To observe the power conversion efficiency of the PSCs and the impact of the copper concentration after the characterization, the samples will be integrated as an ETL in an efficient and famous perovskite solar cell, as is shown in Figure 1, where a SCAPS simulation will be used in future work to foretell the PCE of each sample.

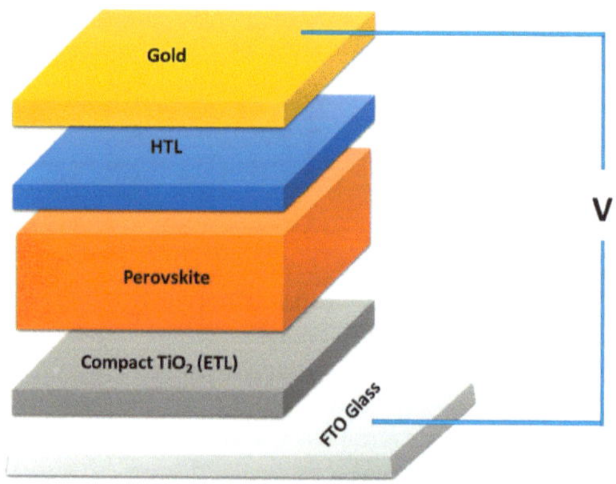

Figure 1. Schematic of perovskite solar cell.

2. Results and Discussion

2.1. Morphological Properties

Figure 2 shows the SEM images of pure and doped TiO_2. The TiO_2 thin films demonstrate small spherical nanoparticles at a range of 200–500 nm that are orderly dispersed on the surface of the FTO glass to form a homogenous film. It seems that the addition of copper dopant to the TiO_2 nanoparticles has no effect on the morphology of the TiO_2 nanoparticles; all films depict a granular nanostructure, and with the increase in dopant concentration percentage, some regions show a microstructural defect with the presence of bumps and holes.

Figure 3 shows the cross-sectional SEM image of the TiO_2 thin films with a thickness of 350 nm deposited on FTO glass. The thickness of the sample remains the same since the deposition parameters of the spin coating were sustained with same speed, solution quantity and duration. Moreover, the thickness seems to range between 340 and 380 nm for the different dopant concentrations.

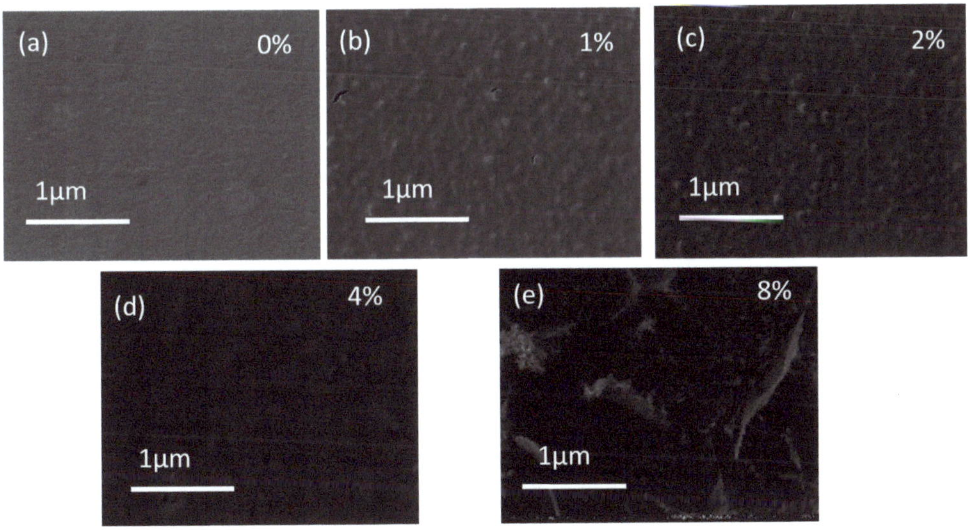

Figure 2. SEM images of the (**a**) undoped and (**b**) 1%, (**c**) 2%, (**d**) 4% and (**e**) 8% Cu-doped TiO_2 nanoparticle thin films.

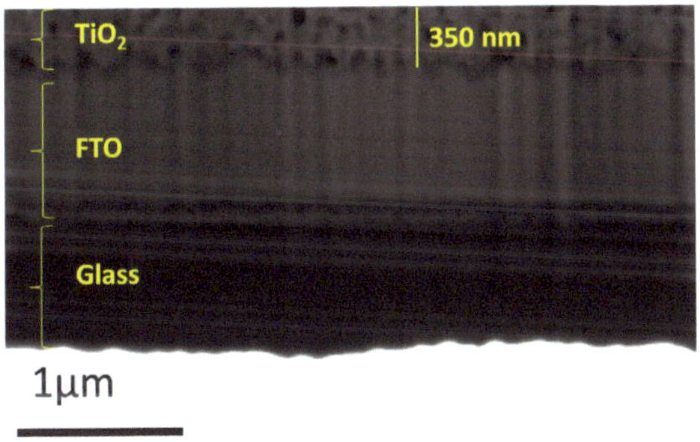

Figure 3. Cross-sectional SEM image of TiO_2 thin films on FTO glass.

2.2. Structural Properties

The X-ray diffraction can provide information about the structural properties of the as-prepared samples. Figure 4a depicts a diffractogram with the visible crystalline patterns of undoped and Cu-doped TiO_2 nanoparticles deposited on FTO glass with a preferred orientation at $2\theta = 25.389°$ that correspond to anatase TiO_2 (101), according to the reference card (JCPDS Card 21-1272) [40], and a small peak, with the other intense peaks corresponding to FTO glass (110), (101), (200) and (211); however, no Cu-related peaks were observable, which could mean the substitution with the Cu ion was successfully carried out without interstitial growth [41]. Moreover, the intensity of the anatase (101) peak increases with the increase in the dopant percentages. This could be due to the high incorporation of copper into the TiO_2 structure, which in turn induces physical stressing on the lattice. From the used precursors, we can confirm the formation of Cu^{2+} in the solution and according to the literature the ionic radius of Cu^{2+} is (86 pm) and Ti^{4+} is (74.5 pm) [42]. According to Figure 4b, it seems that as we increase the percentage of the dopant the TiO_2 (101) peak vaguely shifts to lower angles, especially for 8% sample. This could be better interpreted by the calculation of the crystallite size, lattice strain and microstrain, which are presented below.

The Scherrer equation for crystallite size and lattice strain determination is [43]

$$D = \frac{K\lambda}{\beta \cos\theta} \tag{1}$$

$$\frac{1}{D} \tag{2}$$

The Wilson equation for microstrain determination is [43]

$$\varepsilon = \frac{\beta}{4\tan(\theta)} \tag{3}$$

According to Figure 5 and the calculation in Table 1, it seems that the crystallite size is at its lowest values for the 2% and 4% samples and according to proportionality these samples also exhibits a high lattice strain and microstrain, which provides information on the degree of distortion, dislocations and defects present in the crystalline lattice; this could affect the photosensitivity of TiO_2 nanoparticle thin films by inducing the density of charge carriers [44].

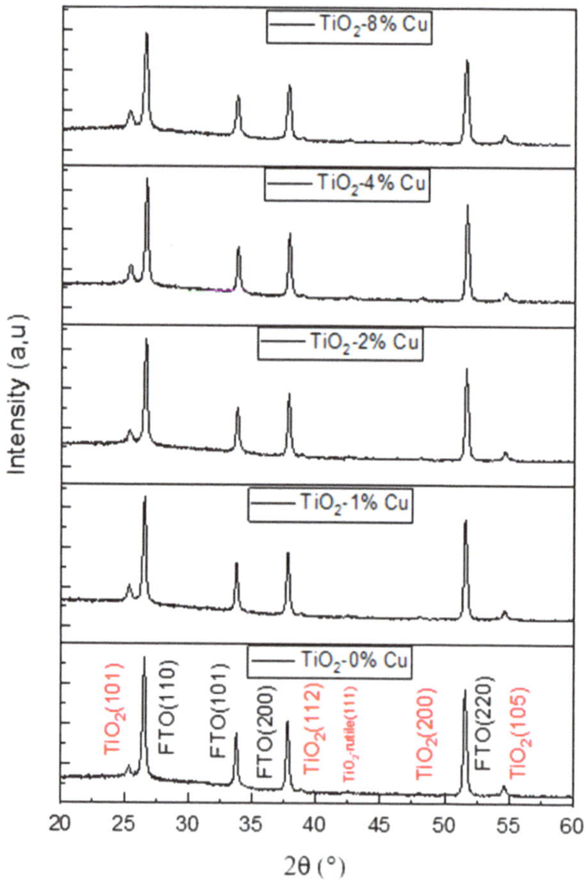

(**a**)

Figure 4. *Cont.*

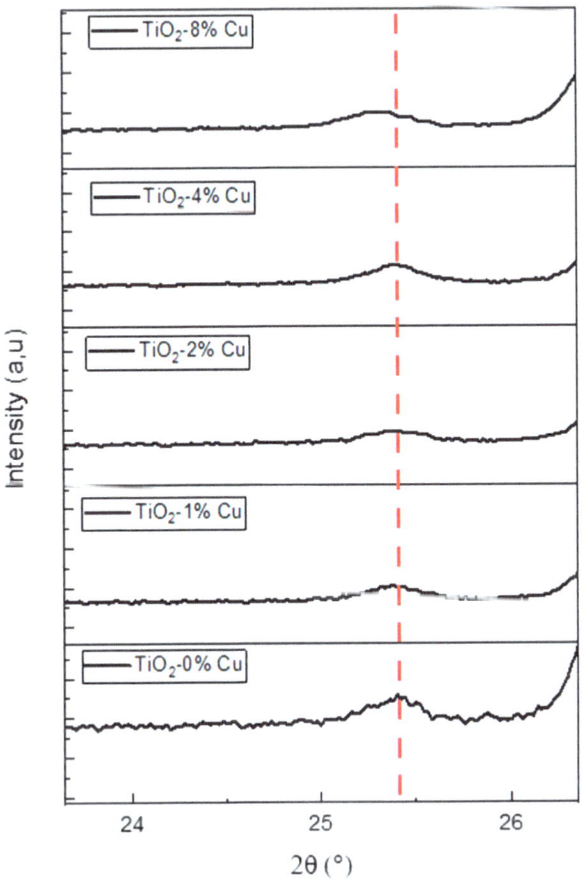

(**b**)

Figure 4. (**a**) XRD diffractogram of the undoped and 1%, 2%, 4% and 8% Cu-doped TiO_2 nanoparticle thin films. (**b**) A close-up of the TiO_2 (101) peak.

(**a**)

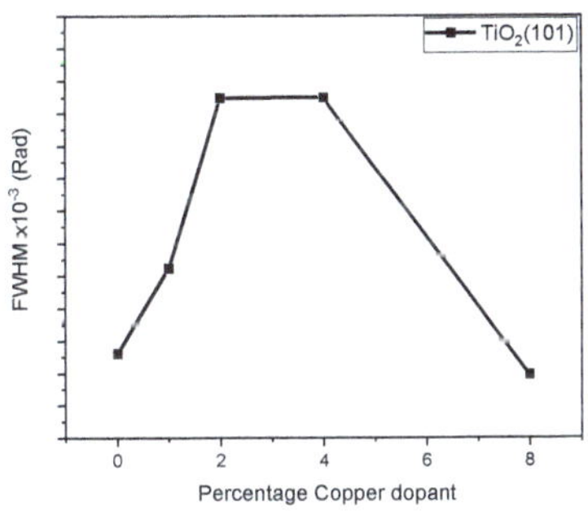

(**b**)

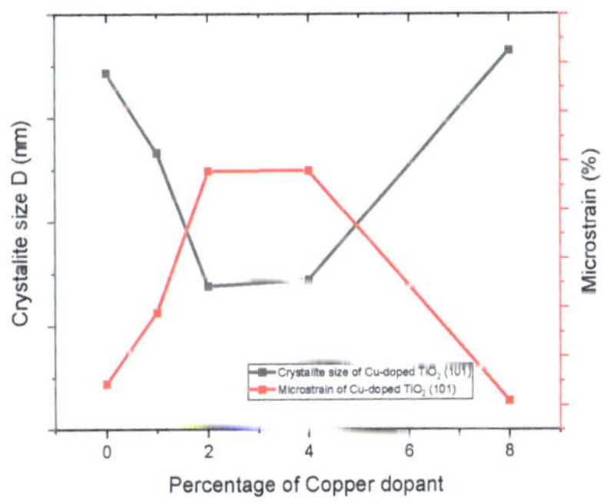

Figure 5. Variation in (**a**) FWHM and (**b**) crystallite size and microstrain of the undoped and 1%, 2%, 4% and 8% Cu-doped TiO_2 nanoparticle thin films.

Table 1. Crystallite size, lattice strain and microstrain calculations of the undoped and 1%, 2%, 4% and 8% Cu-doped TiO_2 at the preferred orientation (101).

Cu-Doped TiO_2 (101)	2θ (deg)	d(101) (Å)	FWHM (rad)	D (nm)	1/D (nm^{-1})	ε(%) × 10^{-3}
0% Cu	25.389	3.506934	0.002962	47.16	0.0212	3.289
1% Cu	25.388	3.515507	0.003223	43.33	0.023	3.579
2% Cu	25.397	3.472665	0.003748	36.9	0.0271	4.16
4% Cu	25.40	3.513734	0.003749	37.2	0.0268	4.161
8% Cu	25.32	3.508971	0.002896	48.23	0.0207	3.22

2.3. Optical Properties

To acknowledge the effect of Cu doping on the optical properties of the TiO_2 nanoparticle thin film, we used UV–Vis spectroscopy and obtained transmittance and absorbance spectra (Figure 6). Figure 6a depicts the transmittance of each sample, and it seems that the 4% copper dopant TiO_2 presents more light transmission than the other samples, followed by the 2% copper dopant TiO_2; this could positively affect the charge mobility and the performance of TiO_2 as an electron transport layer [45]. Figure 6b shows the absorption spectra of the samples, and it seems that at the range 300–400 nm, the sample with 8% copper dopant has a high absorbance value compared to the other samples; as for the 2% and 4% samples, they have a low absorbance value, even lower than the undoped TiO_2, but unexpectedly the 1% sample presents a higher light absorption than the undoped and 2% and 4% Cu-doped TiO_2.

(**a**) (**b**)

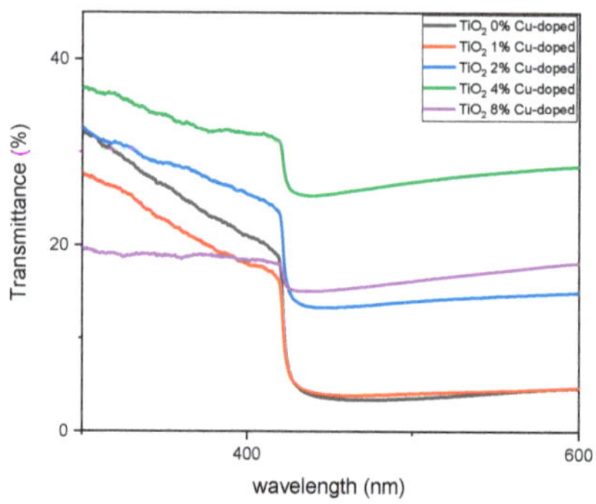

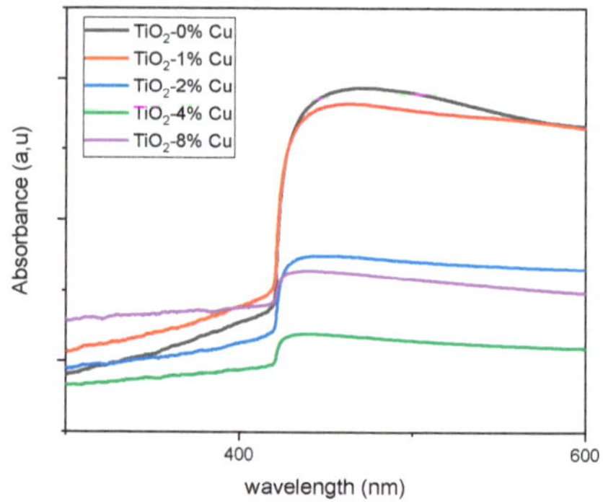

Figure 6. (**a**) Transmittance and (**b**) absorbance spectra of the undoped and 1%, 2%, 4% and 8% Cu-doped TiO_2 nanoparticle thin films.

To better understand the effect of doping with copper on the optical band gap of the samples, we used the Tauc formula below [46]

$$(Ah\upsilon)^{1/n} = C(h\upsilon - Eg) \tag{4}$$

where A is the absorption coefficient, C is a constant, Eg is the average band gap of the material and n depends on the type of the transition. After examination of the obtained Tauc plots $(Ah\upsilon)^2$ vs. hυ (Figure 7), the difference in the optical band gap is clear. If we exclude the 2% sample, it is noticeable that by increasing the copper doping percentage (the concentration of dopant) the optical band gap increases from 3.01 eV for undoped TiO_2 to 3.1 eV for 8% copper-doped TiO_2. But it is obvious to state that the optical band gap for the sample of 2% Cu-doped TiO_2 decreases in comparison with undoped and the other doped samples to a value of 2.96 eV. This is in accordance with Nair et al., who studied the effect of doping and crystalline properties on the band gap of nanoparticles and concluded that clearly there is a correlation between the particle size, strain and band gap [47]. Considering the calculated values in Table 1 and the obtained results from the Figure 5, the 2% Cu-doped TiO_2 nanoparticle sample presents the lowest crystallite size, the highest microstrain and the lowest bang gap. So, in short, and as they reported, when the particle is extremely small, pressure surface increases, meaning lattice strain increases and thus the band gap decreases [48]. So, as reported, Red shift or a decrease in optical band gap with a decrease in particle size arises due to the surface and interface effect and Blue shift or an increase in energy is due to the quantum size effect, which is the case for 1%, 4% and 8% copper-doped TiO_2.

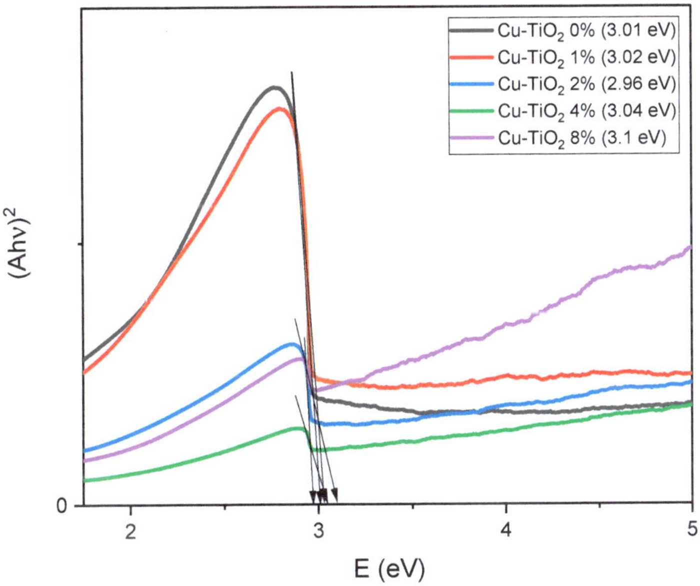

Figure 7. Tauc plots of the undoped and 1%, 2%, 4% and 8% Cu-doped TiO_2 nanoparticle thin films.

2.4. Photoluminescence

We investigated the recombination of charge carriers for undoped and Cu-doped TiO_2 nanoparticles. All samples were optically characterized with photoluminescence spectroscopy. As Figure 8 depicts, TiO_2 nanoparticles are not photoluminescent, since the PL spectra did not show any peak around the absorption wavelength that corresponds to the calculated band gap ranging from 2.96 to 3.1 eV (420–401 nm). The observed peak collection between 450 and 700 nm is related to the oxygen vacancies in the TiO_2 structure [49]. The peak appears to be decreasing with the increase in dopant concentration, which means a decrease in the recombination rate of charge carriers e^-/h^+, hence fast electronic transport and better photoactivity, except for the 8% and 2% samples, which give a higher PL intensity which could be due the high formation of defects, while the 4% sample presents the lowest PL intensity; also, it is clear that the doped samples present the same spectra shape as the undoped one, which means that Cu doping does not induce any PL signals. The significant peak of the 8% Cu-doped TiO_2 nanoparticles at 780 nm corresponds to the band gap of CuO (1.6 eV) according to the literature. But this hypothesis cannot be right since the XRD analysis did not show any peaks related to copper oxide structure, unless it also means that the formed copper oxide nanoparticles are extremely well dispersed on TiO_2 nanoparticles and that is why they are undetectable by the X-ray diffractometer.

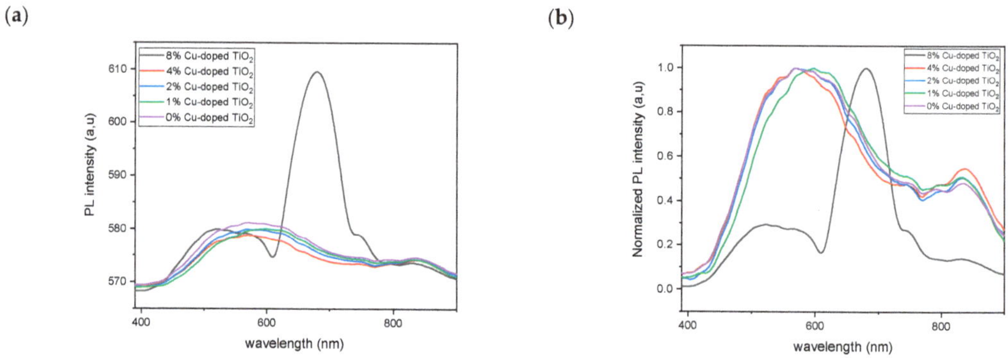

Figure 8. (**a**) Photoluminescence and (**b**) normalized PL spectra of the undoped and 1%, 2%, 4% and 8% Cu-doped TiO_2 nanoparticle thin films.

3. Experimental Details Materials and Methods

3.1. Materials

All chemicals and reagents were used directly without further purification as received. Absolute ethanol, isopropyl alcohol, acetone, FTO, Titanium diisopropoxide bis(acetylacetonante) 75 wt.% in isopropanol ($C_{16}H_{32}O_6Ti$) and Cupric Acetate Monohydrate (Copper (II) Acetate Monohydrate) pure, 98% ($C_4H_8CuO_5$). All the products were bought from Sigma-Aldrich (St. Louis, MO, USA).

3.2. Pure and Cu-Doped TiO_2 Film Elaboration

Firstly, the FTO glass was cleaned with acetone, ethanol and distilled water. Then, to deposit the TiO_2 nanoparticles on the surface of the FTO glass, a simple cost-free method was used: spin coating, which consists of coating the FTO glass surface with a solution of titanium precursor. To study the effect of doping, we added a portion of copper precursor, while varying the amount of copper from 0%, 1%, 2%, 4% to 8%. After the deposition of the Cu dopant TiO_2 thin film with spin coating, the as-prepared samples were annealed at 450 °C for 1 h and then naturally cooled down for morphological, structural and optical characterization. All of the steps are illustrated in Figure 9.

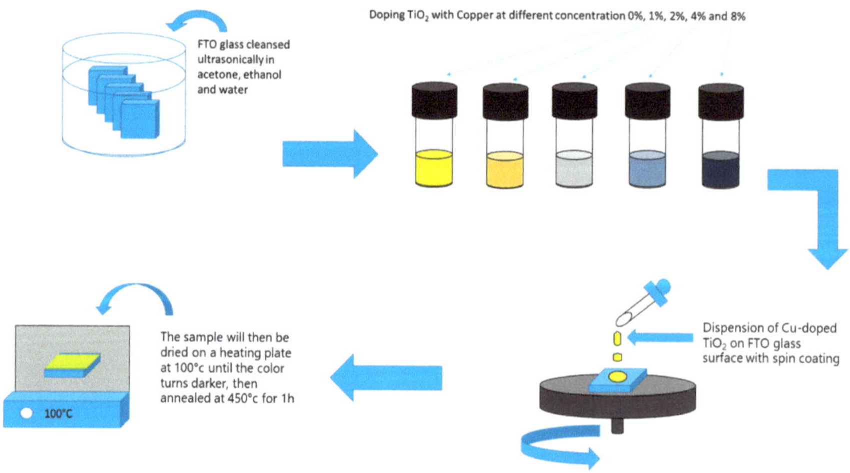

Figure 9. Illustration of the steps for sample preparation of 0%, 1%, 2%, 4% and 8% Cu-doped TiO_2 nanoparticle thin films.

3.3. Characterization Techniques

The samples were morphologically characterized using scanning electron microscopy (SEM, FEI XL30 ESEM Company, Hillsboro, OR, USA). As for the structural stability and crystallographic formation of the phase, X-ray diffraction (XRD) was performed using a Philips X'PERT-MPD diffractometer equipped with CuKα radiation (λ = 1.5406 Å), with the diffraction patterns in the range of 20–80°. To determine the gap energy of each sample and its transmittance and absorbance properties, we used a UV–Vis spectrophotometer and photoluminescence (PL) spectroscopy in the range of 200 and 1200 nm by equipping a PerkinElmer Lambda 950.

4. Conclusions

TiO_2 nanoparticles doped with copper were synthesized with a simple method of spin coating, which consists of dispersing the prepared solution on FTO glass at high speed, and these samples of Cu-doped TiO_2 at 0%, 1%, 2%, 4% and 8% were characterized morphologically, structurally and optically and the acquired results reveal that the 4% sample has the highest transmittance, while the 2% is the second highest; these samples could be perfect candidates for electron transport. The 2% sample exhibited the lowest band gap energy of 2.96 eV, whereas the 8% samples exhibit the highest at 3.05 eV; the 8% sample also showed a new PL peak compared to the other samples, which could be related to the copper oxide.

Author Contributions: Conceptualization, A.H. and B.B.; Methodology, L.K.; Software, S.S. and L.K.; Validation, A.B. and B.M.S.; Formal analysis, S.S. and L.K.; Investigation, A.B.; Resources, B.B.; Data curation, S.S.; Writing—original draft, S.S.; Writing—review & editing, A.B., A.H., L.K. and B.M.S.; Visualization, A.B.; Supervision, A.H., B.B. and B.M.S.; Funding acquisition, B.B. and B.M.S. All authors have read and agreed to the published version of the manuscript.

Funding: Syrine Sassi acknowledged Erasmus plus for the grant, the author Amal Bouich acknowledges MCIN for funding support through MS Fellowship (MCIN/AEI/10.13039/501100011033) and UE NextGeneration. This research was funded by the Ministerio de Ciencia e Innovación (Spain) and by the Spanish Agencia Estatal de Investigación through projects BESTMAT PID2019-107137RB-C21/AEI/10.13039/5011000 11033 and PID2019-107137RB-C22/AEI/10.13039/501100011033 and by ERDF under the funding "A way of making Europe".

Data Availability Statement: The original contributions presented in the study are included in the article, further inquiries can be directed to the corresponding authors.

Conflicts of Interest: The authors declare no conflict of interest.

References

1. Hsu, A.; Rosengarten, C.; Weinfurter, A.; Xie, Y. *Renewable Energy and Energy Efficiency in Developing Countries: Contributions to Reducing Global Emissions*; United Nations Environment Programme: New York, NY, USA, 2017.
2. Green, M.A.; Emery, K.; Hishikawa, Y.; Warta, W.; Dunlop, E.D. Solar cell efficiency tables (Version 45). *Prog. Photovolt. Res. Appl.* **2015**, *23*, 1–9. [CrossRef]
3. Saïdi, H.; Boujmil, M.F.; Durand, B.; Lazzari, J.L.; Bouaïcha, M. Elaboration and characterization of $CuInSe_2$ thin films using one-step electrodeposition method on silicon substrate for photovoltaic application. *Mater. Res. Express* **2018**, *5*, 016414. [CrossRef]
4. Rathore, N.; Panwar, N.L.; Yettou, F.; Gama, A. A comprehensive review of different types of solar photovoltaic cells and their applications. *Int. J. Ambient. Energy* **2021**, *42*, 1200–1217. [CrossRef]
5. Dong, Y.; Yang, Y.; Qiu, L.; Dong, G.; Xia, D.; Liu, X.; Li, M.; Fan, R. Polyoxometalate Based Inorganic-Organic Hybrid [Cu(phen)2]2[(α-Mo8O26)]: A New Additive to Spiro-OMeTAD for Efficient and Stable Perovskite Solar Cells. *ACS Appl. Energy Mater.* **2019**, *2*, 4224–4233. [CrossRef]
6. Wu, M.C.; Chan, S.H.; Lee, K.M.; Chen, S.H.; Jao, M.H.; Chen, Y.F.; Su, W.F. Enhancing the efficiency of perovskite solar cells using mesoscopic zinc-doped TiO_2 as the electron extraction layer through band alignment. *Mater. Chem. A* **2018**, *6*, 16920–16931. [CrossRef]
7. Teimouri, R.; Heydari, Z.; Ghaziani, M.P.; Madani, M.; Abdy, H.; Kolahdouz, M.; Asl-Soleimani, E. Synthesizing Li doped TiO_2 electron transport layers for highly efficient planar perovskite solar cell. *Superlattices Microstruct.* **2020**, *145*, 106627. [CrossRef]
8. Mohammad, A.; Mahjabeen, F. Promises and Challenges of Perovskite Solar Cells: A Comprehensive Review. *BULLET J. Multidisiplin. Ilmu* **2023**, *2*, 1147–1157.

9. Ma, T.; Wang, S.; Zhang, Y.; Zhang, K. Yi, The development of all-inorganic $CsPbX_3$ perovskite solar cells. *J. Mater. Sci.* **2020**, *55*, 464–479. [CrossRef]
10. Bouich, A.; Torres, J.C.; Khattak, Y.H.; Baig, F.; Marí-Guaita, J.; Soucase, B.M.; Mendez-Blas, A.; Palacios, P. Bright future by controlling α/δ phase junction of formamidinium lead iodide doped by imidazolium for solar cells: Insight from experimental, DFT calculations and SCAPS simulation. *Surf. Interfaces* **2023**, *40*, 103159. [CrossRef]
11. Ding, M.; Sun, L.; Chen, X.; Luo, T.; Ye, T.; Zhao, C.; Zhang, W.; Chang, H. Airprocessed, large grain perovskite films with low trap density from perovskite crystal engineering for high-performance perovskite solar cells with improved ambient stability. *J. Mater. Sci.* **2019**, *54*, 12000–12011. [CrossRef]
12. Bouich, A.; Marí-Guaita, J.; Soucase, B.M.; Palacios, P. Bright future by enhancing the stability of methylammonium lead triiodideperovskites thin films through Rb, Cs and Li as dopants. *Mater. Res. Bull.* **2023**, *163*, 112213. [CrossRef]
13. Hossain, M.K.; Samajdar, D.P.; Das, R.C.; Arnab, A.A.; Rahman, M.F.; Rubel, M.H.; Islam, M.R.; Bencherif, H.; Pandey, R.; Madan, J.; et al. Design and simulation of Cs_2BiAgI_6 double perovskite solar cells with different electron transport layers for efficiency enhancement. *Energy Fuels* **2023**, *37*, 3957–3979. [CrossRef]
14. Mukametkali, T.M.; Ilyassov, B.R.; Aimukhanov, A.K.; Serikov, T.M.; Baltabekov, A.S.; Aldasheva, L.S.; Zeinidenov, A.K. Effect of the TiO_2 electron transport layer thickness on charge transfer processes in perovskite solar cells. *Phys. B Condens. Matter* **2023**, *659*, 414784. [CrossRef]
15. Cheng, M.; Zuo, C.; Wu, Y.; Li, Z.; Xu, B.; Hua, Y.; Ding, L. Charge-transport layer engineering in perovskite solar cells. *Sci. Bull.* **2020**, *65*, 1237–1241. [CrossRef] [PubMed]
16. Lin, L.; Jones, T.W.; Yang, T.C.; Duffy, N.W.; Li, J.; Zhao, L.; Chi, B.; Wang, X.; Wilson, G.J. Inorganic electron transport materials in perovskite solar cells. *Adv. Funct. Mater.* **2021**, *31*, 2008300. [CrossRef]
17. Li, F.; Shen, Z.; Weng, Y.; Lou, Q.; Chen, C.; Shen, L.; Guo, W.; Li, G. Novel electron transport layer material for perovskite solar cells with over 22% efficiency and long-term stability. *Adv. Funct. Mater.* **2020**, *30*, 2004933. [CrossRef]
18. Valadi, K.; Gharibi, S.; Taheri-Ledari, R.; Akin, S.; Maleki, A.; Shalan, A.E. Metal oxide electron transport materials for perovskite solar cells: A review. *Environ. Chem. Lett.* **2021**, *19*, 2185–2207. [CrossRef]
19. Zhen, C.; Wu, T.; Chen, R.; Wang, L.; Liu, G.; Cheng, H.M. Strategies for Modifying TiO_2 Based Electron Transport Layers to Boost Perovskite Solar Cells. *ACS Sustain. Chem. Eng.* **2019**, *7*, 4586–4618. [CrossRef]
20. Kaewprajak, A.; Kumnorkaew, P.; Lohawet, K.; Duong, B.; Chonsut, T.; Kayunkid, N.; Saranrom, N.; Promarak, V. An unconventional blade coating for low-cost fabrication of PCDTBT: $PC_{70}BM$ polymer and $CH_3NH_3PbI_xCl_{3-x}$ perovskite solar cells. *Surf. Interfaces* **2021**, *23*, 100969. [CrossRef]
21. Stranks, S.D.; Eperon, G.E.; Grancini, G.; Menelaou, C.; Alcocer, M.J.; Leijtens, T.; Herz, L.M.; Petrozza, A.; Snaith, H.J. Electron-hole diffusion lengths exceeding 1 micrometer in an organometal trihalide perovskite absorber. *Science* **2013**, *342*, 341–344. [CrossRef]
22. Nishat, S.S.; Hossain, M.J.; Mullick, F.E.; Kabir, A.; Chowdhury, S.; Islam, S.; Hossain, M. Performance analysis of perovskite solar cells using DFT-extracted parameters of metal-doped TiO_2 electron transport layer. *J. Phys. Chem. C* **2021**, *125*, 13158–13166. [CrossRef]
23. Che Halin, D.S.; Azhari, A.W.; Mohd Salleh, M.A.; Muhammad Nadzri, N.I.; Vizureanu, P.; Abdullah, M.M.; Wahab, J.A.; Sandu, A.V. Metal-Doped TiO_2 Thin Film as an Electron Transfer Layer for Perovskite Solar Cells: A Review. *Coatings* **2022**, *13*, 4. [CrossRef]
24. Chen, S.H.; Ho, C.M.; Chang, Y.H.; Lee, K.M.; Wu, M.C. Efficient perovskite solar cells with low J-V hysteretic behavior based on mesoporous Sn-doped TiO_2 electron extraction layer. *Chem. Eng. J.* **2022**, *445*, 136761. [CrossRef]
25. Chen, S.H.; Chan, S.H.; Lin, Y.T.; Wu, M.C. Enhanced power conversion efficiency of perovskite solar cells based on mesoscopic Ag-doped TiO_2 electron transport layer. *Appl. Surf. Sci.* **2019**, *469*, 18–26.
26. Dahlan, D.; Saad, S.K.; Berli, A.U.; Bajili, A.; Umar, A.A. Synthesis of two-dimensional nanowall of Cu-Doped TiO_2 and its application as photoanode in DSSCs. *Phys. E Low-Dimens. Syst. Nanostructures* **2017**, *91*, 185–189. [CrossRef]
27. Abbas, M.M.; Rasheed, M. Investigation of structural, Mechanical, Thermal and Optical Properties of Cu Doped TiO_2. *Iraqi J. Phys.* **2021**, *19*, 1–9. [CrossRef]
28. Bhattacharyya, K.; Mane, G.P.; Rane, V.; Tripathi, A.K.; Tyagi, A.K. Selective CO_2 photoreduction with Cu-doped TiO_2 photocatalyst: Delineating the crucial role of Cu-oxidation state and oxygen vacancies. *J. Phys. Chem. C* **2021**, *125*, 1793–1810. [CrossRef]
29. Dhonde, M.; Sahu, K.; Murty, V.V. Cu-doped TiO_2 nanoparticles/graphene composites for efficient dye-sensitized solar cells. *Sol. Energy* **2021**, *220*, 418–424. [CrossRef]
30. Deng, Z.; Li, L.; Ren, Y.; Ma, C.; Liang, J.; Dong, K.; Liu, Q.; Luo, Y.; Li, T.; Tang, B.; et al. Highly efficient two-electron electroreduction of oxygen into hydrogen peroxide over Cu-doped TiO_2. *Nano Res.* **2022**, *15*, 3880–3885. [CrossRef]
31. Raguram, T.; Rajni, K.S. Synthesis and characterisation of Cu-Doped TiO_2 nanoparticles for DSSC and photocatalytic applications. *Int. J. Hydrog. Energy* **2022**, *47*, 4674–4689. [CrossRef]
32. He, Z.; Zhang, S.; Yin, L.; Hayat, M.D.; Cao, P. Cu–TiO_2 nanocomposite coatings prepared from sol-enhanced electrodeposition. *Int. J. Mod. Phys. B* **2020**, *34*, 2040038. [CrossRef]
33. Reda, S.M.; Khairy, M.; Mousa, M.A. Photocatalytic activity of nitrogen and copper doped TiO_2 nanoparticles prepared by microwave-assisted sol-gel process. *Arab. J. Chem.* **2020**, *13*, 86–95. [CrossRef]

34. Farzaneh, A.; Javidani, M.; Esrafili, M.D.; Mermer, O. Optical and photocatalytic characteristics of Al and Cu doped TiO_2: Experimental assessments and DFT calculations. *J. Phys. Chem. Solids* **2022**, *161*, 110404. [CrossRef]
35. Sahu, K.; Dhonde, M.; Murty, V.V. Microwave-assisted hydrothermal synthesis of Cu-doped TiO_2 nanoparticles for efficient dye-sensitized solar cell with improved open-circuit voltage. *Int. J. Energy Res.* **2021**, *45*, 5423–5432. [CrossRef]
36. Jasima, F.H.; Shakirb, H.R.; Chiada, S.S.; Habubic, N.F.; Mosad, Z.S.; Kadhime, Y.H.; Jadanf, M. Characterizations of sprayed TiO_2 and Cu doped TiO_2 thin films prepared by spray pyrolysis method. *Dig. J. Nanomater. Biostructures (DJNB)* **2023**, *18*, 1385–1393. [CrossRef]
37. Lukong, V.T.; Ukoba, K.; Jen, T.C. Review of self-cleaning TiO_2 thin films deposited with spin coating. *Int. J. Adv. Manuf. Technol.* **2022**, *122*, 3525–3546. [CrossRef]
38. Liu, J.; Yang, Z.; Gong, Z.; Shen, Z.; Ye, Y.; Yang, B.; Qiu, Y.; Ye, B.; Xu, L.; Guo, T.; et al. Weak light-stimulated synaptic hybrid phototransistors based on islandlike perovskite films prepared by spin coating. *ACS Appl. Mater. Interfaces* **2021**, *13*, 13362–13371. [CrossRef]
39. Liao, Y.H.; Chang, Y.H.; Lin, T.H.; Chan, S.H.; Lee, K.M.; Hsu, K.H.; Hsu, J.F.; Wu, M.C. Boosting the power conversion efficiency of perovskite solar cells based on Sn doped TiO_2 electron extraction layer via modification the TiO_2 phase junction. *Sol. Energy* **2020**, *205*, 390–398. [CrossRef]
40. Hajjaji, M.A.; Missaoui, K.; Trabelsi, K.; Bouzaza, A.; Bessais, B.; Hajjaji, A.; Assadi, A.A. Electrodeposited Platinum Nanoparticles on Highly Ordered Titanium Dioxide Nanotubes for Photocatalytic Application: Enhancement of Photocatalytic Degradation of Amido Black Dye. *Catal. Lett.* **2023**, *154*, 1242–1254. [CrossRef]
41. Nguyen, T.M.; Bark, C.W. Synthesis of Cobalt-Doped TiO_2 Based on Metal-Organic Frameworks as an Effective Electron Transport Material in Perovskite Solar Cells. *ACS Omega* **2020**, *5*, 2280–2286. [CrossRef] [PubMed]
42. Bensouici, F.; Bououdina, M.; Dakhel, A.A.; Tala-Ighil, R.; Tounane, M.; Iratni, A.; Souier, T.; Liu, S.; Cai, W.J. Optical, structural and photocatalysis properties of Cu-doped TiO_2 thin films. *J. Appl. Sci.* **2017**, *395*, 110–116.
43. Sassi, S.; Trabelsi, K.; El Jery, A.; Abidi, M.; Hajjaji, A.; Khezami, L.; Karrech, A.; Gaidi, M.; Soucase, B.M.; Bessais, B. Synergistic effect of CuxOy-NPs/TiO_2-NTs heterostructure on the photodegradation of amido black staining. *Optik* **2023**, *272*, 170234. [CrossRef]
44. Trabelsi, K.; Jemai, S.; El Jery, A.; Sassi, S.; Guesmi, A.; Khezami, L.; Hajjaji, A.; Gaidi, M.; Bessais, B. Ag-NPs coating influence on TiO_2-NTs photocatalytic performances on Amido Black staining. *Res. Sq.* **2022**. [CrossRef]
45. Sławek, A.; Starowicz, Z.; Lipiński, M. The influence of the thickness of compact TiO_2 electron transport layer on the performance of planar $CH_3NH_3PbI_3$ perovskite solar cells. *Materials* **2021**, *14*, 3295. [CrossRef]
46. Bouich, A.; Torres, J.C.; Chfii, H.; Marí-Guaita, J.; Khattak, Y.H.; Baig, F.; Soucase, B.M.; Palacios, P. Delafossite as hole transport layer a new pathway for efficient perovskite-based solar cells: Insight from experimental, DFT and numerical analysis. *Sol. Energy* **2023**, *250*, 18–32. [CrossRef]
47. Nair, S.S.; Mathews, M.; Anantharaman, M.R. Anantharaman. Evidence for blueshift by weak exciton confinement and tuning of bandgap in superparamagnetic nanocomposites. *Chem. Phys. Lett.* **2005**, *406*, 398–403. [CrossRef]
48. Deotale, A.J.; Nandedkar, R.V. Correlation between particle size, strain and band gap of iron oxide nanoparticles. *Mater. Today Proc.* **2016**, *3*, 2069–2076. [CrossRef]
49. Komaraiah, D.; Radha, E.; Kalarikkal, N.; Sivakumar, J.; Reddy, M.R.; Sayanna, R. Structural, optical and photoluminescence studies of sol-gel synthesized pure and iron doped TiO_2 photocatalysts. *Ceram. Int.* **2019**, *45*, 25060–25068. [CrossRef]

inorganics

MDPI

Article

Thermal Shock Resistance of Commercial Oxide-Bonded Silicon Carbide Reticulated Foams under Concentrated Solar Radiation at PSA: A Feasibility Study

Fernando de Almeida Costa Oliveira [1,*], José Galindo [2], José Rodríguez [2], Inmaculada Cañadas [2] and Jorge Cruz Fernandes [3]

[1] LNEG—Laboratório Nacional de Energia e Geologia I.P., LEN—Laboratório de Energia, UME—Unidade de Materiais para a Energia, Estrada do Paço do Lumiar, 22, 1649-038 Lisboa, Portugal

[2] PSA—Plataforma Solar de Almería, CIEMAT—Centro de Investigaciones Energéticas, Medioambientales y Tecnológicas, Apartado 22, E-04200 Tabernas, Spain; jose.galindo@psa.es (J.G.); jose.rodriguez@psa.es (J.R.); inmaculada.canadas@psa.es (I.C.)

[3] IDMEC—Instituto de Engenharia Mecânica, Instituto Superior Técnico, Universidade de Lisboa, Av. Rovisco Pais, 1049-001 Lisboa, Portugal; cruz.fernandes@tecnico.ulisboa.pt

* Correspondence: fernando.oliveira@lneg.pt

Abstract: Volumetric ceramic receivers can be regarded as a promising technology to heat air above 1000 °C for solar thermal electricity production. In this study, the thermal shock behavior of commercial 10 ppi (A) and 20 ppi (B) oxide-bonded silicon carbide (ob-SiC) reticulated porous ceramic (RPC) foams was evaluated using the SF60 solar furnace at Plataforma Solar de Almería. The foams were subjected to well-controlled temperature cycles ranging from 800 to 1000, 1200, 1300 or 1400 °C, for 25, 100, and 150 cycles. The extent of the damage after thermal shock was determined by crushing tests. The damage was found to be critically dependent on both the bulk density and cell size. Decreasing both the bulk density and cell size resulted in better thermal shock resistance. The B foam exhibited approximately half the stress degradation compared to the A foam when exposed to a temperature difference of 600 K (in the range of 800 to 1400 °C) and subjected to 150 cycles.

Keywords: silicon carbide foams; solar furnace; concentrated solar radiation; thermal shock; crushing strength; open volumetric receivers

Citation: Costa Oliveira, F.d.A.; Galindo, J.; Rodríguez, J.; Cañadas, I.; Cruz Fernandes, J. Thermal Shock Resistance of Commercial Oxide-Bonded Silicon Carbide Reticulated Foams under Concentrated Solar Radiation at PSA: A Feasibility Study. *Inorganics* **2024**, *12*, 246. https://doi.org/10.3390/inorganics12090246

Academic Editor: Roberto Nisticò

Received: 24 July 2024
Revised: 27 August 2024
Accepted: 4 September 2024
Published: 11 September 2024

1. Introduction

The usage of solar energy is one way to increase the sustainability of production processes using a renewable and carbon free heating source [1–3]. Among the solar approaches available, concentrated solar power (CSP) tower technology is well established [4]. It relies on concentrating light through mirrors onto a receiver (also called an absorber) [5,6]. Many studies were conducted on suitable materials for solar receivers, including SiC fiber mesh Ceramat® FN, supplied by Schott Glass (Mainz, Germany), SiC monoliths manufactured by HelioTech (Svenstrup, Denmark), and SiC foams [7,8]. Wang et al. [9] performed thermal analysis of the porous media receiver by combining the Monte Carlo Ray Tracing method with FLUENT software but did not mention the morphology of the SiC foam tested on a solar dish collector. A similar approach has been employed by Kribus et al. [10] for an open receiver made of SiC foam, suggesting that the available commercial materials offered a limited range of features that are still not optimal for the application. Optimization of geometry (porosity and pore size) appears to be insufficient to reach a high efficiency, whilst a double layer SiC foam holds great potential to enhance efficiency. This is attributed to the fact that the thickness of the first porous layer has a significant effect on the temperature distribution and pressure drop.

On the other hand, Lidor et al. [11] showed that there is a significant temperature gradient across large scale porous absorbers. Development of new thick, porous materials for volumetric air receivers is therefore important. In this respect, one must bear in mind that the extinction coefficient is another key parameter for estimating radiation properties in porous media, as it enables the determination of the radiative thermal conductivity [12]. It has been shown that the extinction coefficient increases with both decreasing porosity and pore size [13]. Furthermore, high optical thicknesses (larger pore sizes) favor more homogenous temperature gradients across thicker ceramic foams, as the concentrated solar beam can penetrate deep inside the porous structure [14].

A 5-kW solar receiver prototype proved to deliver a high-temperature air flow (>1000 °C) with a reasonably high thermal efficiency (>0.65) using 10 ppi (pores per inch) SiSiC foams [15]. Thence, there is plenty of room for improvement on both the geometry and the material properties for large scale absorbers. Indeed, challenges associated with durability, receiver efficiency, and the specific cost remain to be sorted out. The inherent advantages of air receivers, such as availability of the fluid, no trace heating necessary, non-toxic, and 3–5 h of thermal storage, allow higher-efficiency thermodynamic cycles, and the high receiver thermal efficiency (>75% due to the volumetric effect which reduces thermal radiation losses) makes this technology simpler, cheaper, and more efficient than other currently available technologies [16]. However, the material's suitability and its durability still need to be addressed in a suitable fashion. Among the current solar receivers, three major types exist, namely: (a) Surface receivers (tubular, external, cavity), (b) Porous receivers (wire mesh, ceramic/metallic foams, honeycombs), and (c) Particle receivers (falling curtain, entrained particles), according to Mey-Coltier et al. [17]. Exploiting the 3D structure and properties of ceramic foams in this kind of application can open new routes for producing heat from solar irradiation. First used in liquid metal filtration, reticulated porous ceramic (RPC) foams are currently being used as gas (particulate) filters, kiln furniture, catalyst substrates, porous burner substrates, scaffolds for bone regeneration, advanced thermal management, heat exchangers, and bacteria/cell immobilization, among others [18,19].

The present study aimed at evaluating the damage imposed on two commercial oxide-bonded silicon carbide reticulated porous ceramic foams, hereafter referred to as ob-SiC RPC foams, having pore densities of 10 and 20 ppi (pores per inch), through exposure to concentrated solar radiation, under drastic thermal shock conditions, replicating those experienced by real solar open volumetric air receivers. The development of solar thermal energy conversion processes faces challenges related to receiver/reactor materials and radiative energy distribution technology. Recently, innovative approaches to efficient solar energy conversion have been developed using complex coupled numerical models applied to a high-flux solar reactor and a simulator [20]. Before widespread use, the suitability and cyclic stability of the materials must be demonstrated. To the best of the author's knowledge, such data has not yet been reported in the open literature for commercial ob-SiC RPC foams, which would be useful for modelling purposes.

The simplest way to assess thermal shock resistance is by rapidly transferring the heated samples from a resistance furnace to a quenching bath containing distilled water at room temperature. Thus, for comparison purposes, the ob-SiC RPC foams under investigation were kept at a soaking temperature (namely 420 and 620 °C) for 1 h prior to quenching. The damage was assessed by measuring the retained crushing strength after water quenching tests at various temperature differences, ΔT, which are defined as the pre-set temperature of the samples minus the temperature of the bath, typically at 20 °C. The resistance to thermal shock of the dense ceramics is found to fall in an abrupt fashion, whereas in the case of RPC foams it was found that the crushing strength retained undergoes a gradual decrease with increasing quench temperature because of an increase in mechanical damage throughout the material. A similar trend was observed for thermal shock tests carried out on cordierite foams [21].

Regarding the current state of the art, thermal shock was assessed by means of an innovative experimental setup, developed at PSA, including a guillotine system, a secondary concentrator, and a temperature control system, which allowed the performance of thermal shock cycles under conditions that cannot be realized in conventional electrical or gas furnaces [22].

Thermal cycling of the ob-SiC RPC foams under direct solar irradiation is regarded as an interesting alternative to the current honeycombs based on siliconized silicon carbide commonly used in pilot power plants, not only in terms of cost, but also performance [23–25]. After cycling, the retained crushing strength was measured, and the results are quite promising.

2. Results

2.1. Materials Characterization

Table 1 summarizes the bulk densities (ρ_b) and porosities of the tested materials under investigation. The true density of the ob-SiC RPC strut material (ρ_r) was determined by He pycnometry to be 3.01 ± 0.01 Mg m^{-3}. Therefore, the maximum porosity was of about 87%, bearing in mind that $P = 1 - (\rho_b/\rho_r) = 1 - (0.38/3.01)$.

Table 1. Samples characteristics.

Pore Size (ppi)	Bulk Density [1] (Mg m^{-3})	Relative Density	Porosity (%)
10	0.46 ± 0.01	0.15	85
20	0.38 ± 0.01	0.13	87

[1] mean ± SD of 10 samples.

The macrostructure of the as-received foams is shown in Figure 1. Typically, the mean thickness of the struts ranged from 0.75 ± 0.20 mm and 0.45 ± 0.15 mm, for the 10 ppi and 20 ppi foams, respectively. The linear intercept method was used to determine the cell size and the values obtained match the specification.

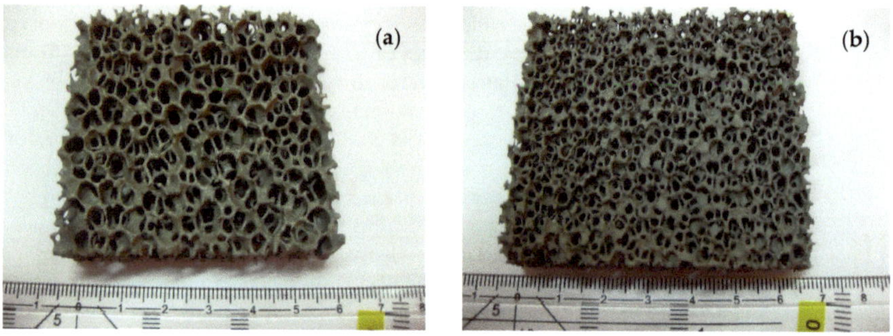

Figure 1. Top view of as-received 10 ppi (**a**) and 20 ppi (**b**) ob-SiC RPC foams.

The X-ray diffraction (XRD) profiles of the as-received and exposed ob-SiC RPC foams, after grinding into powder form using an agate mortar, are shown in Figure 2. The diffraction patterns, in the range of 20° < 2θ < 75°, show peaks corresponding to the planes, which match with the pattern of the ICDD data base, having the card no. 01-072-0018 (corresponding to the phase Moissanite-6H-SiC/•/) and traces of Moissanite-4H-SiC/□/(ICDD card no. 029-1127), SiO_2/◊/(ICDD card no. 039-1425) and Al_2O_3/○/(ICDD card no. 01-070-7346).

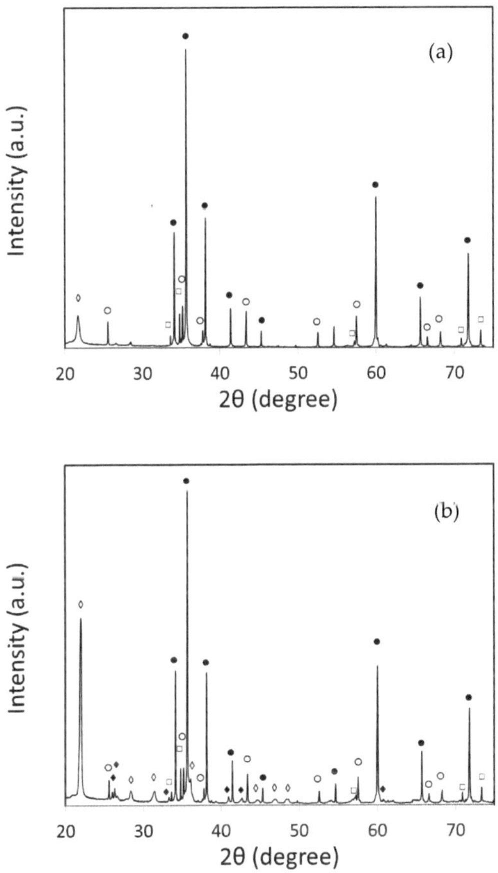

Figure 2. Typical XRD patterns of ob-SiC RPC foam before (**a**) and after exposure at 1400 °C for 150 cycles (**b**).

After exposure to 1400 °C for 150 cycles, oxidation of the ob-SiC RPC foams took place as indicated by the increase in the intensity of the peaks matching the SiO_2 phase and the formation of minor amounts of mullite ($Al_6Si_2O_{13}$)/♦/(ICDD card no. 015-0776).

2.2. Thermal Shock

To evaluate the damage imposed on reticulated porous ceramic (RPC) materials, namely commercial silicon carbide foams having pore densities of 10 and 20 ppi, thermal shock cycling was performed under conditions replicating those experienced by real solar open volumetric air receivers through direct exposure to concentrated solar radiation.

Most thermal performance data available refer to tests performed using xenon lamps simulators, so information regarding the effect of the entire wavelengths spectrum of solar radiation reaching the materials surface on their degradation is not known with certainty. Five new porous morphologies suitable for volumetric solar receivers have been analyzed experimentally in a laboratory-scale solar simulator in terms of the efficiency of the absorber's thermal performance, but no data have been provided on their mechanical performance [26]. It is therefore pertinent to address this issue, particularly resistance to thermal shock. In this study, ten thermal shock tests were undertaken in the SF60 solar furnace of PSA under well-controlled conditions listed in Table 4.

Figure 3 shows the temperature distribution over the top surface exposed area of the foams. The maximum deviation observed between the target value (point 1) and the temperature measured at the middle of each of the four exposed samples was around 30 °C. The temperature measured by the IR camera was similar to that measured by the B1 thermocouple.

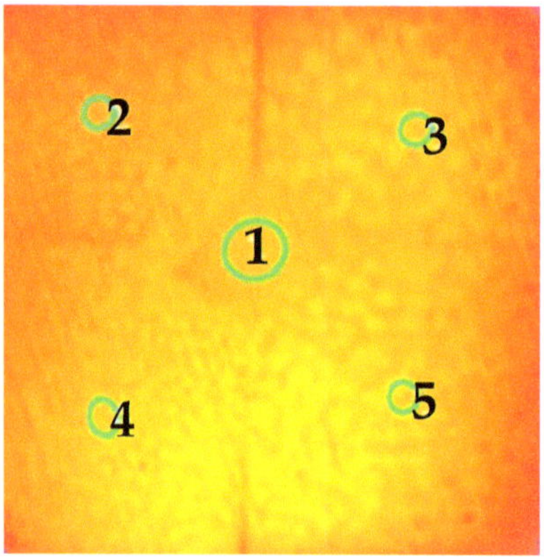

Point	T (°C)
1	1402 ± 13
2	1376 ± 20
3	1426 ± 17
4	1421 ± 18
5	1430 ± 17

Figure 3. Temperature distribution over the exposed top surface area showing typical temperatures measured using the IR camera when the maximum temperature was set at 1400 °C.

The temperature measured by the thermocouple B1 was therefore used to control the shutter opening/closing for performing the thermal cycles shown in Figure 4.

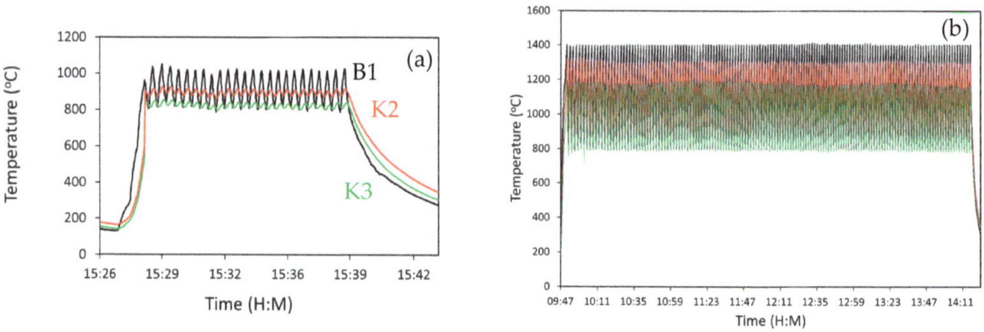

Figure 4. Temperature versus time graphs at ΔT = 600 K after 25 cycles (**a**) and 150 cycles (**b**) for 10 ppi ob-SiC RPC foam showing profiles measured by thermocouples B1 (black), K2 (red) and K3 (green).

2.3. Oxidation Data

Table 2 lists the mass changes recorded before and after the thermal shock tests. Oxidation is more evident on the top surface than on the bottom one, as denoted by the darker surface appearance (Figure 6d), particularly at temperature above 1200 °C. At ΔT = 200 K, no mass change was observed after 25 cycles.

Table 2. Mass changes (in %) of the exposed 10 ppi (**A**) and 20 ppi (**B**) ob-SiC RPC foams.

(A)				
N°. Cycles	**ΔT (K)**			
	200	400	500	600
25	0	0.04 ± 0.01	0.36 ± 0.17	1.77 ± 0.65
100		0.28 ± 0.03	1.37 ± 0.32	2.68 ± 0.39
150		0.99 ± 0.20	1.47 ± 0.06	3.30 ± 0.67
(B)				
N°. Cycles	**ΔT (K)**			
	200	400	500	600
25	0	0.25 ± 0.03	0.61 ± 0.08	1.23 ± 0.49
100		0.38 ± 0.03	1.69 ± 0.51	2.26 ± 0.08
150		0.89 ± 0.05	1.72 ± 0.52	2.99 ± 0.81

2.4. Crushing Data

Subjecting the top surface of the ob-SiC RPC foams to rapid change in temperature produces a thermal stress distribution, the magnitude of which depends on the heat transfer coefficient at the surface relative to the rate at which heat can be conducted to or from the surface through the bulk.

Representative curves of crushing strength versus strain are shown in Figure 5. It was observed that a macrocrack propagated through the compressed foam when the slope of the curve changed. For this reason, the value corresponding to such change was considered as the rupture stress.

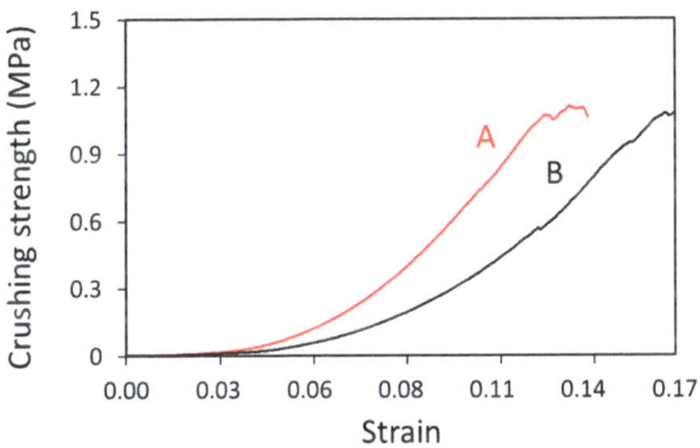

Figure 5. Typical crushing strength curves for foams A and B tested at ΔT = 600 K for 150 cycles.

The mean values of crushing strength are listed in Table 3 highlighting the effect of both oxidation and thermal stresses on their retained structural integrity.

Table 3. Crushing strength data obtained for the different testing conditions.

Foam		**Number of Cycles**				
A (10 ppi)	**ΔT (K)**	**0**	**1**	**25**	**100**	**150**
	200			0.70 ± 0.01		
	400			0.59 ± 0.04	0.56 ± 0.02	0.58 ± 0.01
	500			0.60 ± 0.02	0.55 ± 0.01	0.52 ± 0.01
	600			0.57 ± 0.02	0.56 ± 0.02	0.53 ± 0.02
	As-received	0.70 ± 0.01				
	400 (H_2O)		0.52 ± 0.02			
	600 (H_2O)		0.50 ± 0.02			

Table 3. *Cont.*

B (20 ppi)	ΔT (K)	Number of Cycles				
		0	1	25	100	150
	200			0.67 ± 0.01		
	400			0.65 ± 0.01	0.59 ± 0.07	0.60 ± 0.01
	500			0.64 ± 0.01	0.59 ± 0.04	0.60 ± 0.04
	600			0.61 ± 0.02	0.58 ± 0.01	0.59 ± 0.01
	As-received	0.67 ± 0.01				
	400 (H_2O)		0.62 ± 0.01			
	600 (H_2O)		0.56 ± 0.02			

3. Discussion

The damage associated with oxidation was determined by the mass change upon exposure to the concentrated solar beam, owing to the oxidation of silicon carbide into silica, according to the reaction:

$$SiC\ (s) + 3/2\ O_2\ (g) \rightarrow SiO_2\ (s) + CO\ (g) \quad (1)$$

as confirmed by XRD analysis (Figure 2). This results in the formation of a surface SiO_2 (cristobalite) layer, which protects the substrate and alters the surface condition [27]. In addition, SiO_2 reacts with Al_2O_3 leading to the formation of mullite ($3Al_2O_3{\cdot}2SiO_2$) [28]. The mass change increased with increasing either ΔT or the number of cycles, as expected, mainly due to the oxidation of the SiC. In the same way as silicon nitride, it is expected that thermal cycling will have a minimal effect on the rate of oxidation. However, it could affect the characteristics of the oxidation products on the surface, depending on how often the cycling occurs. Whilst the oxide layers develop cracks upon cooling, they quickly heal when exposed to high temperatures again. There was no observable change in the mechanism controlling the rate of oxidation during the investigated exposure time [29].

Beside degradation caused by oxidation, the thermal stresses imposed on the foams upon thermal cycling are expected to damage the structural integrity of the materials.

The current setup permitted to accomplish a more or less homogenous temperature distribution (±30 °C) over the irradiated surface (Figure 3). However, when the control temperature (B1) varied from 800 to 1000 °C, the thermocouples placed underneath K2 and K3 (Figure 6e) fell in the range of 880 ± 5 °C to 920 ± 10 °C (difference of about 80 °C compared to B1) and 800 ± 5 °C to 850 ± 10 °C, respectively, as can be seen in Figure 4. The maximum temperature difference between the top and bottom surfaces' temperatures set and those actually measured was about 150 °C.

In the case of the temperature (B1) ranged from 800 to 1400 °C, the thermocouples K2 and K3 ranged from 870 ± 5 °C to 1300 ± 20 °C (difference of about 100 °C compared to B1) and 785 ± 5 °C to 1175 ± 15 °C (indicating that the difference from the top surface temperature was around 200 °C), respectively. This shows that the actual temperature gradients across each sample varied from 600 K (on the radiated surface) to about 400 K (non-radiated one). As expected, the samples were subjected to several temperature gradients, namely across their thickness and their width, resulting in unknown thermal stress fields.

The crushing strength of the RPC foams is known to depend on their relative porosity [30]. In the present case, the difference in porosity is small (85 and 87%), so its effect is low: the as-received crushing strength decreased from 0.70 ± 0.01 to 0.67 ± 0.01 MPa, going from 10 ppi to 20 ppi ob-SiC RPC foams, respectively. Using the bending micromechanical model developed by Gibson and Ashby [31], one can estimate the compressive strength of an open-cell ceramic foam, σ_{cf}, through the relationship:

$$\sigma_{cf} = C_1 \sigma_{fs} \left(\frac{\rho_{cf}}{\rho_s} \right)^{\frac{3}{2}} \quad (2)$$

where σ_{fs} is the flexural strength of the strut material (assumed to be 40 MPa [32]), ρ_{cf} is the bulk density of the ceramic foam, ρ_s is the density of the solid struts (3.01 Mg m^{-3}, measured by pycnometry), and C_1 is a geometric constant characteristic of the unit cell shape. The value of C_1 was found to be equal to 0.16, by Zhang and Ashby, for brittle open-cell foams assuming tetrakaidecahedral unit cell geometry [33]. For the 10 ppi foam, the ρ_{cf} = 0.46 Mg m^{-3} and the calculated σ_{cf} is 0.38 MPa (i.e., about 50% lower than that measured experimentally). This discrepancy is attributed to the fact that the tested foams contained four external solid walls (Figure 6d). For A foams (10 ppi), increasing the ΔT from 200 to 600 K resulted in a crushing strength decrease of 18.6% after 25 cycles and 24.3% after 150 cycles. In contrast, the decrease in crushing strength for B foams (20 ppi) was around 9% and ≈12% after 150 cycles under the same conditions (Table 3). At ΔT = 600 K, for A foams the decrease in strength was around 7% when increasing the number of cycles from 25 to 150. In the case of the B-type foams, under the same conditions, the decrease was roughly half (3.3%).

The thermal shock stresses imposed on the foams when testing them in water baths is more severe than that experienced upon solar exposure. For ΔT = 400 K, the retained crushing strength dropped ≈26% and ≈8% for type A and B foams, respectively. In the case of ΔT = 600 K, the decrease was respectively ≈29% and ≈16% for type A and B foams. One can therefore assume that ΔTc for the A foams is 600 K. The results obtained showed that the thermal shock resistance of the B foam is better than their A foam counterparts (both in air and water media). Experiments using ceramic foams showed that very high porosities (>70%) result in thermal shock resistance improvements, which is attributed to penetration of the foam structure by the cooling medium [34]. Indeed, such open-cell foams are characterized by a very high permeability, with larger cell sizes facilitating the infiltration of the structure by the quenching media. At least two sources of thermal stress can thus be considered: one associated with the heating of the quenching medium as it infiltrates the cellular structure; and the other the temperature gradient across the microscopic struts [35]. The former temperature gradient seems to be more pronounced than the latter one in the present work, as the thermal shock resistance was found to increase with decreasing in cell size, which was also observed for cordierite foams [21].

The experimental data obtained are in good agreement with those published for similar SiC foams, where foams having high ppi resulted in the best solar-to-thermal performances [17,36].

4. Materials and Methods

4.1. Materials

The two commercial ob-SiC RPC foams manufactured by Vesuvius GmbH (Borken, Germany) were SEDEX 65 × 65 × 18 mm, 20 ppi (pores per inch) and SEDEX 60 × 60 × 18, 10 ppi, respectively. Their crystal structure was analyzed by XRD on a PANalytical's X'Pert PRO MPD diffractometer with a step size of 0.02° and the accelerating voltage of 40 kV. Phase identification was carried out with the DIFFRAC.EVA V7 software and the ICDD (International Centre for Diffraction Data). The measured density of the foams was determined using their mass and geometrical dimensions. The true density of the foams was measured by Helium pycnometry (Accupyc 1330, Micromeritics Inc., Norcross, GA, USA).

4.2 Thermal Cycling

Experiments were carried out at the renewed SF60 solar furnace of PSA described by Rodríguez et al. [37] shown in Figure 6.

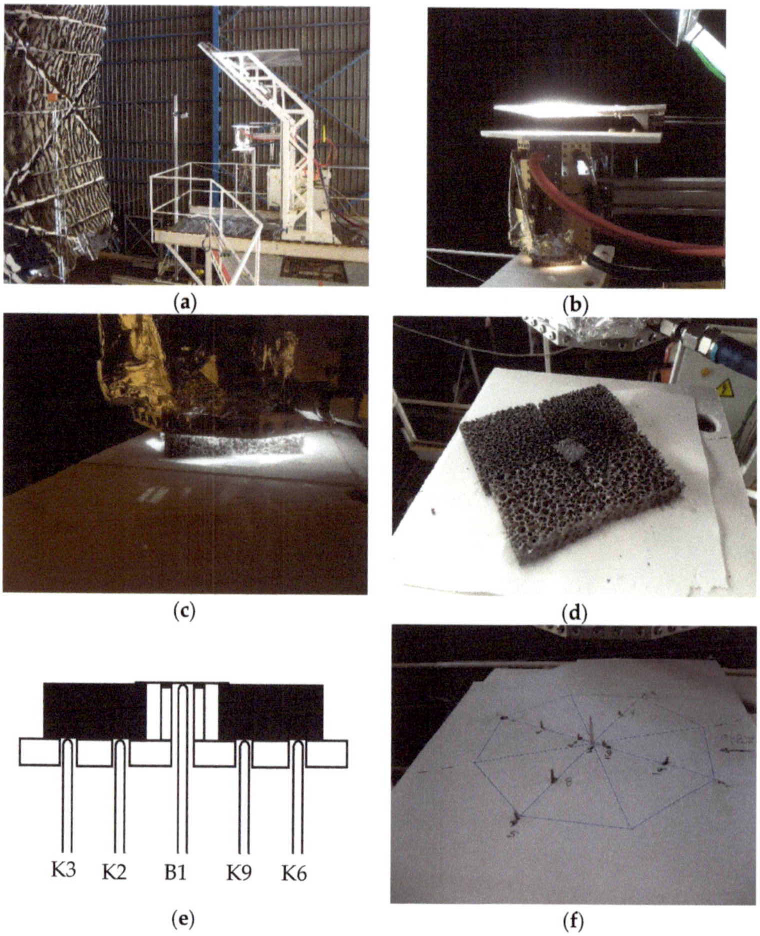

Figure 6. Experimental system showing the uncut (**a**) and in cut (**b**) guillotine positions; ob-SiC RPC samples being exposed (**c**); samples after the test at $\Delta T = 500$ K for 100 cycles (**d**); scheme of the B- and K-type thermocouples positions (**e**); actual (nine) thermocouples positioning (**f**).

Briefly, this facility reaches a peak flux of 6722 kW m^{-2} in a focus diameter of 22 cm. The optical axis of the furnace is horizontal, so that the HT130 heliostat, and the parabolic concentrator, allow the solar beam to be aligned on the optical axis of the parabola. The amount of incident sunlight was controlled through the angular motion of attenuator slats. Furthermore, a mirror placed at 45° was used to carry out the tests in the horizontal plane.

Owing to the Gaussian energy distribution of the concentrated solar beam, a secondary concentrator of octagonal shape, hereafter referred to as homogenizer, was placed in line to obtain a more homogeneous temperature distribution over the horizontal plane.

Ten thermal shock tests were carried out during the 10-working day campaign (from the 5th to the 16th of September 2022), according to the conditions listed in Table 4. In order to control the guillotine's on/off cycle, and consequently, the heating and cooling times, a specially designed LabView program was implemented. Cooling was achieved under natural conditions.

Table 4. Thermal cycling test conditions.

ΔT (°C)	Heating Time (s)	Cooling Time (s)	Number of Cycles		
			25	100	150
800–1000	15	15	x	-	-
800–1200	35	45	x	x	x
800–1300	40	45	x	x	x
800–1400	45	50	x	x	x

For every test condition, two samples of each type of foam were exposed to the direct concentrated solar beam. The temperature recorded for the B-type thermocouple was used as the control reference for performing the thermal cycles. Underneath each exposed foam, two K-type thermocouples were placed as shown in Figure 6f. The measurements of the temperature distribution across the surface of the exposed foams were also carried out using an infra-red (IR) camera (model Equus 327k SM PRO), manufactured by IRCam GmbH, Erlangen, Germany. A flat piece of silicon carbide wall circa 1 mm thick was placed above the B-type thermocouple to measure the samples' surface temperature and compare it with both the solar-blind pyrometer (Infratherm IGA 5LO, IMPAC Electronic GmbH, Frankfurt, Germany) and the IR camera measurements (Figure 6d).

For comparison, water quenching tests were performed by rapid heat transfer of a test sample at an elevated temperature in a distilled water bath at room temperature according to the ASTM C1525-04 standard [38]. Two temperature differences were set: 400 and 600 K. The thermal shock resistance is defined as a critical temperature interval (ΔTc), which corresponds to at least 30% reduction in the mean retained strength.

To assess oxidation damage, prior and after each test the samples were weighed for mass change measurements.

4.3. *Crushing Testing*

After cycling, the retained crushing strength was determined at room temperature using an Instron Corp., Norwood, MA, USA (model 4302) testing machine with the compressive plates closing at a crosshead speed of 0.5 mm min-1 with a load cell of 10 kN. Since the foams have four external walls, the procedure used to calculate the "crushing strength" was the same as described elsewhere [39]. Briefly, a compliant 1 mm thick rubber spacer was inserted between the loading plates and the foams to ensure uniform loading. On the other hand, the nominal load value required to calculate the "crushing strength" was determined as the point in the chart curve were the slope changed.

5. Conclusions

The purpose of the present study was to evaluate the ability of commercial ob-SiC RPC foams as solar receivers for new CSP plants. For this purpose, a dedicated experimental setup was developed allowing a ±30 °C temperature distribution over the entire irradiated surface plane to be achieved. The following conclusions could be drawn:

- The ob-SiC RPC foams showed good performance to thermal shock under the conditions investigated;
- The B foam having both lower bulk density and cell size showed better thermal shock behavior;
- Under the most extreme conditions (ΔT = 600 K, 150 cycles), the B foam (20 ppi) showed a crushing strength degradation of nearly half of the A foam (10 ppi) counterparts;
- Exposure of the foams to the thermal cycling resulted in the formation of a SiO_2-rich surface layer containing traces of mullite;
- The cooling medium effect on the retained crushing strength was found to be ΔTc = 600 K for A foam;

- The experimental crushing strengths were found to be about double of the predicted by the Gibson-Ashby model, which can be attributed to the fact that outer solid walls are present.
- Future work should focus on evaluating long-term durability, conducting a higher number of thermal cycles, and introducing an air stream during cooling to simulate real appliance conditions.

Author Contributions: Conceptualization, F.d.A.C.O. and J.C.F.; methodology, J.G., J.R. and I.C.; software, J.R.; validation, F.d.A.C.O. and J.C.F.; formal analysis, F.d.A.C.O. and J.C.F.; investigation, F.d.A.C.O., I.C. and J.C.F.; resources, J.C.F.; data curation, F.d.A.C.O. and J.C.F.; writing—original draft preparation, F.d.A.C.O. and J.C.F.; writing—review and editing, F.d.A.C.O., J.G., J.R., I.C. and J.C.F.; visualization, F.d.A.C.O. and J.C.F.; supervision, F.d.A.C.O. and J.C.F.; project administration, F.d.A.C.O.; funding acquisition, F.d.A.C.O. and J.C.F. All authors have read and agreed to the published version of the manuscript.

Funding: The funding provided by the European Commission through the SFERA-III project (Grant Agreement No. 823802) is acknowledged. The study was also financially supported by Fundação para a Ciência e a Tecnologia (FCT) via the project LAETA Base Funding (DOI: 10.54499/UIDB/50022/2020), and the INIESC-National Research Infrastructure for Concentrated Solar Energy through contract ALT20-03-0145-FEDER-022113.

Data Availability Statement: The datasets generated during and/or analyzed during the current study are available from the corresponding author upon reasonable request.

Acknowledgments: F.d.A.C.O. and J.C.F. gratefully acknowledge the CIEMAT-PSA Laboratory for providing access to its installations. The administrative support of Marta Ruiz is also appreciated. SEDEX filters were kindly donated by Foseco Portugal—Produtos para Fundição, Lda., Porto, Portugal.

Conflicts of Interest: The authors declare no conflicts of interest.

References

1. Farjana, S.H.; Huda, N.; Parvez Mahmud, M.A.; Saidur, R. Solar process heat in industrial systems—A global review. *Renew. Sust. Energy Rev.* **2018**, *82 Pt 3*, 2270–2286. [CrossRef]
2. Ravi Kumar, K.; Krishna Chaitanya, N.V.V.; Sendhil Kumar, N. Solar thermal energy technologies and its applications for process heating and power generation—A review. *J. Clean. Prod.* **2021**, *282*, 125296. [CrossRef]
3. Kumar, L.; Ahmed, J.; El Haj Assad, M.; Hasanuzzaman, M. Prospects and challenges of solar thermal for process heating: A comprehensive review. *Energies* **2022**, *15*, 8501. [CrossRef]
4. Aprà, F.M.; Smit, S.; Sterling, R.; Loureiro, T. Overview of the enablers and barriers for a wider deployment of CSP tower technology in Europe. *Clean Technol.* **2021**, *3*, 377–394. [CrossRef]
5. Ávila-Marín, A.L. Volumetric receivers in solar thermal power plants with central receiver system technology: A review. *Sol. Energy* **2011**, *85*, 891–910. [CrossRef]
6. Ho, C.K. Advances in central receivers for concentrating solar applications. *Sol. Energy* **2017**, *152*, 38–56. [CrossRef]
7. Fend, T.; Hoffschmidt, B.; Pitz-Paal, R.; Reutter, O.; Rietbrock, P. Porous materials as open volumetric solar receivers: Experimental determination of thermophysical and heat transfer properties. *Energy* **2004**, *29*, 823–833. [CrossRef]
8. Roldán, M.I.; Smirnova, O.; Fend, T.; Casas, J.; Zarza, E. Thermal analysis and design of a volumetric solar absorber depending on the porosity. *Renew. Energy* **2014**, *62*, 116–128. [CrossRef]
9. Wang, F.; Shuai, Y.; Tan, H.; Yu, C. Thermal performance analysis of porous media receiver with concentrated solar irradiation. *Int. J. Heat Mass Transf.* **2013**, *62*, 247–254. [CrossRef]
10. Kribus, A.; Gray, Y.; Grijnevich, M.; Mittelman, G.; Mey-Cloutier, S.; Caliot, C. The promise and challenge of solar volumetric absorbers. *Sol. Energy* **2014**, *110*, 463–481. [CrossRef]
11. Lidor, A.; Fend, T.; Roeb, M.; Sattler, C. Parametric investigation of a volumetric solar receiver-reactor. *Sol. Energy* **2020**, *204*, 256–269. [CrossRef]
12. Hajimirza, S.; Sharadga, H. Learning thermal radiative properties of porous media from engineered geometric features. *Int. J. Heat Mass Transf.* **2021**, *179*, 121668. [CrossRef]
13. Zhao, Y.; Tang, G.H. Monte Carlo study on extinction coefficient of silicon carbide porous media used for solar receiver. *Int. J. Heat Mass Transf.* **2016**, *92*, 1061–1065. [CrossRef]
14. Nimvari, M.E.; Jouybari, N.F.; Esmaili, Q. A new approach to mitigate intense temperature gradients in ceramic foam solar receivers. *Renew. Energy* **2018**, *122*, 206–215. [CrossRef]

15. Patil, V.R.; Kiener, F.; Grylka, A.; Steinfeld, A. Experimental testing of a solar air cavity-receiver with reticulated porous ceramic absorbers for thermal processing at above 1000 °C. *Sol. Energy* **2021**, *214*, 72–85. [CrossRef]
16. Ávila-Marín, A.L.; Caliot, C.; Flamant, G.; Alvarez de Lara, M.; Fernandez-Reche, J. Numerical determination of the heat transfer coefficient for volumetric air receivers with wire meshes. *Sol. Energy* **2018**, *162*, 317–329. [CrossRef]
17. Mey-Cloutier, S.; Caliot, C.; Kribus, A.; Gray, Y.; Flamant, G. Experimental study of ceramic foams used as high temperature volumetric solar absorber. *Sol. Energy* **2016**, *136*, 226–235. [CrossRef]
18. Fey, T.; Betke, U.; Rannabauer, S.; Scheffler, M. Reticulated replica ceramic foams: Processing, functionalization, and characterization. *Adv. Eng. Mater.* **2017**, *19*, 1700369. [CrossRef]
19. Hammel, E.C.; Ighodaro, O.L.-R.; Okoli, O.I. Processing and properties of advanced porous ceramics: An application based review. *Ceram. Int.* **2014**, *40*, 15351–15370. [CrossRef]
20. Lougou, B.G.; Wu, L.; Ma, D.; Geng, B.; Jiang, B.; Han, D.; Zhang, H.; Łapka, P.; Shuai, Y. Efficient conversion of solar energy through a macroporous ceramic receiver coupling heat transfer and thermochemical reactions. *Energy* **2023**, *271*, 126989. [CrossRef]
21. Costa Oliveira, F.A.; Dias, S.; Fernandes, J.C. Thermal shock behaviour of open-cell cordierite foams. *Mater. Sci. Forum* **2006**, *514–516*, 764–767. [CrossRef]
22. Costa Oliveira, F.A.; Fernandes, J.C.; Galindo, J.; Rodríguez, J.; Canãdas, I.; Rosa, L.G. Thermal resistance of solar volumetric absorbers made of mullite, brown alumina and ceria foams under concentrated solar radiation. *Sol. Energy Mater Sol. Cells* **2019**, *194*, 121–129. [CrossRef]
23. Hoffschmidt, B.; Tellez, F.M.; Valverde, A.; Fernandez, J.; Fernandez, V. Performance evaluation of the 200-kWth HiTRec-II open volumetric air receiver. *J. Sol. Energy Eng.* **2003**, *125*, 87–94. [CrossRef]
24. Fend, T.; Pitz-Paal, R.; Reutter, O.; Bauerb, J.; Hoffschmidt, B. Two novel high-porosity materials as volumetric receivers for concentrated solar radiation. *Sol. Energy Mater. Sol. Cells* **2004**, *84*, 291–304. [CrossRef]
25. Behar, O.; Khellaf, A.; Mohammedi, K. A review of studies on central receiver solar thermal power plants. *Renew. Sustain. Energy Rev.* **2013**, *23*, 12–39. [CrossRef]
26. Ávila-Marín, A.L.; Alvarez-Lara, M.; Fernandez-Reche, J. Experimental results of gradual porosity wire mesh absorber for volumetric receivers. *Energy Procedia* **2014**, *49*, 275–283. [CrossRef]
27. Narushima, T.; Goto, T.; Hirai, T.; Iguchi, Y. High-temperature oxidation of silicon carbide and silicon nitride. *Mater. Trans.* **1997**, *38*, 821–835. [CrossRef]
28. Ding, S.; Zhu, S.; Zeng, Y.; Jiang, D. Fabrication of mullite-bonded porous silicon carbide ceramics by in situ reaction bonding. *J. Eur. Ceram. Soc.* **2007**, *27*, 2095–2102. [CrossRef]
29. Costa Oliveira, F.A.; Baxter, D.J. The effect of thermal cycling conditions on oxide scale structure and morphology on a hot-pressed silicon nitride. *J. Eur. Ceram. Soc.* **1999**, *19*, 2747–2756. [CrossRef]
30. Seuba, J.; Deville, S.; Guizard, C.; Stevenson, A.J. Mechanical properties and failure behavior of unidirectional porous ceramics. *Sci. Rep.* **2016**, *6*, 24326. [CrossRef]
31. Gibson, L.G.; Ashby, M.F. *Cellular Solids Structure & Properties*, 2nd ed.; Cambridge University Press: Cambridge, UK, 1997.
32. Baitalik, S.; Kayal, N. Thermal shock and chemical corrosion resistance of oxide bonded porous SiC ceramics prepared by infiltration technique. *J. Alloys Compd.* **2019**, *781*, 289–301. [CrossRef]
33. Brezny, R.; Green, D.J. Uniaxial strength behavior of brittle cellular materials. *J. Am. Ceram. Soc.* **1993**, *76*, 2185–2192. [CrossRef]
34. Orenstein, R.M.; Green, D.J. Thermal shock behavior of open-cell ceramic foams. *J. Am. Ceram. Soc.* **1992**, *75*, 1899–1905. [CrossRef]
35. Vedula, V.R.; Green, D.J.; Hellman, J.R. Thermal shock resistance of ceramic foams. *J. Am. Ceram. Soc.* **1999**, *82*, 649–656. [CrossRef]
36. Zaversky, F.; Sánchez, M.; Roldán, M.I.; Ávila-Marín, A.L.; Füssel, A.; Adler, J.; Knoch, M.; Dreitz, A. Experimental evaluation of volumetric solar absorbers—Ceramic foam vs. an innovative rotary disc absorber concept. *AIP Conf. Proc.* **2018**, *2033*, 040044.
37. Rodríguez, J.; Galindo, J.; Cañadas, I.; Monterreal, R.; Fernández-Reche, J. Desing and characterization of the new FAHEX100 concentrator of PSA's SF60 Solar Furnace. In Proceedings of the SolarPACES 2022, 28th International Conference on Concentrating Solar Power and Chemical Energy Systems, Albuquerque, NM, USA, 27–30 September 2022; Volume 1. [CrossRef]
38. *ASTM C1525-04*; Standard Test Method for Determination of Thermal Shock Resistance for Advanced Ceramics by Water Quenching. ASTM International: West Conshohocken, PA, USA, 2004.
39. Costa Oliveira, F.A.; Sardinha, M.; Galindo, J.; Rodríguez, J.; Cañadas, I.; Leite, M.; Fernandes, J.C. Manufacturing and thermal shock resistance of 3D-printed porous black zirconia for concentrated solar applications. *Crystals* **2023**, *13*, 1323. [CrossRef]

MDPI AG
Grosspeteranlage 5
4052 Basel
Switzerland
Tel.: +41 61 683 77 34

Inorganics Editorial Office
E-mail: inorganics@mdpi.com
www.mdpi.com/journal/inorganics

www.ingramcontent.com/pod-product-compliance
Lightning Source LLC
Chambersburg PA
CBHW041356210926
51850CB00013B/3100
* 9 7 8 3 7 2 5 8 3 8 5 4 7 *